Behavioural Genetics for Education

Behavioural Genetics for Education

Edited by

Yulia Kovas
Goldsmiths, University of London, UK

Sergey Malykh
Russian Academy of Education, Russia

and

Darya Gaysina
University of Sussex, UK

palgrave
macmillan

BEHAVIOURAL GENETICS FOR EDUCATION

First published 2016 by
PALGRAVE MACMILLAN

Palgrave Macmillan in the UK is an imprint of Macmillan Publishers Limited, registered in England, company number 785998, of Houndmills, Basingstoke, Hampshire RG21 6XS.

Palgrave Macmillan in the US is a division of Nature America, Inc., One New York Plaza, Suite 4500 New York, NY 10004–1562.

Palgrave Macmillan is the global academic imprint of the above companies and has companies and representatives throughout the world.

Hardback ISBN: 978–1–137–43731–0
E-PUB ISBN: 978–1–137–43733–4
E-PDF ISBN: 978–1–137–43732–7
DOI: 10.1057/9781137437327

Distribution in the UK, Europe and the rest of the world is by Palgrave Macmillan®, a division of Macmillan Publishers Limited, registered in England, company number 785998, of Houndmills, Basingstoke, Hampshire RG21 6XS.

Library of Congress Cataloging-in-Publication Data

Names: Kovas, Yulia, editor. | Malykh, S. B. (Sergey Borisovich), editor. | Gaysina, Darya, 1978– editor.

Title: Behavioural genetics for education / edited by Yulia Kovas, Sergey Malykh, Darya Gaysina.

Other titles: Behavioral genetics for education

Description: Houndmills, Basingstoke, Hampshire ; New York, NY : Palgrave Macmillan, 2016. | Includes index.

Identifiers: LCCN 2015040547 | ISBN 9781137437310 (hardback)

Subjects: LCSH: Behavioral assessment. | Educational psychology. | Behavior genetics. | Learning ability—Genetic aspects. | Cognition in children. |

BISAC: EDUCATION / Educational Psychology. | EDUCATION / Educational Policy & Reform / General. | PSYCHOLOGY / Developmental / Child. | SCIENCE / Life Sciences / Genetics & Genomics.

Classification: LCC LB1060.2 .B446 2016 | DDC 370.15—dc23

LC record available at http://lccn.loc.gov/2015040547

A catalog record for this book is available from the Library of Congress

A catalogue record for the book is available from the British Library

Contents

List of Illustrations

Boxes

Figures

Table

Foreword

This book is dedicated to the links between genetics and education. Despite the overwhelming evidence for the importance of genetics in cognitive development and other educationally relevant traits, genetics is rarely considered in relation to education (Asbury & Plomin, 2014). This book is timely and provides an important contribution to the emerging field of educational genomics. The chapters, written by collaborative international teams, present a comprehensive overview of the advances in quantitative and molecular genetic methodology applied to education.

Quantitative genetic research has moved beyond simply demonstrating whether and how much genetics influences individual differences in behaviour to understanding *how* genes influence behaviour. One example of going beyond heritability estimates to ask questions about how genes influence behaviour is developmental behavioural genetics: rather than studying samples of a wide age range cross-sectionally, genetically sensitive longitudinal cohort studies throughout the lifespan can investigate the crucial question of how genotypes become phenotypes in development. A second example is the use of multivariate genetic strategies to understand the general effects of genes as well as specific effects across traits at the behavioural level (e.g., genetic effects on learning skills such as reading and mathematics), as well as between the behavioural level and other levels of investigation, most notably brain structure and function.

A third example of going beyond heritability estimates is the investigation of the interactions and correlations between nature and nurture. Much of the behavioural genetic research reported in this book is focused on investigations of genetic and environmental interactions and correlations in relation to the most massive environmental intervention of all: education. A fourth direction is opening up a completely new area of research on the interplay between nature and nurture by going beyond individual differences within one culture to comparisons between cultures.

Although quantitative genetics will continue to make important contributions to understanding the origins of individual differences in behaviour, the future of behavioural genetics research lies with molecular genetics, which aims to identify DNA differences responsible for heritability. We now know that the heritability of behavioural traits is caused by many genes of very small effect, which means that we need very large samples to detect these small effects. Throughout the life sciences, molecular geneticists are grappling with the "missing heritability" problem, the gap between heritability and the small amount of heritability that can be accounted for by replicated DNA associations (Plomin, 2013). Until the missing heritability gap is narrowed, instead of joining the flood of underpowered studies reporting results of a few candidate genes – studies that seldom replicate – behavioural geneticists from different countries can join efforts to collect behavioural data and DNA on large samples and wait until relevant gene associations are found, probably using strategies that have not yet been developed. Then they can use many genes – perhaps hundreds or thousands – in polygenic composites indexing genetic risk and resilience in order to investigate questions about development, multivariate links, and gene–environment interplay at the level of DNA.

The field of molecular genetics is advancing so quickly that it is advantageous to wait until we can harvest the next big crop, which will be whole-genome sequencing. Sequencing all three billion pairs of DNA is the end of the story in terms of DNA sequence variation – once a child's genome is sequenced, there is no need to do any more genotyping. The most exciting possibility is that, as the cost of whole-genome sequencing declines dramatically, DNA of all newborns will be sequenced, which has been predicted by the head of the US National Institutes of Health (Collins, 2010). This would mean that we would not even need to collect DNA in order to use DNA in behavioural genetics research.

One hallmark of today's behavioural genetic research, clearly on display in this book, is its international flavour, moving away from the insularity of earlier research. For me, the most wonderful aspect of science is its ability to ignore international boundaries and, especially with modern technology such as the internet, to feel that the scientific community is truly worldwide. A case study demonstrating this international nature of science can be seen in the development of twin registers in many countries (Hur & Craig, 2013), which can be combined to answer many important questions. The synergy that comes from bringing together the energy of scientists throughout the world is awesome, particularly for scientists who value collaborative rather than competitive science.

Robert Plomin
Kings College London

References

Asbury, K., & Plomin, R. (2014). *G is for genes: What genetics can teach us about how we teach our children*. Oxford: Wiley Blackwell.

Collins, F. S. (2010). *The language of life: DNA and the revolution in personalised medicine*. New York: Harper Collins.

Hur, Y.-M., & Craig, J. M. (2013). Twin registries worldwide: An important resource for scientific research. *Twin Research and Human Genetics, 16*, 1–12.

Plomin, R. (2013). Child development and molecular genetics: 14 years later. *Child Development, 84*, 104–120.

Preface

To date, no educational methods exist which are universal and effective for each child. Educational environments interact with children's unique genetic profiles, leading to wide individual differences in learning ability, motivation, and achievement in different academic subjects – even when children study with the same teacher, attend the same school, and follow the same curriculum. The most effective educational programmes will take into account each student's individual psychological profile with their complex genetic and environmental aetiology. In contemporary science, many important findings come from interdisciplinary research which combines methods and approaches of different disciplines to study the same phenomenon. Education, one of the most important aspects of human life, is no exception. Today, all theoretical, methodological, and technological foundations exist for applying interdisciplinary approaches to educationally relevant research. This book provides an overview of the implications of combining different scientific approaches (psychology, quantitative genetics, molecular genetics, neuroscience, and others) for the understanding of complex educational processes.

The book begins with a Foreword by the world leading expert in behavioural genetics, Robert Plomin. The book includes chapters written by international experts whose research interests span a wide range of topics: cognitive abilities and disabilities, aspects of motivation and personality, behavioural and emotional problems, aspects of social functioning, such as peer and parent–child relations, well-being, and academic achievement. Although the chapters of the book can be read in any order, we recommend that the reader begins with Chapters 1 and 2. In order to help the reader to orient in the complex material, we included a Glossary of specialized terms and concepts at the end of the book.

Chapter 1 highlights the benefits of applying behavioural genetic methods to research into individual differences in educationally relevant traits and reviews the relevant recent research. The aim of Chapter 2 is to evaluate the role of environmental factors and their complex interplay with genetic factors (gene–environment co-action). Chapter 3 continues with the topic of gene–environment interplay and presents a review of investigations into gene–environment correlations in the context of the role of parent–child and peer relationships in socio-emotional development. Chapter 4 presents a review of research into reading and mathematical abilities and disabilities, as well as into comorbidity between the two disabilities. The chapter also presents advantages of studying neurobiological markers for understanding genetic and environmental underpinnings of reading and mathematical abilities and disabilities. Chapter 5 is dedicated to molecular genetic investigations of educationally relevant traits. The chapter presents a new approach to the study of molecular genetic foundations of complex traits informed by the knowledge of rare genetic disorders. The following three chapters present reviews of behavioural genetic research into self-concept and self-esteem (Chapter 6), psychological wellbeing (Chapter 7), and personality (Chapter 8) – complex traits that are linked to each other and to academic achievement. Chapters 9, 10, and 11 focus on the aetiology of individual differences in psychological problems: psychopathology in childhood and adolescence (Chapter 9), sleep disorders (Chapter 10), and suicidal behaviour (Chapter 11). Chapter 12 presents an overview of genetically informative investigations of neurophysiological traits that are important for education. The Concluding chapter summarizes the main achievements of behavioural genetics in the field of education to date, and looks into the future of educational genomics.

This book will be of interest to teachers, parents, educational and developmental psychologists, and researchers in different fields working on educationally relevant issues. We hope that this book will provide a platform for further international and truly interdisciplinary research, including disciplines and cultures that are still to join this quest, leading to progress in education worldwide.

Acknowledgements

The writing of this book was supported by Megagrant funding from the Government of the Russian Federation (grant no 11.G34.31.0043). We thank all the chapter authors for their invaluable contribution. In addition, we would like to thank Ivan Voronin, Ilya Zakharov, Margherita Malanchini, Irina Voronina and Robert Chapman for their assistance with the creation of the book. Finally, we thank all the families, teachers, administrators and everyone else involved in educational process, whose participation makes educational behavioural genetic research possible.

Notes on Contributors

Editors

Darya Gaysina is Lecturer in Psychology at the University of Sussex, UK and leads the interdisciplinary research programme in Genetics and Psychopathology (EDGE Lab). She is also a member of the Rudd Centre for Adoption Research and Practice. Darya holds a PhD in Human Genetics from the Russian Academy of Sciences. Her research focuses on exploring environmental and genetic factors and their interplay, in relation to psychopathology across the life course. She is the author of more than 40 peer-reviewed papers and book chapters.

Yulia Kovas is Professor of Genetics and Psychology at Goldsmiths, University of London, where she directs InLab – International Laboratory for Interdisciplinary Investigations into Individual Differences in Learning. She is also Professor at New York University in London. Yulia is the Director of the Laboratory for Cognitive Investigations and Behavioural Genetics, Tomsk State University (TSU); Co-Director of the International Centre for Research in Human Development (TSU); and Co-Director of the Russian–British Laboratory for Behavioural Genetics, Psychological Institute of the Russian Academy of Education. She directs the Russian School Twin Register (RSTR) and leads the mathematics research direction in the Twins Early Development Study (TEDS) at King's College London. She is the author of more than 100 peer-reviewed papers and book chapters.

Sergey Malykh is a professor at the Psychological Institute of the Russian Academy of Education, where he directs the Laboratory for Developmental Behavioural Genetics and co-directs the Russian–British Laboratory for Behavioural Genetics. He is also the Head (Academician-Secretary) of the Department of Psychology and Developmental Physiology of the Russian Academy of Education and Co-Director of the International Centre for Research in Human Development, Tomsk State University (TSU). The focus of his research is currently on the genetic and environmental origins of variation in psychological and psychophysiological traits. He is the author of more than 300 publications, including *Foundation of Behavioural Genetics* (1998), the first textbook on this topic in Russia.

Contributors

Nicola L. Barclay is Senior Lecturer in Psychology and Associate Director of the Northumbria Centre for Sleep Research at Northumbria University, Newcastle upon Tyne, UK. She completed her PhD at Goldsmiths, University of London, in 2011 under the supervision of Dr Alice Gregory. She specializes in the genetic and neurobiological basis of sleep and circadian rhythms, with particular interest in insomnia, circadian rhythm disorders, and physiological and behavioural effects of sleep deprivation. She is the author of more than 20 peer-reviewed journal articles and has contributed to a number of handbooks on sleep since 2009.

Philipp Barsky has been a researcher in the Developmental Behavioural Genetics Laboratory at the Psychological Institute of Russian Academy of Education, Moscow, since 2005. He holds a PhD from the Russian Academy of Education and a Master's in Psychology from Lomonosov Moscow State University. His scientific interests include the problems of specific environmental influences and gene–environment interactions in a variety of phenotypes, such as personality, intelligence, and family relationships; he has also taught and published on qualitative research methods and narratives.

Michel Boivin is Canada Research Chair in Child Development and Director of the Research Unit on Children's Psychosocial Maladjustment (GRIP) at Université Laval, Quebec City, Canada where he leads a research programme on the interplay of biological and social factors in child development. This research focuses on large longitudinal follow-ups of children, including the Quebec Newborn Twin Study and the Quebec Longitudinal Study of Child Development. His programme of research has led to more than 200 peer-reviewed publications. He also leads a knowledge mobilization initiative on early childhood development (ECD), and is one of the editors of the Encyclopedia on Early Childhood Development. He is a member of the Royal Society of Canada.

Mara Brendgen is Professor of Psychology at the Université du Québec à Montréal and a Senior Research Scientist at the Ste Justine Hospital Research Centre in Montréal. Her research focuses on the interplay between individual, family-related, and peer-related factors in the etiology of aggression and victimization in youth utilizing a genetically informed longitudinal twin design. She is the author of more than 120 peer-reviewed papers and book chapters, and her research has been consistently funded by the Canadian Institutes of Health Sciences, the Social Sciences and Humanities Research Council and the Fonds Québécois de Recherche sur la Société et la Culture.

Huajian Cai is a professor at the Institute of Psychology, Chinese Academy of Sciences, Beijing, China. He received his PhD in Psychology from East China Normal University in 2002. In recent years, he has examined the genetic basis of self-positivity, including narcissism, explicit and implicit self-esteem. His articles have appeared in the *Journal of Personality and Social Psychology, Psychological Science, Social Psychological and Personality Science, Social Cognition, Emotion, European Journal of Personality, Journal of Research in Personality, Journal of Cross-Cultural Psychology, Plos One, Personality and Individual Differences,* and *Frontiers in Psychology.*

Sophia Docherty received her PhD in 2011 from the SGDP Centre, Institute of Psychiatry, King's College London. Her thesis explored the aetiology of mathematical ability and disability at the molecular genetic level. Since her PhD, she has continued to investigate the genetic basis of cognitive abilities, working on gene–environment interaction projects and exploring variation in learning abilities in the context of serious genetic disorders.

Philip Gehrman is Assistant Professor of Psychology in the Department of Psychiatry, University of Pennsylvania School of Medicine USA, where he directs the Sleep and Traumatic Stress programme. He completed his graduate training in clinical psychology at the University of California, San Diego, including a predoctoral internship at the Durham VA Medical Center and a postdoctoral fellowship in sleep medicine at Penn. He has an active research programme exploring the mechanisms and treatment of insomnia in the context of mental illness. His work includes studies on the genetics of insomnia for which he leads an international consortium of genetics researches.

Elena Gindina is a senior scientist in the Laboratory of Developmental Behavioural Genetics at the Psychological Institute of the Russian Academy of Education, Moscow. She holds an MA in Psychology (Honors) from Lomonosov Moscow State University and an MA in Applied Psychology from Columbia University, New York. Elena received her PhD in 2005 from the Psychological Institute of the Russian Academy of Education. Elena's thesis focused on the role of environmental and genetic factors in the aetiology of emotional and behavioural problems in adolescents. Her current major area of interest is the aetiology of mental health problems in children. Elena is the author of more than 35 scientific works published in Russian and in international journals. In 2006 she was awarded the Gold Medal for young scientists by the Russian Academy of Education.

Alice M. Gregory is a professor in the Department of Psychology, Goldsmiths, University of London. She first developed an interest in sleep research as an undergraduate student at Oxford University. Following completion of her degree, she studied in Japan for a year collecting cross-cultural data for her first publication in the field of sleep research. Upon her return she commenced her PhD at the Institute of Psychiatry using epidemiological and twin samples to learn more about the associations between sleep disturbances, anxiety, and depression concurrently and longitudinally. She is the author of more than 100 publications, many of which focus on sleep and associated traits.

Claire M. A. Haworth is Reader in Behavioural Genetics at the MRC Integrative Epidemiology Unit, University of Bristol UK, and Co-Director of the Dynamic Genetics Lab. She studied Experimental Psychology at the University of Oxford and then spent a year working as an assistant psychologist. Funded by an MRC studentship, she completed an MSc and PhD in Social Genetic and Developmental Psychiatry at the Institute of Psychiatry, King's College London with Professor Robert Plomin. Since her PhD she has been funded by fellowships and grants from the MRC, ESRC, British Academy and the Wellcome Trust.

Jeffrey Henry is a graduate student at Université Laval, Quebec City, Canada. His work is funded by the Canadian Health Institutes for Research, on the Frederick Banting and Charles Best Canada Graduate Scholarship Doctoral Award. As a researcher, his primary interest is in uncovering the various gene–environment processes involved in the development of callous-unemotional traits and persistent conduct problems. In the context of his doctorate, he benefits from a large, normative twin sample (Quebec Newborn Twin Study; QNTS). His work is supervised by Professors Michel Boivin and Ginette Dionne.

Anastasiya Kazantseva is a senior researcher at the Institute of Biochemistry and Genetics, Ufa Scientific Center of the Russian Academy of Sciences, Russia. Her research focuses on the investigation of personality traits, cognitive abilities, addiction and depression and is supported by the Russian Foundation for Humanities and the Russian Foundation for Basic Research. The author of more than 40 publications in molecular, medical and psychiatric genetics, she is a board member of Vavilov's Society of Geneticists and Breeders and of the Russian Society of Medical Geneticists.

Elza Khusnutdinova is a professor and Head of the Genomics Department, Institute of Biochemistry and Genetics, Ufa Scientific Center of the Russian Academy of Sciences, Russia. She is also the Head of the Department of Genetics and Fundamental Medicine of the Faculty of Biology of Bashkir State University, and a former member of the Council for Science and Education under the President of the Russian Federation (2012–2013). The author of more than 700 publications in molecular, medical and population genetics, she is a board member of the Scientific Council of Russian Academy of Sciences on genetics and breeding.

Yu L. L. Luo is an associate professor at the Institute of Psychology, Chinese Academy of Sciences, Beijing, China. She received her PhD in Psychology from Beijing Normal University in 2011. She is interested in self-positivity (e.g., narcissism, self-esteem), particularly the genetic basis and neural mechanism underlying such constructs. Her research methods include twin study and functional MRI. She has presented her studies in the *Journal of Research in Personality, Journal of Neuroscience, Journal of Cognitive Neuroscience, Plos One, Personality and Individual Differences, Learning and Individual Differences, Frontiers in Psychology* and other journals.

Rachael E. O'Leary currently works in Insights for a large healthcare company. She developed an interest in sleep research whilst on a year placement at Goldsmiths University assisting Alice Gregory. She was involved in several projects focusing on sleep, including a twin study and a study focusing on cognitive aspects of sleep in adolescents. Rachael holds a BSc in Psychology from the University of Kent.

Michael J. Parsons works at MRC Harwell. His research combines both human psychiatric genetics and the use of genetic mouse models to determine the effects of sleep and circadian disruption on psychiatric illness. His research aims to (1) determine the unique and shared genetics of circadian rhythms/sleep and psychiatric disorders in human; (2) generate genetic mouse models to validate and further characterize these associations; and (3) identify novel biomarkers and targets for treatment of sleep and psychiatric disorders.

Stephen A. Petrill received his PhD from Case Western Reserve University in 1995 and completed postdoctoral studies in 1997 at the Social, Genetic and Developmental Research Centre in the UK. Since that time, he has held faculty positions at Wesleyan University (1997–2001), Pennsylvania State University (2001–2006), and Ohio State University (2006–present). He currently directs one of four Learning Disabilities Innovation Hubs in the United States. Funded by NICHD in 2012, the Ohio State HUB conducts research at the intersection of neuroimaging, behavioural genetics, and learning disabilities. Petrill has more than 150 peer-reviewed papers, is a member of the National Advisory Child Health and Human Development Council, and was awarded a Fulbright Research Scholar Award (Russia).

Robert Plomin has been MRC Research Professor of Behavioural Genetics at the Institute of Psychiatry, Psychology and Neuroscience, King's College London since 1994, when he co-founded and subsequently directed the MRC Social, Genetic and Developmental Psychiatry Centre, whose goal is to bring together genetic and environmental strategies to study behavioural development. In 1995, he launched the Twins Early Development Study (TEDS) of all twins born in England and Wales during 1994–1996, which focuses on developmental problems in cognition and behaviour. He is the author of more than 500 papers and more than a dozen books and has received lifetime research achievement awards from the three major international associations in his field.

Kaili Rimfeld is a doctoral candidate at the Social, Genetic and Developmental Psychiatry Centre, King's College London, working under the supervision of Professors Robert Plomin and Yulia Kovas. She holds a BSc in Psychology and an MSc in Developmental Psychology from Birkbeck College, University of London. Her research explores the role of gene–environment interplay in educational achievement. She is also interested in early predictors of educational achievement, as well as associated life outcomes, such as quality of life and health outcomes.

Maja Rodic received her PhD in 2015 from Goldsmiths, University of London. The principal goal of her PhD research was to increase the knowledge of the links between early (untaught) numerical development and later (taught) mathematical skills, using cross-cultural and genetically informative designs. She is currently a postdoctoral research fellow at the University of Sussex working on an evaluation of the AdOpt parenting programme, specifically designed for adoptive parents to help promote parenting techniques and supports that address particular difficulties that adopted children may experience.

Fatos Selita is a Barrister of England and Wales, and an Attorney of the State of New York, USA. He is a consultant at Emerging Law in London, the Director of AIR Courses at Goldsmiths, University of London and of the Department of Intellectual Property Transfers at Tomsk State University (TSU), Russia. A visiting lecturer at TSU and Higher School of Economics in Moscow, he collaborates with scientists on issues related to implications and dissemination of research, and chairs the Council on Ethical, Societal and Legal Implications of Genetic Research in Child Development and Education at TSU.

Tatiana Tikhomirova is a senior researcher at the Psychological Institute, Russian Academy of Education. She is also the scientific director of the Experimental Platform on the basis of the Secondary Public School at Moscow region. She holds an MSc in Pedagogic Sciences and a PhD in Psychology from the Russian Academy of Sciences. The current focus of her research is environmental influences on development of individual differences in educationally relevant traits. Tatiana is leading the longitudinal project studying cognitive sources of variation in learning achievement. She is the author of more than 60 scientific works.

Maria G. **Tosto** holds a PhD in Psychology and Behavioural Genetics from Goldsmiths, University of London. She has taught across different branches of the University of London and is now an associate professor in the Department of General and Pedagogical Psychology at Tomsk State University. Her research is centred on learning abilities and disabilities and encompasses the study of genetic and environmental effects on the numerical ability of "number sense". She actively collaborates with different twin studies in the UK, US, Russia, and Canada.

Adelina Valiullina is a researcher at the Institute of Biochemistry and Genetics, Ufa Scientific Center of the Russian Academy of Sciences. Her research interests include genetic predisposition to suicidal and aggressive behaviour. The author of more than 20 publications in molecular, medical and psychiatric genetics, she is a board member of Vavilov's Society of Geneticists and Breeders and of the Russian Society of Medical Geneticists.

Aigul Zainullina is a senior researcher at the Institute of Biochemistry and Genetics, Ufa Scientific Center of the Russian Academy of Sciences. Her field of research includes genetic predisposition to suicidal and aggressive behaviour and schizophrenia. The author of more than 40 publications in molecular, medical and psychiatric genetics, she is a board member of Vavilov's Society of Geneticists and Breeders and of the Russian Society of Medical Geneticists.

1
How Genetics Can Help Education

Yulia Kovas, Tatiana Tikhomirova, Fatos Selita, Maria G. Tosto, and Sergey Malykh

The vast individual differences observed in all educationally relevant psychological traits, such as academic achievement and cognitive, motivational and emotional characteristics, develop through complex processes of gene–environment co-action. Research suggests that educational environment interacts with unique genetic profiles of students, leading to great variability among students (see Kovas, Haworth, Dale, & Plomin, 2007).

One of the main aims of most educational systems is to raise the level of achievement of all students. Behavioural genetic investigations in education provide important insights into individual differences in academic ability and achievement. Better understanding of the aetiology of individual differences can form a foundation for developing personalized educational methods and technologies that can help improve the performance of all children.

Educational research that attempts to explain individual differences in learning abilities, often does not take into account the role of genetic factors. They are often ignored, even when complex multilevel models are applied to development, including multiple approaches and theoretical constructs (see examples in Tommerdahl, 2010). One explanation for this is that genetic research is associated with the mistaken belief that genetic information may lead to selection and increase social inequality (see examples in Pinker, 2002; Malykh, Tikhomirova, & Kovas, 2012).

Moreover, most studies in education are dedicated to the development of universal methods of education for all students, a so called "one-size-fits-all" approach. This is paradoxical, because teachers and other educational specialists know better than anyone that each student has a unique cognitive and emotional profile. Research into the effectiveness of various educational programmes and methods has shown that, irrespective of which curriculum is followed in a particular school, teachers in every class have to adapt the curriculum to the needs of individual students (Krasa & Shunkwiler, 2009). In order to increase students' learning success, teachers are expected to take into account the profiles of individual students, such as aspects of their cognitive and metacognitive functioning. For example, in the USA, the principles and standards of school mathematics curricula, set by the National Council of Teachers of Mathematics (2008), demand that mathematics teachers possess content knowledge, pedagogical methods and the ability to identify strong and weak features of cognitive development of each student, as well as each student's interests and motivation (Krasa & Shunkwiler, 2009).

Despite this, research that aims for development of personalized methods of teaching of academic disciplines, such as mathematics, is still in its infancy. In the USA, where there is no National curriculum, schools try to select the best programmes for their students; but, with limited understanding of the aetiology of individual differences in academic achievement, such attempts to choose and apply educational programmes are in "a state of chaos" (Krasa & Shunkwiler, 2009, p. 185). In countries with more centralized curricula, such as the UK and Russia, "chaos" is in the implementation of these centralized systems, because one programme cannot be appropriate for all students. In 2008, the US National Mathematics Advisory Panel analysed results from a large number of studies in education and concluded that no unified, ideal method, suitable for all students, exists. The results showed that the success of the educational process depends on many interacting factors, such as ability and motivation of students, personal and professional qualities of teachers, educational methods and curricula (Krasa & Shunkwiler, 2009). Therefore, in order to select the most effective methods for a specific student in a specific subject and in specific learning conditions, better understanding of the aetiology of individual differences is needed.

This chapter presents potential benefits of behavioural genetic approaches to the study of cognitive, motivational and emotional factors underlying individual differences in academic achievement in different areas of learning. In a recently published volume "Neuroscience in education. The good, the bad and the ugly." the authors characterize the potential contribution of genetics to education as "good"; the general public's attitude to genetics in education as "bad"; and the slow progress in identification of specific genetic factors involved in academic achievement as "ugly" (Kovas & Plomin, 2012). The present chapter demonstrates the "good": multiple insights from behavioural genetic investigations into the aetiology of individual differences in learning and the potential of these findings for developing effective educational policies. These insights call for the change of the "bad" attitude towards behavioural genetics, by showing that all claims that genetic effects are unchangeable and deterministic are myths. The chapter also presents the results of molecular genetic investigations that attest to the gradual progress in the identification of genetic factors involved in educational processes.

Foundations of behavioural genetic research in education

Behavioural genetics is an interdisciplinary science that uses knowledge and methods of different fields, including psychology, molecular genetics and neuroscience. Behavioural genetics investigates gene–environment interplay and how it affects individual differences in psychological traits. Quantitative genetic investigations are based on comparisons between different types of family members, including twins, adoptive relatives and other types of relatives (for more detail see Plomin, DeFries, Knopik, & Neiderhiser, 2012). Molecular genetics aims to identify specific DNA markers that are involved in individual differences.

Quantitative genetic approach

The twin method is the most commonly used quantitative genetic approach that involves comparisons between monozygotic (MZ, identical) and dizygotic (DZ, fraternal) twins (see Box A.1 for details of the twin method). Twins can be considered an ideal natural experiment as MZ twins are 100% genetically similar, whereas DZ twins share on average only 50% of the segregating (changeable) DNA.

As twins grow up together they objectively share many aspects of their environment, including their parents, socio-economic conditions and often schools, classrooms, teachers and peers (Plomin et al., 2012). The twin method allows for estimation of the relative contribution of two types of environment: shared (common) environment that contributes to the similarity between the two twins in a pair on a particular trait; and non-shared (unique) environment that contributes to the dissimilarity between the two twins. The question of why children in the same family are very different from each other has been extensively researched in behavioural genetics (e.g., Plomin & Daniels, 1987; Plomin, Asbury, & Dunn, 2001). One of the findings is that objectively shared factors more often than not end up as non-shared effects. For example, socio-economic status of the family, objectively shared by the two twins, may actually lead to the differences between the twins if they perceive or respond to this status differently (Plomin, Asbury, & Dunn, 2001). A low socio-economic status may motivate one child to strive for achievement, but lead to lower motivation and achievement in another child. Such subjective reactions may form under the influence of many factors, including genetics.

Beyond the role of environment, the twin method allows for estimation of heritability – the role of genetic factors in individual differences in a particular trait in a specific population. Heritability is estimated as double the difference between MZ and DZ twin correlations for a particular trait (see Box A.1). The role of shared environment is estimated as the difference between MZ correlation and heritability. In other words, it estimates to what extent similarity between MZ twins is greater than would be expected from their genetic relatedness. In addition to comparing twin correlations, modern twin research uses structural model-fitting to provide more precise parameters (Neale, 1997; Neale & Maes, 2003; See Box A.1).

Twin registries represent a unique resource for studying genetic and environmental influences on complex psychological traits in the context of development. Special twin registries exist in many countries, including those that involve thousands of participants and those that follow the same participants for many years. In many scientific institutions of Europe, America and Asia, twin registries support interdisciplinary investigations into gene–environment interplay and its role in the variation in psychological traits, including cognitive characteristics and learning abilities and disabilities. Here we provide several examples of large-scale twin registries that have contributed to the study of psychological processes important for learning and education (also see an excellent review of twin research in Polderman et al., 2015 and a web resource Meta-Analysis of twin correlations and heritability http://match.ctglab.nl/#/home).

Twins Early Development Study (TEDS), UK. TEDS is one of the most impressive examples of a developmental twin study – a representative longitudinal study of more than 10000 twin pairs in the UK followed for 20 years, from birth to date. The main aim of this project is the study of different aspects of children's psychological development, such as cognitive abilities, behaviour, and learning abilities and disabilities (Haworth et al., 2013).

The Quebec Newborn Twin Study (QNTS), Canada. This study is another large-scale longitudinal cohort project that has followed 662 pairs of twins, born between 1995 and 1998 in Montréal (and greater Montréal area). The aims of this twin project are: the study of individual differences in cognitive, behavioural and socio-emotional aspects of development; identification of early biosocial factors and understanding their role in the further socio-emotional adaptation to school education (Boivin et al., 2013).

The Danish Twin Registry (DTR), Denmark. This twin registry is the oldest in the world and includes 86398 twin pairs, born in Denmark between 1870 and 1996. The project studies cognitive development, aging and the aetiology of various disorders and diseases (Skytthe et al., 2013).

The German Twin Study on Cognitive Ability, Self-Reported Motivation, and School Achievement (CoSMoS), Germany. This longitudinal project follows 408 pairs of twins, investigating the nature of the links between cognitive abilities, motivation and school achievement (Hahn et al., 2013).

The Young Netherlands Twin Register (YNTR), Netherlands. This twin registry includes more than 70000 pairs of twins born in the Netherlands since 1985. The project investigates cognitive development, neurophysiological processes, physical health, learning abilities and school achievement (van Beijsterveldt et al., 2013).

The South Korean Twin Registry (SKTR), South Korea. The registry was created in 2001 and today includes approximately 10000 pairs of twins of up to

30 years of age. The research focuses on gene–environment interplay and its role in variation in psychological health, personality and cognitive abilities (Hur et al., 2013).

The Colorado Twin Registry (CTR), USA. This project includes 17136 pairs of twins, born in Colorado since 1968. The registry combines four twin samples: (1) general twin sample; (2) newborn twin sample; (3) longitudinal twin sample; and (4) early reading development study sample. The project investigates early cognitive development, early reading ability, cognitive processes and antisocial behaviour (Rhea et al., 2013).

Russian School Twin Register (RSTR), Russia. The work on the establishment of RSTR began in Russia in 2011 and the register is continuously growing (http://www.protwins.ru/). The aim of the project is to study the role of co-action between genetic and environmental factors in the development of individual differences in achievement in different academic disciplines; and in cognitive, emotional and motivational characteristics of learners (Kovas et al., 2013; Malykh, Tikhomirova, & Kovas, 2012).

These and other twin registries are an invaluable source of theoretical and practical knowledge for teachers, psychologists, parents and learners themselves. Collaborative studies on the basis of these registries form a promising new research direction – cross-cultural behavioural genetics. The use of comparable research methods and measures in twin samples from different countries allows for studying the role of different aspects of culture in the aetiology of individual differences in educationally relevant traits (Kovas, Tikhomirova, & Malykh, 2011).

Molecular genetic approach

In recent years, significant progress has been achieved in our understanding of the complex processes through which genetic factors contribute to individual differences. In this chapter we provide a brief description of how genetic variability is related to observed individual differences among people. Later in this book we provide more detailed information about molecular genetic foundations of individual differences (see Chapter 5). Here we also present several examples from molecular genetic research that demonstrate how incorrect are the myths about genetic effects as deterministic and static.

Every person has a unique genetic profile which contributes to the formation of his or her individual-specific psychological profile. A person's unique genetic profile includes a unique DNA sequence, as well as a unique pattern of genetic expression and, consequently, a unique pattern of gene–environment co-action.

The human genome consists of 3 billion nucleotide base pairs, with only approximately 2% of the sequence constituting genes. The size of the genome, the number of genes and the sequence of base pairs are practically identical for all people, with less than 1% of the sequence varying among people. However, even this small amount of variable DNA means that there are many regions in the genome that differ across people. For example, approximately one in every thousand letters in the genomic "text" differs by one nucleotide (Single Nucleotide Polymorphism, SNP); and if rare polymorphisms are taken into account, one in every hundred letters differs across people. In addition, other types of variation occur in the genome, such as substitutions, insertions and deletions of genomic "text" (Plomin et al., 2012). This DNA variation may contribute to the observed individual differences across people.

Any of the millions of polymorphisms (variants) in the human genome may contribute to individual differences in a particular human trait. In addition, every polymorphism may contribute to a large number of psychological traits. This fact complicates the study of relations between genes and observed psychological traits. However, modern technologies make this research possible. Today, many research centres conduct molecular genetic investigations aimed at understanding the role of such genomic variation in learning abilities (Plomin et al., 2012).

In the future, the results of this research will aid implementation of early diagnosis of learning disabilities – because the unique DNA sequence of an individual does not change under the influence of development, behaviour or experience (of course, gene expression changes, but that is a different matter). Identification of relevant genes might help with finding the most suitable teaching and learning approaches for children's specific cognitive and motivational profiles. However, progress with genetic identification has turned out to be slower than initially expected, since, as widely accepted today, learning abilities and disabilities are influenced by multiple genetic variants of small effects (a phenomenon called polygenicity).

Finding answers to the questions "How many genetic variants are involved in each trait?" and "How small are their effects?" requires significant research efforts as small effects are very difficult to detect reliably. If learning disabilities were influenced by just one genetic variant, it would be easy to find a place on a chromosome, and then its precise genomic address. Indeed, specific genetic mutations have already been identified for hundreds of monogenic disorders that are usually extremely debilitating and very rare, occurring in 1 in 10000 individuals and not in 1 in 10 or 20, as is the case for most learning disabilities. Symptoms of more than 250 monogenic disorders include cognitive difficulties (Inlow & Restifo, 2004). For example, Huntington's disease is determined by a mutation in just one gene, irrespective of other genetic or environmental effects. It is likely that other monogenic disorders associated with cognitive impairments will be identified (Raymond & Tarpey, 2006). However, all of the currently known monogenic disorders, as well as many chromosomal anomalies, explain cognitive deficits in less than 1% of children who suffer from learning disabilities (Plomin et al., 2012). Since prevalence of learning disabilities is 5–10%, the majority of such cases cannot be explained by known single gene or chromosomal mutations.

In theory, it is possible that other, as yet unidentified single gene mutations and chromosomal anomalies will explain some of the observed learning difficulties. However, as previously mentioned, most behavioural geneticists have reached the conclusion that common medical and behavioural disabilities result from a large number of genetic factors of small effect. These factors are referred to as Quantitative Trait Loci (QTLs). If a trait is influenced by several genetic factors, it is quantitatively distributed, irrespective of where on this continuum we place the threshold for extreme values (e.g., threshold for disability) (Plomin, Haworth, & Davis, 2009). Therefore, disability can be viewed as a low (or high) end of the continuum, and QTLs have probabilistic rather than deterministic effects.

Identifying genetic factors linked with quantitative traits is carried out with the use of DNA markers, such as single nucleotide polymorphism (SNP) – a change of just one nucleotide in the DNA sequence. If one allele (e.g., one form of a particular SNP) increases a particular cognitive ability, and another allele decreases this ability, the frequency of the two alleles is likely to be different in high and low ability groups. Although most SNPs have very small effects on phenotypes, tens and hundreds of individual SNPs can be aggregated into SNP composites (polygenic index), which might be used in the future to determine a degree of genetic risk for developing a particular disorder, such as dyslexia, and to plan the most suitable prevention or intervention programme.

One of the latest developments in molecular genetics is the Genome-Wide Association Study (GWAS) approach, in which hundreds of thousands of DNA markers are tested simultaneously for association with a particular trait (Hirschhorn & Daly, 2005). The use of this method with large samples (thousands of participants) allows for identification of multiple DNA markers of small effect. By 2010, GWAS had already identified 779 significant associations between common genetic variants (polymorphisms) and 148 traits (Hindorff et al., 2010) and, more recently, even more polymorphisms have been identified (see http://www.genome.gov/gwastudies/ for a continuously updated catalogue). Since many identified associations have not been replicated in follow up studies (McCarthy et al., 2008), any newly established association requires replication on independent samples. Inconsistencies in results of different studies may stem from both false-positive associations and objective factors that mask associations, such as lack of statistical power, age or ethnic differences between samples, and the use of different methods of phenotypic assessment (Manolio et al., 2009).

Genome-Wide Association Study became possible with the development of DNA microarrays (chips) which contain multiple segments of single-stranded polynucleotide sequence, each containing a variable locus. One DNA chip can hold millions of different segments, which makes genotyping relatively inexpensive and fast (Plomin & Schalkwyk, 2007). This method is necessary because hundreds, or even thousands, of DNA markers are probably linked with different aspects of learning abilities and disabilities. Contemporary technologies have already allowed for the creation of specialized DNA chips that can be used to genotype individuals for all known associations with specific traits (e.g., cardiovascular disease) relatively inexpensively and quickly. This opens the possibility of creating specialized DNA chips for educationally relevant traits, such as "cognitive" chip, "dyslexia" chip, etc.

Establishing the sequence of all 3 billion base pairs of an individual is becoming progressively cheaper (currently, the cost of sequencing a genome is approximately $1000; Wetterstrand, 2015, www.genome.gov/sequencing-costs). In the near future, it will be possible to sequence the whole genome of all people. This development may lead to unprecedented opportunities for scientific discoveries and for applying these discoveries for the benefit of all people.

Gene–environment co-action

The influence of genetic and environmental factors on psychological development is realized through complex processes of their co-action. Two main types of gene–environment co-action have been identified: gene–environment interaction and gene–environment correlation (co-variation).

Gene–environment interaction is a phenomenon reflecting the joint effects of genes and environment. The essence of this phenomenon is that phenotypic differences might not emerge among carriers of different genotypes in the absence of specific environmental conditions. For example, genotypic differences among people may express only in the presence of certain socio-cultural (educational, family, etc.) conditions. And vice versa, a particular environmental condition may have an effect on a particular behaviour only in the presence of certain genetic variants. This type of gene–environment co-action may partially explain the widely different academic achievement results of children in the same class (Tikhomirova & Kovas, 2013; Kovas, Tikhomirova, & Malykh, 2011; Malykh, Egorova, & Meshkova, 1998). One recent study in a sample of 2000 school children investigated whether the link between ten specific genetic polymorphisms and mathematical achievement depends on home and school educational conditions. The results suggested that the 10 polymorphisms were more strongly associated with mathematical achievement in less favourable home environments (characterized by a high level of chaos in the home and negative parenting; Docherty et al., 2011).

Gene–environment correlation (rGE) is the phenomenon reflecting an uneven distribution of genotypes in different environments. The gene–environment correlation reflects the extent to which people's environment is linked (correlated to) their genotype (see Glossary for information on rGE). Such processes might be at play in education, for example, when gifted children are selected for enriched educational environment (e.g., such as specialized mathematical education), and this environment, in turn, facilitates further development. In such circumstances, a child can also actively select and organize suitable environmental conditions (Kovas, Tikhomirova, & Malykh, 2011). Three types of gene–environment (rGE) correlation have been described (Plomin, DeFries, & Loehlin, 1977):

passive rGE – members of the same family partially share both genetic factors and environmental conditions; for example, mathematically able parents may provide a more mathematically stimulating environment to their children;

evocative (reactive) rGE – people's (partly heritable) behaviour evokes a particular response from their environment; for example, a pupil who exhibits aggressive behaviour (partially under the influence of genetic factors) evokes a particular attitude from teachers and classmates;

active rGE – people actively seek or create environments that are most suitable to their genetic profiles; for example mathematically gifted children may actively enrol in mathematical competitions and seek mathematically stimulating environments, such as suitable peer groups or online resources.

The role of each type of gene–environment correlation may be stronger (or dominant) at different times in the course of development. For example, it has been shown that the first type of correlation is dominant in early childhood (Malykh, 2009). Further chapters in this book provide more information and examples of different types of gene–environment co-action (e.g., Chapters 2 and 3).

To summarize, the research has already led to some important conclusions about genetic and environmental effects. It is clear that neither genetic nor environmental effects are deterministic. The same genes can have different effects, depending on environmental conditions. For example, the same trait (e.g., intelligence) may be under a stronger genetic influence in one socio-cultural environment, whereas in another environment it may be under a stronger influence of environmental factors (e.g., access to quality education). In turn, genetic effects can mediate environmental effects, for example, when a particular environment has an effect only on carriers of a particular genotype. It follows that complex psychological traits cannot be explained by any single factor.

It is also clear that genetic effects are not static. Research has shown that the same psychological traits may be under stronger or weaker genetic influence at different stages of development; and that some genetic factors become active only at a certain age (Kovas et al., 2007). For example, individual differences in intelligence in early childhood largely stem from environmental factors; on the contrary, in adulthood, individual differences largely stem from genetic factors (Haworth et al., 2010; Kovas et al., 2013; Plomin & Deary, 2014; Davies et al., 2011).

Interdisciplinary approach: genetics, neuroscience and education

Most investigations in educational neuroscience (see Box A.2 for the description of the neuroscience methods) focus on group comparisons and aim to identify brain mechanisms involved in particular psychological functions. For example, research into neural underpinnings of mathematical ability suggests that certain types of mathematical tasks (e.g., operations with symbolic and non-symbolic numerosities), are correlated with enhanced activation in the intraparietal sulcus and superior frontal gyrus of the brain (e.g., Stanescu-Cosson et al., 2000; Venkatraman, Ansari, & Chee, 2005; Dehaene, Piazza, Pinel, & Cohen, 2003; Piazza, Pinel, Le Bihan, & Dehaene, 2007; Cantlon et al., 2006). These regions have also been linked with differences between high and low mathematical ability groups. Therefore, it has been suggested that disability may result from a particular lesion/lesions or dysfunction in these areas. However, interpretation of these results in terms of the direction of causality requires caution. Indeed, any "abnormal" brain activity in a particular area may be the result rather than the cause of low mathematical ability. Unfortunately, incorrect conclusions from neuroscience can misinform educational programmes. For example, in educational practice in the USA, educational methods based on information about brain processes ("brain-based learning") were initially enthusiastically received by all participants of educational processes – from students at schools and universities to their parents and teachers (Alferink & Farmer-Dougan, 2010). However, recently, the US National Mathematics Advisory Panel has warned that attempts to use conclusions from neuroscience research with the aim of raising student mathematical achievement are premature. It is necessary to evaluate the effectiveness of educational programmes that, according to the authors, are based on the latest neuroscience knowledge (Krasa & Shunkwiler, 2009). Moreover, it is necessary to study how different educational methods and programmes in different disciplines are associated with individual differences among people in their neuronal processing.

Only recently, educational neuroscience has begun to address the neural underpinnings of the observed individual differences (rather than group differences) in educationally relevant traits (see Box A.3 for the discussion of individual vs. group differences). Moreover, a shift is taking place from studying separate brain areas towards studying the brain as a unified functional system. An example of this approach is the network analysis of structural and functional connectivity that aims to characterize the organization of brain networks – path lengths, clustering, hierarchy and regional inter-connectivity. Such methods require large numbers of participants, advanced technologies and expertise – making them extremely costly. Nevertheless, the future of neuroscience requires this shift from group analyses of regions of interest to identifying individual-specific neuronal networks and processes (Bressler & Menon, 2010).

The complex genetic and environmental aetiology of individual differences in educationally relevant traits suggests a similarly complex aetiology at the level of the neuronal processes (Kovas & Plomin, 2006). Identifying genetic polymorphisms involved in individual differences in learning will lead to the development of educational neurogenomics. Educational neurogenomics, in turn, may lead to developments in early diagnostics and identification of individual developmental profiles and will help with selecting the most suitable educational trajectories for all learners (e.g., Gabrieli, 2009). In addition, the use of genetic data will help establish the causal links between the brain processes and behaviour. Therefore, integration of educational, genetic and neuroscientific investigations can lead to faster progress in optimizing education.

However, for societies to reap the benefits of neurogenomics, they need to be prepared, and the relevant stakeholders need to be ready to accept these findings. For this, researchers need to disseminate neurogenomic findings in a way comprehensible to the society at large. This work would include scientists taking great care when explaining findings to journalists in order to avoid misinterpretation of findings. In addition, relevant materials need to be distributed to all teachers, and seminars on research findings should be held in schools. Most importantly, collaboration between scientists, teachers and policy makers is essential for faster progress in implementing neurogenomic findings towards improving educational achievement for all children. As an immediate step, policy makers should include training in the field of educational neurogenomics in teachers' professional development. Such training will be particularly important once most children's DNA is sequenced at birth and the impact of gene–environment co-action on educational achievement has become apparent. Furthermore, clear protection, and guidance on the use of genetic data need be in place prior to these data becoming widely available, which is very likely to be in the next few years. With the right steps taken by policy makers at this stage, in the near future all children, regardless of socio-demographic characteristics, will benefit enormously from educational neurogenomic findings.

Insights from behavioural genetic research in education

A wealth of behavioural genetic investigations have been conducted into educationally relevant traits, such as learning abilities and academic motivation and achievement. In such research, learning abilities are measured through specific cognitive tests. Academic achievement in different domains (e.g., reading, language and mathematical ability) is measured through teacher reports and grades, as well as test results (Kovas et al., 2007; Malykh, Tikhomirova, & Kovas, 2012).

Aetiology of individual differences in learning, motivation and achievement

The relative contribution of genetic and environmental factors to individual differences in academic achievement has been studied in several recent behavioural genetic investigations. For example, one study of academic achievement of 6–12 year old children found that heritability and shared environment each explained approximately 40% of the observed variance (Thompson, Detterman, & Plomin, 1991). Similar estimates of genetic and shared environmental factors were found in other samples from the USA (Loehlin & Nichols, 1976), Germany (Bartels, Rietveld, van Baal, & Boomsma, 2002) and Australia (Wainwright, Wright, Luciano, Geffen, & Martin, 2005).

Several behavioural genetic studies explored the aetiology of individual differences in specific learning abilities and academic achievement in specific school subjects. For example, for reading ability, approximately 40% of the variance was explained by genetic factors, with shared environment explaining 25–45% (Light, DeFries, & Olson, 1998; Stromswold, 2001). For mathematics, genetic effects explained more variance – approximately 70%, with only modest contributions of shared environment (Light, DeFries, & Olson, 1998).

In TEDS, the following estimates were obtained of the contribution of genetic and shared environmental factors, respectively, to variation in academic achievement: 64% and 15% for language; 68% and 10% for mathematics (Kovas et al., 2007). In TEDS, data on general intelligence and school achievement (measured through teacher reports and tests) were collected. Interestingly, in primary school the contribution of genetic factors was stronger for academic achievement than for general intelligence (Kovas et al., 2007, 2013).

The relative contribution of genetic and environmental factors can differ across populations with different socio-demographic characteristics, such as age or educational opportunity. For example, a unified National Curriculum may lead to a situation where individual differences in academic achievement are largely explained by genetic factors. This could be a potential explanation for the higher heritability estimates for learning abilities in the UK than, for example, in the USA. The system of education in the UK is strictly regulated, it is based on the unified National Curriculum, and is therefore relatively homogeneous. The system of education in the USA is one of the most decentralized in the world. This heterogeneity of educational environment may lead to a higher contribution of shared environment to individual differences in learning (Kovas et al., 2007).

Education is the best example of environmental influence on psychological development. School curricula, qualifications of teachers, educational methods, class set-ups, and schools' educational and cultural ethos are all environmental factors that may influence students' academic achievement in different disciplines. Recent behavioural genetic investigations have focused on aspects of education such as teacher and classroom effect on individual differences in motivation and achievement.

For example, several studies explored whether teacher effect is a class-wide or individual-specific effect. A class-wide positive teacher effect implies that all children in a particular class show better performance than they would show if they were taught by a different teacher (Kovas, Haworth, Dale, & Plomin, 2007; Byrne et al., 2010). An individual-specific teacher effect implies that the same teacher, potentially, can increase and decrease performance of different children in the same class.

Teacher effect was assessed in TEDS, using the classical twin method, taking the teacher/class factor into account. In the UK, parents of twins can choose whether their children will attend the same or different class (or even school). Analysis of the TEDS data showed that being taught by different teachers (in different classes) does not increase dissimilarities between twins in any of the academic characteristics, such as reading, science and mathematical ability, general cognitive ability and being interested in particular subjects (e.g., Kovas et al., 2007; Coventry et al., 2009).

Behavioural genetic research often leads to unexpected results. For example, it has been shown that individual differences in mathematical self-evaluation (self-perceived ability) and interest (motivation) result from genetic and non-shared environmental factors, rather than from shared environment (Luo, Kovas, Haworth, & Plomin, 2011). This result is not specific to mathematics. Research in TEDS showed that individual differences in motivation and self-perceived ability in different school subjects were explained by genetic and non-shared environmental factors. Interestingly, the link of motivation and self-perceived ability with academic achievement was largely explained by genetic factors (e.g., Greven, Harlaar, Kovas, Chamorro-Premuzic, & Plomin, 2009).

A recent large-scale investigation provided further insights into the aetiology of individual differences in academic motivation (Kovas et al., 2015). Enjoyment and self-perceived ability in different school subjects were measured in a sample of 13000 twins, aged 9–16 years, from six countries: UK, Canada, Japan, Germany, USA and Russia. In addition, in the UK sample, information was available on whether twins attended the same or different classes at ages 7 and 9. The results showed striking consistency in the aetiology of individual differences in enjoyment and self-evaluation across all school subjects, ages and countries. All measures of enjoyment and self-evaluations were moderately heritable. Non-shared environment explained most of the variation in these constructs. In other words, the results suggest that similarity among family members in academic motivation is largely explained by their genetic similarity, whereas differences among family members are largely explained by non-shared environmental factors.

Moreover, no class-wide teacher/classroom effect was found in this study: differences between co-twins in the level of enjoyment and self-evaluation of school subjects were not greater for twins attending different classes than those between twins studying in the same class. These results suggest that, contrary to common opinion, genetic effects on enjoyment of school subjects and self-evaluation of ability in these subjects are not weaker than genetic effects on cognitive ability. The results also emphasize the role of non-shared (rather than shared) environmental factors in the development of academic motivation (Kovas et al., 2015).

These results require careful consideration. An erroneous conclusion would be that the role of school and teacher in the development of academic motivation is minimal. However, the absence of shared environmental influence suggests that the role of parents, teachers, class and school should be viewed as non-shared environment. On one hand, parents, teachers, siblings and peers may treat children in the same family/class differently (Harris & Morgan, 1991; Babad, 1993); on the other hand, children may perceive their parents, teachers, peers and school differently (Zhou, Lam, & Chan, 2012). Further investigations are needed in order to identify specific non-shared environments involved in individual differences in academic motivation and achievement.

Molecular genetic investigations have only just begun, but several genetic variants have already been identified as potentially important for educationally relevant traits. For example, one recent investigation identified a link between a SNP on chromosome 3q29 and a General Certificate of Secondary Education (GCSE) exam, taken by school children at the age of 16 in the UK (Baron-Cohen et al., 2014). A different study, based on the TEDS sample and using DNA chips, found an association between 10 polymorphisms and mathematical performance at age 10 (Docherty et al., 2010).

Aetiology of inter-relations among different academic subjects in the context of development

Behavioural genetic investigations go beyond estimating the relative contribution of genetic and environmental factors to specific learning abilities to studying the aetiology of inter-relations among different learning abilities in the context of development. Multivariate genetic analysis (See Box A.1) can determine the extent to which genetic and environmental factors that contribute to, for example, reading ability, also influence other abilities, such as language or mathematics. Genetic and environmental correlations, estimated in this analysis reflect the extent to which the same genetic and environmental factors affect two traits.

Research found high genetic correlations for educationally relevant traits: 0.67–1.00 between reading and learning abilities (Stromswold, 2001); 0.47–0.98 between reading and mathematics (Kovas et al., 2007); 0.59–0.98 between language and mathematics (Plomin & Kovas, 2005). One recent study investigated genetic and environmental influences on the links between compulsory examinations in different academic subjects taken by UK pupils at age 16. The results showed substantial genetic correlations across all subjects, even after controlling for IQ (Rimfeld, Kovas, Dale, & Plomin, 2015).

Such strong genetic correlations suggest that, to a large extent, the same set of "generalist" genes affects a wide range of abilities and academic performance (the generalist genes hypothesis; Plomin & Kovas, 2005; Kovas & Plomin, 2006). In addition, genetic correlations among learning abilities are higher than between learning abilities and general intelligence (*g*). For example, one study showed that 1/3 of genetic effects on mathematical ability were also involved in individual differences in reading and *g*; 1/3 were common for reading and mathematics but not shared with *g*; and the remaining 1/3 were specific for mathematics (Kovas et al., 2007). Generalist genes are also likely to be involved in comorbidity – the presence of two or more cognitive/behavioural disabilities in one person (Butterworth & Kovas, 2013).

Similar to genetic correlations, shared environmental correlations are also usually strong, such as a correlation of 0.74 between reading and mathematical abilities (Kovas, Harlaar, Petrill, & Plomin, 2005), 0.86 among different aspects of mathematical ability (Kovas et al., 2007). In other words, shared environmental factors are also largely generalist, influencing a wide range of cognitive abilities and achievement in different academic subjects. This is not surprising as shared environment, possibly including socio-economic status and different aspects of family and educational environment, are likely to influence learning in general, rather than specific aspects of learning. Further large-scale cross-cultural research is needed to achieve better understanding of these effects.

Non-shared environmental factors are shown to be largely specific to each ability. In other words, non-shared environmental factors might contribute to the differentiation at the level of cognitive abilities and achievement in different school subjects in one child. For example, a child's achievement level can be lower than his or her cognitive ability; or the level of mathematical achievement may significantly exceed achievement in other subjects (Kovas & Plomin, 2007).

Multivariate analysis also allows an estimate of the extent to which genetic and environmental factors explain the observed (phenotypic) co-variation between two traits (biviariate heritability and environmentalities; see Box A.1). In TEDS, genetic factors explain most of the co-variation among different learning abilities; shared environment explains approximately

16% of the co-variation on average; and contribution of non-shared environment is even smaller (Kovas et al., 2007). These results suggest that non-shared environmental factors may play an important role in differentiation of abilities in a child – which should be taken into account in educational research and practice.

Behavioural genetic investigations also allow for estimation of the relative contribution of genetic and environmental factors to individual differences in a particular trait at different stages of development. Genetic and environmental effects on learning abilities and academic achievement may differ across the lifespan for multiple reasons. First, the content of a subject area may change conceptually (e.g., mathematics that children study at age 7 differs substantially from mathematics studied by 16 year-olds). Second, different cognitive and social processes may be involved in learning ability at different ages (e.g., a change in interests and social influences in adolescence). These cognitive and social processes may be linked to different genetic and environmental factors. The actual strength of the relative contribution of genetic and environmental factors may be similar or different across development. In order to systematically evaluate genetic and environmental effects in the course of development, more large-scale genetically informative longitudinal investigations are needed.

One consistently replicated example of the dynamic processes of genetic and environmental effects is a gradual increase in heritability of intelligence – from approximately 20% in infancy, to 40% in childhood reaching 60% (or more) in adulthood (Haworth et al., 2010). Moreover, a study of separated monozygotic twins showed that heritability of intelligence can be as high as 80% in adulthood (Plomin, 2003). This research has led to the formulation of the "amplification theory", according to which genetic influences increase in the course of ontogenesis (e.g., Plomin et al., 2012; Plomin, 2003). This dynamic nature of heritability across development is interesting because it contradicts the commonly expected increase of environmental influences. The exact reasons for the increase in heritability are not known, but it is likely that gene–environment correlation is a contributing factor: people actively form their environment, choosing and changing its elements in accordance with their genetically influenced individual characteristics (Plomin, 2003).

In TEDS, the stability of cognitive abilities across development is largely explained by genetic factors, whereas developmental changes are largely explained by non-shared environment (Kovas et al., 2007). For example, genetic correlations of 0.60–0.72 have been found for intelligence and learning abilities at different ages. These strong genetic correlations mean that, to a large extent, genetic factors that influence a particular ability at one age continue to influence this ability at other ages. Longitudinal correlations for non-shared environment are weaker (0.03–0.26). In other words, the dynamics of learning and achievement may be largely explained by non-shared environmental factors that contribute to dissimilarities between the children in the same family, and make the same child perform differently at different ages (Kovas et al., 2007).

Genetic links among different cognitive abilities, and among measures of a particular ability at different ages may stem from pleiotropy – a phenomenon by which one genetic factor affects multiple traits (Kovas & Plomin, 2006; Ligthart & Boomsma, 2012). For example, the same genes may contribute to the ability to estimate numerosity without counting (i.e., number sense) and calculation ability, possibly through affecting the same brain processes. Pleiotropic explanation for genetic effects implies that improving one cognitive ability or skill (e.g., number sense) does not necessarily lead to the improvement in another correlated trait (e.g., mathematical performance). Instead, any improvement in the ability targeted by a particular intervention may not generalize to the other (linked) ability, leading to a decreased correlation between the two traits.

Another possible explanation for genetic correlations is the mechanism of causality, by which one trait directly influences another at the cognitive, emotional or some other level (Ligthart & Boomsma, 2012). For example, a particular set of genes may contribute to individual differences in basic estimation ability. In turn, estimation ability may underlie the development of exact calculations and other mathematical skills. Therefore, genetic factors affecting estimation would be indirectly involved in mathematical ability.

Current understanding of the observed links between different abilities at the behavioural, cognitive, neurophysiological and gene–environment levels does not allow one to disentangle between pleiotropy and causality, and other possible mechanisms underlying such links. In the absence of good understanding of the causal processes, many assumptions about development require proper evaluation. For example, different basic skills are often viewed as a foundation for successful acquisition of more advanced skills and knowledge later in life. Educational programmes are developed to raise the level of future performance through increasing the level of early cognitive development and basic skills. Some educational systems, for instance, emphasize the importance of "school readiness". For mathematical ability, school readiness refers to acquisition of adequate levels of such basic skills as estimation (number sense), working memory, verbal ability, knowledge of numerals, symbolic expression of magnitudes and counting (Mazzocco, Feigenson, & Halberda, 2011; Krasa & Shunkwiler, 2009). However, much research is still needed to fully understand which early abilities causally underlie further development.

Molecular genetic investigations can contribute to this work through identifying specific genetic markers that are involved in the links between different traits. Only a few examples of such research into educationally relevant traits are currently available. For example, identification of several polymorphisms associated with mathematical ability at age 10, described earlier (Docherty et al., 2010), led to further analyses of the longitudinal multivariate data to test the generalist genes hypothesis (Plomin & Kovas, 2005). The 10 polymorphisms were tested for association with different mathematical phenotypes at different ages, using genotypic information on 5000 TEDS children. Consistent with the generalist genes hypothesis, the results showed that polymorphisms associated with the overall mathematical performance score at age 10, were also associated with different aspects of mathematical performance (e.g., calculation vs. non-numerical mathematical operations). The same investigation showed that the same polymorphisms were significantly (but more weakly) associated with mathematical performance at ages 7, 9 and 12. This finding suggests that the same factors contribute to mathematical achievement in the whole course of development, but new genetic effects may also be present at each age. Furthermore, the same set of polymorphisms was also significantly (and with similar strength) associated with reading ability and general cognitive ability at age 10 – which is also consistent with the generalist genes hypothesis (Docherty et al., 2010).

Once the generalist and specialist genes associated with educationally relevant traits are identified, behavioural genomic work will begin to specify the actual causal mechanisms through which each gene influences behaviour: from its expression under the influence of environment, to neurophysiological processes, and to the functioning at the level of behaviour.

Understanding these mechanisms will lead to effective interventions for early learning and behavioural problems. It will also enable identification of genetic and environmental factors that can be used as early warning systems and targets for prevention of early problems. This is particularly important because by the time the strengths and weaknesses in a child's cognitive development become apparent, the development might already be following an abnormal trajectory.

Conclusion

Genetically informative research into educational processes is a powerful tool to aid our attempts to raise the level of academic achievement of all students. Behavioural genetic research suggests that individual differences in educationally relevant traits form through complex processes of gene–environment co-action. The results, discussed in this chapter, demonstrate that genetic differences can interact with environmental factors, leading to situations where the same educational environment may lead to vastly different outcomes for different children. Genomic investigations have begun to identify specific genes and environments that are involved in specific educational processes (Baron-Cohen et al., 2014; Docherty et al., 2010; Meaburn et al., 2008). Research has also begun into the aetiology of the links between different learning abilities and into the causal mechanisms underlying these links.

In the future, we may be able to evaluate a person's DNA for specific genetic markers, associated with learning abilities and other educationally relevant traits. This information can be used to select the most suitable educational methods and compensatory strategies, as is already happening in personalized medicine (Collins, 2010). Therefore, the role of schools will shift towards developing and supporting individualized educational plans that take into account each student's strengths and weaknesses. Strong cooperation among scientists, practitioners and policy makers is needed to make this possible.

Quantitative and molecular genetic findings raise some ethical concerns, as they could potentially be used to increase inequality. However, knowledge itself does not lead to any social or political decisions. Moreover, both fair and unfair decisions can be made with or without scientific foundation. Understanding the aetiology of individual differences in learning ability better equips us to optimize the use of resources and to reduce unfairness in their distribution. In the context of humanistic values, genetic knowledge enables us to optimize education for all children, from high achieving to those with learning difficulties, from all socio-economic backgrounds. Genetic research can lead to a better understanding of individual differences. Instead of using simplistic explanations, for example that learning disabilities are a product of bad teaching or parenting, this research can help to establish complex causal mechanisms, enabling prediction and prevention of learning disabilities. Genomic investigations, informed by the needs of educational practitioners, will help avoid erroneous assumptions when developing educational policies desssigned to optimize education for all learners.

References

Alferink, L. A., & Farmer-Dougan, V. (2010). Brain-(not) based education: dangers of misunderstanding and misapplication of neuroscience research. *Exceptionality, 18*(1), 42–52.

Babad, E. (1993). Teachers' differential behavior. *Educational Psychology Review, 5*(4), 347–376.

Baron-Cohen, S., Murphy, L., Chakrabarti, B., Craig, I., Mallya, U., Lakatošová, S., ... & Warrier, V. (2014). A genome wide association study of mathematical ability reveals an association at chromosome 3q29, a locus associated with autism and learning difficulties: a preliminary study. *PloS One, 5*, e96374.

Bartels, M., Rietveld, M. J., Van Baal, G. C. M., & Boomsma, D. I. (2002). Heritability of educational achievement in 12-year-olds and the overlap with cognitive ability. *Twin Research, 5*(06), 544–553.

Boivin, M., Brendgen, M., Dionne, G., Dubois, L., Pérusse, D., Robaey, P., ... & Vitaro, F. (2013). The Quebec newborn twin study into adolescence: 15 years later. *Twin Research and Human Genetics, 16*(01), 64–69.

Bressler, S. L., & Menon, V. (2010). Large-scale brain networks in cognition: emerging methods and principles, *Trends in Cognitive Sciences, 14*(6), 277–290.

Butterworth, B., & Kovas, Y. (2013). Understanding neurocognitive developmental disorders can improve education for all. *Science, 340*(6130), 300–305.

Byrne, B., Coventry, W. L., Olson, R. K., Wadsworth, S. J., Samuelsson, S., Petrill, S. A., ... & Corley, R. (2010). "Teacher effects" in early literacy development: evidence from a study of twins. *Journal of educational psychology, 102*(1), 32.

Cantlon, J. F., Brannon, E. M., Carter, E. J., & Pelphrey, K. A. (2006). Functional imaging of numerical processing in adults and 4-y-old children. *PLOS Biology, 4*(5).

Collins, F. (2010). Has the revolution arrived? *Nature, 464*(7289), 674–675.

Coventry, W. L., Byrne, B., Coleman, M., Olson, R. K., Corley, R., Willcutt, E., & Samuelsson, S. (2009). Does classroom separation affect twins' reading ability in the early years of school? *Twin Research and Human Genetics, 12*(05), 455–461.

Davies, G., Tenesa, A., Payton, A., Yang, J., Harris, S. E., Liewald, D., ... & Deary, I. J. (2011). Genome-wide association studies establish that human intelligence is highly heritable and polygenic. *Molecular psychiatry, 16*(10), 996–1005.

Dehaene, S., Piazza, M., Pinel, P., & Cohen, L. (2003). Three parietal circuits for number processing. *Cognitive neuropsychology, 20*(3–6), 487–506.

Docherty, S. J., Davis, O. S. P., Kovas, Y., Meaburn, E. L., Dale, P. S., Petrill, S. A., ... & Plomin, R. (2010). A genome-wide association study identifies multiple loci associated with mathematics ability and disability. *Genes, Brain and Behavior, 9*(2), 234–247.

Docherty, S. J., Kovas, Y., Petrill, S. A., & Plomin, R. (2010). Generalist genes analysis of DNA markers associated with mathematical ability and disability reveals shared influence across ages and abilities. *BMC genetics, 11*(1), 61.

Docherty, S. J., Kovas, Y., & Plomin, R. (2011). Gene-environment interaction in the etiology of mathematical ability using SNP sets. *Behavior genetics, 41*(1), 141–154.

Gabrieli, J. D. (2009). Dyslexia: a new synergy between education and cognitive neuroscience. *Science, 325*(5938), 280–283.

Greven, C. U., Harlaar, N., Kovas, Y., Chamorro-Premuzic, T., & Plomin, R. (2009). More than just IQ school achievement is predicted by self-perceived abilities.– but for genetic rather than environmental reasons. *Psychological Science, 20*(6), 753–762.

Hahn, E., Gottschling, J., & Spinath, F. M. (2013). Current twin studies in Germany: report on CoSMoS, SOEP, and ChronoS. *Twin Research and Human Genetics, 16*(01), 173–178.

Harris, K. M., & Morgan, S. P. (1991). Fathers, sons, and daughters: differential paternal involvement in parenting. *Journal of Marriage and the Family,* 531–544.

Haworth, C. M. A., Wright, M. J., Luciano, M., Martin, N. G., De Geus, E. J. C., Van Beijsterveldt, C. E. M., ... & Plomin, R. (2010). The heritability of general cognitive ability increases linearly from childhood to young adulthood. *Molecular Psychiatry, 15*(11), 1112–1120.

Haworth, C., Davis, O. S., & Plomin, R. (2013). Twins Early Development Study (TEDS): a genetically sensitive investigation of cognitive and behavioral development from childhood to young adulthood. *Twin Research and Human Genetics, 16*(01), 117–125.

Hindorff, L. A., Junkins, H. A., Hall, P. N., Mehta, J. P., & Manolio, T. A. (2010). A catalog of published genome-wide association studies. *National Human Genome Research Institute.*

Hirschhorn, J. N., & Daly, M. J. (2005). Genome-wide association studies for common diseases and complex traits. *Nature Reviews Genetics, 6*(2), 95–108.

Hur, Y. M., Jeong, H. U., Chung, K. W., Shin, J. S., & Song, T. B. (2013). The South Korean twin registry: an update. *Twin Research and Human Genetics, 16*(01), 237–240.

Inlow, J. K., & Restifo, L. L. (2004). Molecular and comparative genetics of mental retardation. *Genetics, 166*(2), 835–881.

Kovas, Y., Galajinsky, E. V., Boivin, M., Harold, G. T., Jones, A., Lemelin, J. P., ... & Malykh, S. (2013). The Russian School Twin Registry (RSTR): project progress. *Twin Research and Human Genetics, 16*(01), 126–133.

Kovas, Y., Garon-Carrier, G., Boivin, M., Petrill, S. A., Plomin, R., Malykh, S. B., ... & Vitaro, F. (2015). Why children differ in motivation to learn: insights from over 13,000 twins from 6 countries. *Personality and Individual Differences, 80*, 51–63.

Kovas, Y., Harlaar, N., Petrill, S. A., & Plomin, R. (2005). "Generalist genes" and mathematics in 7-year-old twins. *Intelligence, 33*(5), 473–489.

Kovas, Y., Haworth, C. M., Dale, P. S., Plomin, R., Weinberg, R. A., Thomson, J. M., & Fischer, K. W. (2007). The genetic and environmental origins of learning abilities and disabilities in the early school years. *Monographs of the Society for Research in Child Development*, i–156.

Kovas, Y., & Plomin, R. (2006). Generalist genes: implications for the cognitive sciences. *Trends in cognitive sciences, 10*(5), 198–203.

Kovas, Y., & Plomin, R. (2007). Learning abilities and disabilities generalist genes, specialist environments. *Current Directions in Psychological Science, 16*(5), 284–288.

Kovas, Y., & Plomin, R. (2012). Genetics and genomics: good, bad and ugly. In S. Della Salla & M. Anderson (Eds), *Neuroscience in Education: The Good, The Bad and The Ugly* (pp. 155–173). Oxford University Press.

Kovas, Y., Tikhomirova, T. N., & Malykh, S. B. (2011). Problema stabil'nosti i izmenchivosti obshchih sposobnostej v psihogenetike [The problem of stability and variability of general abilities in behavioural genetics]. *Voprosy psihologii, 6*, 44–54.

Kovas, Y., Voronin, I., Kaydalov, A., Malykh, S. B., Dale, P. S., & Plomin, R. (2013). Literacy and numeracy are more heritable than intelligence in primary school. *Psychological science, 24*(10), 2048–2056.

Krasa, N., & Shunkwiler, S. (2009). *Number Sense and Number Nonsense: Understanding the Challenges of Learning Math*. Brookes Publishing Company.

Light, J. G., Defries, J. C., & Olson, R. K. (1998). Multivariate behavioral genetic analysis of achievement and cognitive measures in reading-disabled and control twin pairs. *Human Biology*, 215–237.

Ligthart, L., & Boomsma, D. I. (2012). Causes of comorbidity: pleiotropy or causality? Shared genetic and environmental influences on migraine and neuroticism. *Twin Research and Human Genetics, 15*(02), 158–165.

Loehlin, J. C., & Nichols, R. C. (1976). *Heredity, Environment, and Personality: A Study of 850 Sets of Twins*. University of Texas Press.

Luo, Y. L., Kovas, Y., Haworth, C. M., & Plomin, R. (2011). The etiology of mathematical self-evaluation and mathematicS achievement: understanding the relationship using a cross-lagged twin study from ages 9 to 12. *Learning and Individual Differences, 21*(6), 710–718.

Malykh, S. B. (2009). Problema razvitiya intellekta v psihogenetike [The problem of intelligence development in behaviour genetics]. *Kognitivnye issledovaniya: Problema razvitiya. Sbornik nauchnyh trudov, 3*, 287–302.

Malykh, S. B., Egorova, M. S., & Meshkova, T. A. (1998). Osnovy psihogenetiki [The basics of behavior genetics]. Moscow: Epidavr.

Malykh, S. B., Tikhomirova, T. N., & Kovas, Y. V. (2012). Individual'nye razlichiya v sposobnostyah k obucheniyu: vozmozhnosti i perspektivy psihogeneticheskih issledovanij [Individual differences in learning abilities: opportunities and perspectives of behaviour genetic studies]. *Voprosy obrazovaniya, 4*, 186–199.

Manolio, T. A., Collins, F. S., Cox, N. J., Goldstein, D. B., Hindorff, L. A., Hunter, D. J., ... & Visscher, P. M. (2009). Finding the missing heritability of complex diseases. *Nature, 461*(7265), 747–753.

Mazzocco, M. M., Feigenson, L., & Halberda, J. (2011). Preschoolers' precision of the approximate number system predicts later school mathematics performance. *PLoS one, 6*(9), e23749.

McCarthy, M. I., Abecasis, G. R., Cardon, L. R., Goldstein, D. B., Little, J., Ioannidis, J. P., & Hirschhorn, J. N. (2008). Genome-wide association studies for complex traits: consensus, uncertainty and challenges. *Nature Reviews Genetics, 9*(5), 356–369.

Meaburn, E. L., Harlaar, N., Craig, I. W., Schalkwyk, L. C., & Plomin, R. (2008). Quantitative trait locus association scan of early reading disability and ability using pooled DNA and 100K SNP microarrays in a sample of 5760 children. *Molecular psychiatry, 13*(7), 729–740.

Meta-Analysis of twin correlations and heritability; http://match.ctglab.nl/#/home

Neale, M. C. (1997). Mx: Statistical Modeling (Version 1.1) (4th ed.). Richmond (VA): Department of Psychiatry, University of Virginia.

Neale, M. C., & Maes, H. H. M. (2003). *Methodology for Genetic Studies of Twins and Families*. Dordrecht, NL: Kluwer Academic Publishers.

Piazza, M., Pinel, P., Le Bihan, D., & Dehaene, S. (2007). A magnitude code common to numerosities and number symbols in human intraparietal cortex. *Neuron, 53*(2), 293–305.

Pinker, S. (2002). *The Blank Slate: The Modern Denial of Human Nature*. New York: Viking.

Plomin, R. (2003). Genetics, genes, genomics and g. *Molecular Psychiatry, 8*, 1–5.

Plomin, R., Asbury, K., & Dunn, J. F. (2001). Why are children in the same family so different? Nonshared environment a decade later. *Canadian Journal of Psychiatry, 46*, 225–233.

Plomin, R., & Daniels, D. (1987). Why are children in the same family so different from one another? *Behavioral and Brain Sciences, 10*(01), 1–16.

Plomin, R., & Deary, I. J. (2014). Genetics and intelligence differences: five special findings. *Molecular psychiatry.*

Plomin, R., DeFries, J. C., & Loehlin, J. C. (1977). Genotype–environment interaction and correlation in the analysis of human behavior. *Psychological bulletin, 84*(2), 309.

Plomin, R., DeFries, J. C., Knopik, V. S., & Neiderhiser, J. M. (2012). *Behavioral Genetics* (6th edition). New York: Worth.

Plomin, R., Haworth, C. M., & Davis, O. S. (2009). Common disorders are quantitative traits. *Nature Reviews Genetics, 10*(12), 872–878.

Plomin, R., & Kovas, Y. (2005). Generalist genes and learning disabilities. *Psychological bulletin, 131*(4), 592–617.

Plomin, R., & Schalkwyk, L. C. (2007). Microarrays. *Developmental Science, 10*(1), 19–23.

Polderman, T. J., Benyamin, B., de Leeuw, C. A., Sullivan, P. F., van Bochoven, A., Visscher, P. M., & Posthuma, D. (2015). Meta-analysis of the heritability of human traits based on fifty years of twin studies. *Nature Genetics, 47*(7), 702–709.

Raymond, F. L., & Tarpey, P. (2006). The genetics of mental retardation. *Human Molecular Genetics, 15*(suppl 2), 110–116.

Rhea, S. A., Gross, A. A., Haberstick, B. C., & Corley, R. P. (2013). Colorado twin registry: an update. *Twin Research and Human Genetics, 16*(01), 351–357.

Rimfeld, K., Kovas, Y., Dale, P. S., & Plomin, R. (2015). Pleiotropy across academic subjects at the end of compulsory education. *Scientific Reports, 5*, 11713.

Skytthe, A., Christiansen, L., Kyvik, K. O., Bødker, F. L., Hvidberg, L., Petersen, I., ... & Christensen, K. (2013). The Danish Twin Registry: linking surveys, national registers, and biological information. *Twin Research and Human Genetics, 16*(01), 104–111.

Stanescu-Cosson, R., Pinel, P., van de Moortele, P. F., Le Bihan, D., Cohen, L., & Dehaene, S. (2000). Understanding dissociations in dyscalculia. *Brain, 123*(11), 2240–2255.

Stromswold, K. (2001). The heritability of language: a review and metaanalysis of twin, adoption, and linkage studies. *Language, 77*(4), 647–723.

The National Human Genome Research Institute: http://www.genome.gov/gwastudies/.

Thompson, L. A., Detterman, D. K., & Plomin, R. (1991). Associations between cognitive abilities and scholastic achievement: genetic overlap but environmental differences. *Psychological Science, 2*(3), 158–165.

Tikhomirova, T. N., & Kovas, Y. V. (2013). Vzaimosvyaz' kognitivnyh harakteristik uchashchihsya i uspeshnosti resheniya matematicheskih zadanij (na primere starshego shkol'nogo vozrasta) [The association of cognitive characteristics and successful mathematical performance (in high school students)]. *Psihologicheskij zhurnal, 34*(1), 35–45.

Tommerdahl, J. (2010). A model for bridging the gap between neuroscience and education. *Oxford Review of Education, 36*(1), 97–109.

Van Beijsterveldt, C. E., Groen-Blokhuis, M., Hottenga, J. J., Franić, S., Hudziak, J. J., Lamb, D., ... & Boomsma, D. I. (2013). The Young Netherlands Twin Register (YNTR): longitudinal twin and family studies in over 70,000 children. *Twin Research and Human Genetics, 16*(01), 252–267.

Venkatraman, V., Ansari, D., & Chee, M. W. (2005). Neural correlates of symbolic and non-symbolic arithmetic. *Neuropsychologia, 43*(5), 744–753.

Wainwright, M. A., Wright, M. J., Luciano, M., Geffen, G. M., & Martin, N. G. (2005). Multivariate genetic analysis of academic skills of the Queensland core skills test and IQ highlight the importance of genetic g. *Twin Research and Human Genetics, 8*(06), 602–608.

Wetterstrand, K. A. (2013). DNA sequencing costs: data from the NHGRI Genome Sequencing Program (GSP). *National Human Genome Research Institute*, www.genome.gov/sequencingcosts.

Wetterstrand, K. A. (2015). DNA Sequencing Costs: Data from the NHGRI Genome Sequencing Program (GSP). www.genome.gov/sequencingcosts.

Zhou, N., Lam, S. F., & Chan, K. C. (2012). The chinese classroom paradox: a cross-cultural comparison of teacher controlling behaviors. *Journal of Educational Psychology, 104*(4), 1162–1174.

2

Gene–Environment Interplay and Individual Differences in Psychological Traits

Philipp Barsky and Darya Gaysina

Numerous behavioural genetic studies have demonstrated that individual differences in psychological traits develop under the influence of both genetic and environmental factors. These studies have also suggested that genetic factors do not influence a trait in a deterministic way. Moreover, each genetic factor may affect several different neural pathways in the brain, and therefore can affect a wide range of behaviours (see Chapter 1 for more details on the phenomenon of pleiotropy). In addition, effects of genetic factors on complex behavioural traits can depend on environmental factors.

The present chapter focuses on the role of environmental factors and their interplay with genetic factors in the development of complex behavioural traits. Firstly, the key concepts in relation to environment and environmental variance are considered. Secondly, we provide examples of behavioural genetic research into the role of environmental factors in individual differences in psychological traits. Thirdly, we consider the role of two main types of gene–environmental interplay – gene–environmental correlations (rGE) and gene–environmental interactions (GxE). Finally, we discuss the value of genetically informative approaches for gaining new insights into the causal mechanisms underlying normal and abnormal behaviour.

Environment in behavioural genetic research

In behavioural genetics, *environment* is understood as any influence on a specific phenotype that is not genetic (not related to DNA sequence) and that contributes to individual differences in this phenotype in a population. This definition of environment is broader than the one that is usually used in psychology. For example, this definition may include influences in the prenatal and postnatal periods, such as nutrition, illnesses, hormonal levels and social relationships within and outside of the family (Plomin & Caspi, 1999).

Traditionally, in quantitative behavioural genetics, two types of environmental influences are defined: shared (common) and non-shared (unique).

The simplest way to estimate the contribution of environmental factors is by subtracting genetic influences (i.e., heritability) from the total phenotypic variance. The relative contribution of genetic and environmental factors to individual differences can be assessed by comparisons between relatives of different degrees of genetic relatedness (e.g., twins). Environmental influences can lead to either increased resemblance (shared environmental influences; C) or decreased resemblance (non-shared environmental influences; E) between relatives in the same family. These environmental components of the influences on a trait's variance are not directly measured but defined based on the phenotypic resemblance between family members (see Box 1 for more details).

A distinction between *objective* and *effective environment* has been proposed (Goldsmith, 1993; Turkheimer & Waldron, 2000). The *objective environment* stands for factors that can be assessed (observed, measured). The *effective environment* is defined as the result of some environmental influences (e.g., on the children in a family). This way, if an environmental factor is objectively shared by members of one family (e.g., parenting or socio-economic status) but affects them differently, its effective contribution will be included in the unique (non-shared) environmental component. For example, parental divorce is an objective risk factor for low academic performance and psychopathology in children (D'Onofrio et al., 2007; Harold, Aitken, & Shelton, 2007). However, it may be perceived differently by different children in the same family, and in this case it will contribute to differences, not to similarities, between them, and therefore will act as a factor of unique (non-shared) environment (McGuire, 2003; Reiss, Neiderhiser, Hetherington, & Plomin, 2000; Rutter, Pickles, Murray, & Eaves, 2001; Rutter, 2006; Turkheimer & Waldron, 2000).

Overall, the results of a number of large scale twin and adoption studies suggest that the contribution of environmental factors to variation in psychological traits is as significant as that of genetic factors. However, behavioural genetic research has demonstrated that environmental factors contribute mostly to increased differences between family members, while their resemblance is mostly due to the common genetic factors (Hanscombe et al., 2011; Plomin, 2011; Plomin & Daniels, 1987).

One of the most important aims of behavioural genetic research is to identify specific environmental factors that contribute to individual differences in psychological traits (Goldsmith, 1993; Reiss et al., 2000; Turkheimer & Waldron, 2000). Different methods and approaches to achieve this aim have been developed using genetically informative designs, such as twin and adoption designs (Moffitt, Caspi, & Rutter, 2005; Plomin & Bergeman, 1991; Reiss et al., 2000; Rutter et al., 2001; Rutter, 2006; Turkheimer & Waldron, 2000). Many genetically informative studies use multiple informants (children, parents and other family members, peers, teachers, researchers) to measure environmental factors supposedly "unique" for every child, such as parent–child relationships, sibling relationships and relationships with peers and teachers.

The results of a meta-analysis of 43 studies have demonstrated that non-shared environment accounts for ~50% of variability of many behavioural traits. However, none of the measured environmental factors ("objective" environments) has explained more than 2% of phenotypic variance (Turkheimer & Waldron, 2000). Since the publication of this meta-analysis, multiple studies have been conducted searching for objective non-shared environmental factors involved in behaviour of children and adolescents in different contexts (Asbury, Dunn, Pike, & Plomin, 2003; Deater-Deckard et al., 2001; Liang & Eley, 2005). For example, one study using a sample of discordant monozygotic twins has demonstrated that parenting explains 3% of the variance in socio-emotional problems in children. In the presence of adversity (e.g., low family socio-economic status) coupled with extreme differences in parent–child relationships between the co-twins, the effect of parenting accounts for up to 11% of the variance in children's socio-emotional problems (Asbury et al., 2003). Another study, using the children of twins (COT) method, has shown that the effect of parental divorce on offspring substance use problems – but not emotional problems – was significant even when genetic and other environmental risks were taken into account (D'Onofrio et al., 2007).

A large scale longitudinal study, the "NEAD" (Non-shared Environment and Adolescent Development), which includes data from more than 700 families, has aimed: 1) to assess the differences in experiences of children from the same families; 2) to find the links between the differences in experience and behavioural outcomes; and 3) to determine the causal directions of these links (Plomin, Asbury, & Dunn, 2001). The NEAD studies have focused on different behavioural phenotypes, such as psychopathology, adjustment and relationships among family members (Spotts, Neiderhiser, Hetherington, & Reiss, 2001). Different aspects of children's environment have been studied, such as relationships with teachers, peers and close friends. The results have shown significant non-shared environmental influences on many behavioural traits in adolescents. However, specific measures of non-shared environment (e.g., parent–child relationships, asymmetrical sibling relationships, peer relationships) were not significant (Spotts et al., 2001).

It has been demonstrated that contributions of shared and non-shared environmental factors may vary for different psychological traits in children and adolescents (Burt, 2014). Shared environment has been shown to be important for different traits including: antisocial and delinquent behaviour (Marceau et al., 2012; Miles & Carey, 1997; Rhee & Waldman, 2002); attitudes (Abrahamson, Baker, & Caspi, 2002); positive emotionality (Riemann, Angleitner, & Strelau, 1997); personality traits (Horwitz, Videon, Schmitz, & Davis, 2003); and adjustment (Loehlin, Neiderhiser, & Reiss,

2003). As shown in a recent meta-analysis of 490 twin and adoption studies of emotional and behavioural problems in adolescents, the contribution of shared environment for different types of psychopathology is 10%–30%, with the exception of ADHD, which is primarily influenced by genetic factors (Burt, 2009).

Inconsistencies in the estimates of shared environment for different traits may be partially explained by differences in research methods and measurements used in different studies. For example, for aggression, the contribution of genetic factors was smaller and of shared environment – larger in the laboratory studies as compared with the studies using questionnaires (Miles & Carey, 1997). Observational studies of personality have yielded unusually high estimates of shared environment (~25%; Borkenau, Riemann, Spinath, & Angleitner, 2000; Borkenau, Riemann, Angleitner, & Spinath, 2001). A study using data from the NEAD project has found significant shared environmental influences on observational measures of problem solving and antisocial behaviour and on the association between the two (Spotts et al., 2001). These results suggest that laboratory based studies can show higher contributions of shared environment than questionnaire based research. It is possible that genetic effects are more easily detected for more global, or general, measures of behaviour. For more specific and situational-context-dependent behaviours, such as a child's reactions in a laboratory situation, there is a higher probability of detecting a shared environmental effect (Borkenau et al., 2000).

Inconsistencies in the estimated relative contributions of genetic and environmental factors may stem from the differences in age of participants. Research suggests that for some traits (e.g., IQ), the contribution of genetic influences increases, whereas shared environmental influences decreases, with age (Haworth et al., 2010). The results of a study in a sample of 12–14 month old twins, using a laboratory "strange situation", showed that individual differences in attachment at this age were explained by shared environment (50%) and non-shared environment (50%), with no contribution of genetic factors (Bokhorst, Bakermans–kranenburg, Fonagy, & Schuengel, 2003). A similar pattern of results has been reported in a study of attachment in preschool age twins, using the same method of attachment assessment (Deater-Deckard et al., 2001).

The relative contributions of genetic and environmental factors can vary depending on specific circumstances (e.g., different SES; before and after training; Hanscombe et al., 2012; Malykh, 2004; Rowe, Jacobson, & Van den Oord, 1999; Turkheimer et al., 2003). For example, an experimental study of problem-solving in 6–7 year-old twins revealed different patterns of results before and after training: genetic factors (53%) and non-shared environment (47%) were significant in a pre-training series of tests, whereas after the training, shared environment (52%) and non-shared environment (48%) contributed to the same traits (Malykh, 2004).

Quantitative behavioural genetic research has improved our understanding of the ways environmental factors influence individual differences in psychological traits. However, there is a problem with identifying specific environmental factors contributing to the variance in complex traits, the so-called "missing environment" problem. This problem is similar to the "missing heritability" problem – the problem with identifying specific genetic variants contributing to phenotypic variance. It seems that each specific environmental factor has only a small effect on a trait, and that environmental influences act in a cumulative manner. Nevertheless, the fact that specific environmental factors have not been identified does not negate their significance, but simply indicates that causal mechanisms of environmental influences might be more complex than it was previously assumed (Rutter, 2006, 2007; Rutter et al., 2001).

Gene–environment interplay

Genetic and environmental factors do not act separately from each other but co-act. Gene–environment co-action includes gene–environment correlation (rGE) and gene–environment interaction (GxE).

Gene–environment correlation (rGE)

Environment can correlate with one's genotype – this phenomenon is referred to as gene–environment correlation. It has been demonstrated that many environmental factors can be under genetic influence. Genetically informative designs have been used to test this phenomenon by treating environmental factors as phenotypes and estimating their heritability. For example, studies of parenting that used twin and adoption methods have shown that individual differences in perception of parental warmth can be heritable (Rowe, 1981; Bokhorst et al., 2003). Moderate genetic influences on 15 composite measures of family environment and family relationships have been demonstrated using data from the NEAD study (Plomin, Reiss, Hetherington, & Howe, 1994; Spotts et al., 2001). In addition to child reports on perceptions of parenting, retrospective reports on childhood experiences by adults have also been widely used. This strategy was employed in the Swedish Adoption/Twin Study of Aging (Plomin et al., 1990). Findings using data from older MZ and DZ twins raised apart and together, with retrospectively reported experiences, have suggested a possible genetic influence on parental warmth, as well as on other dimensions of family environment. Similar findings have been reported using data from the Minnesota Study of Twins Reared Apart, based on the retrospective reports on rearing family environment obtained from adult twins (Krueger, Markon, & Bouchard Jr, 2003).

Many behavioural genetic studies have been based on self-reported data. Fewer studies have used observations of family relationships, such as the interactions of mothers and infants, siblings, and other dyads of family members. These studies have also reported genetic influences on individual differences in environmental factors (Dunn & Plomin, 1986; Rende, Slomkowski, Stocker, Fulker, & Plomin, 1992; Deater-Deckard et al., 2001; Spotts et al., 2001).

Genetic factors can also influence the probability of experiencing certain life events. A number of genetically informative studies have shown that heritability of life events depends on the extent of personal control over the event: heritability is higher for more controllable events (e.g., divorce, financial difficulties) then for less controllable events (e.g., a relative's illness) for which heritability is low (Billig, Hershberger, Iacono, & McGue, 1996; Kendler, Neale, Kessler, Heath, & Eaves, 1993; Plomin et al., 1990). As demonstrated in a systematic review of 55 studies of 35 environmental influences, including stressful life events, parenting, family environment, social support, peer relationships and marital quality, such experiences are partly influenced by genetic factors, with heritability estimates varying from 7% to 39% (with the weighted heritability of 27% for all environmental measures in all studies; Kendler & Baker, 2007).

How can genes influence environments? The results of behavioural genetic studies suggest that environmental factors are subject to genetic influences since genetic variants can influence behaviour related to change, modification and creation of specific environmental conditions. Therefore, environments less susceptible to behavioural influence are less heritable.

Gene–environment correlations can be divided into three types: passive, reactive and active (Jaffee & Price, 2008; Plomin & Bergeman, 1991; Rutter, 2006, 2007; Rutter et al., 2001).

Passive gene–environment correlation (rGE) refers to the situation when the rearing environment that parents provide for their children is influenced by the parents' own behavioural characteristics, and these characteristics are partly under genetic influence. A positive rGE may provide the situation of dual advantage or dual disadvantage – genetic plus environmental. For example, children of Johann Sebastian Bach inherited musical giftedness and grew up in a musically rich environment, as their father dedicated a lot of time to their musical education. Mathematical giftedness, passed on from one generation to another in the Bernoulli family, may also be influenced by genetic factors as well as by a "mathematically rich" family environment (Copp, 1916; Lick, 1969). Parental psychopathology can serve as an example of dual disadvantage resulting from a passive rGE. Being partly heritable, it can also be related to family conflicts and disagreements (Quinton & Rutter, 1988). Therefore, children of parents with psychopathology may inherit some genetic risk factors for psychopathology and may also be exposed to family conflicts (Rutter, 2006; Rutter et al., 1997).

The presence of passive rGE can be detected using different behavioural genetic approaches (for more detail, see Plomin et al., 2013). For example, a study using the twin-family design demonstrated that family discord and maladaptation were associated with a two-fold increase in risk for conduct disorder symptomatology in children. However, when parental conduct disorder (indicating genetic risk for conduct problems in children) was included in the model, the environmental mediation effect for family maladaptation remained, but that for family discord was lost (Meyer et al., 2000). These findings demonstrated that genetic factors fully explained the association between family discord and conduct disorders in children (indicating the presence of passive rGE). However, the association between family maladaptation and children's conduct problems was shown to be environmental (i.e., free from passive rGE).

Another behavioural genetics method that can be employed to consider the role of rGE is the adoption method. It is based on the fact that adopted children and their adoptive parents are not genetically related and, on average, only share a common postnatal environment. The advantage of this method is that adoptive family environment is free from passive rGE (Leve, Neiderhiser, Scaramella, & Reiss, 2010). For example, one study using data from the Early Growth and Development Study (EGDS) – a longitudinal study of adopted children and their parents (adoptive and biological) – has demonstrated that parental hostility towards their children is a true environmental risk factor (i.e., free from passive rGE) for aggressive behaviour in young children (Stover et al., 2012).

The classical adoption method ("adoption after birth") cannot be used to disentangle prenatal environmental influences (e.g., maternal smoking during pregnancy; maternal depression during pregnancy) from genetic influences, because biological mothers provide both genes and prenatal environment to their children (i.e., passive rGE with prenatal environment). A novel adoption design ("adoption at conception") allows a separation of the influences of prenatal environment from genetic influences on the child's development. The method utilizes a design that is provided by modern *in vitro* fertilization (IVF) technologies, and is based on the comparison of pairs of parents and children with different degrees of genetic relatedness: both parents are genetically related to their child; only father is genetically related to the child (egg donation); only mother is genetically related to the child (sperm donation); none of the parents are genetically related to the child (embryo donation) (Rice et al., 2009; Thapar et al., 2007). If a prenatal factor is associated with the child's trait only in genetically related parent–child pairs, but not in genetically unrelated parent–child pairs, then this factor cannot be considered environmental, as it is likely to involve passive rGE. If a prenatal factor is associated with the child's trait in both genetically related and genetically unrelated parent–child pairs, then this factor is more likely to be environmental (not influenced by common genetic factors). For example, the Cardiff IVF study has found that associations between maternal smoking during pregnancy and a child's antisocial behaviour (Rice et al., 2009) and ADHD (Thapar et al., 2009) are likely to be influenced by genetic factors shared by mothers and children, rather than by maternal smoking itself. At the same time, the effect of the mother smoking during pregnancy on a child's birth weight has been shown to be environmentally mediated (Rice et al., 2009). These and other studies demonstrate that genetically informative research designs are required to study both prenatal and postnatal influences on child development.

Active gene–environment correlation refers to situations when people actively influence their own environment (including relationships with other people) through their behaviours, partly for genetic reasons. **Reactive (evocative) gene–environment correlation** is observed when a person's behaviour that is partly under genetic influence evokes a response from other people. Therefore, different aspects of someone's environment may correlate with his/her genetic propensities (Scarr & McCartney, 1983).

Behavioural genetic research has provided a number of examples of active and reactive rGE. For instance, family conflict is known to be associated with the development of depression in one of the spouses (Weissman, 1987). Behavioural genetic research suggests that family conflict does not act as a causal environmental factor for depression; but rather a spouse with depression susceptibility (that is partly under genetic influence) is likely to evoke negative responses from another spouse leading to family conflict (South & Krueger, 2008; Spotts et al., 2004).

Another study, using data from a longitudinal adoption study of children and their biological and foster (adoptive) parents, investigated influences of inherited characteristics of adopted children on parenting practices of their adoptive parents (Ge et al., 1996). The results showed that symptoms of psychopathology in biological parents were associated with antisocial behaviour of their child (indicating the presence of genetic risk). Moreover, symptoms of psychopathology in biological parents were associated with more negative parenting of adoptive parents. This association was mediated by antisocial behaviour of adopted children, that is, children with higher genetic risk for antisocial behaviour were more likely to evoke more negative parenting from their adoptive parents. Other genetically informative studies have confirmed this finding of the role of evocative rGE in the association between negative parenting and child antisocial behaviour (for review, see Jaffee, Strait, & Odgers, 2012). Recent studies have also provided new evidence for evocative rGE in the link between: parenting and children's ADHD (Harold et al., 2013); parenting and children's peer behaviour problems (Elam et al., 2014); and parenting and children's self-control (Pener-Tessler et al., 2013).

Gene–environment correlations can be investigated using molecular genetic methods examining the effects of specific DNA variants on specific environments. Although molecular genetic studies of gene–environment correlations are relatively rare, several statistically significant associations have already been found, including those between: the dopamine D2 receptor (*DRD2*) gene and parenting (Beaver et al., 2012; Lucht et al., 2006; Mills-Koonce et al., 2007); the dopamine transporter (*DAT1*) gene and parenting (Hayden et al., 2013); the brain derived neurotrophic factor (*BDNF*) gene and childhood adversity (Wichers et al., 2008); the serotonin transporter (*5-HTT*) gene and maternal parenting (Pener-Tessler et al., 2013); the serotonin 2A receptor (*HTR2A*) gene and popularity (Burt, 2008); and the gamma-aminobutyric acid A receptor (*GABRA2*) gene and marital status (Dick et al., 2006). These results are still preliminary and need to be replicated in independent studies; nevertheless they allow us to hypothesize that specific genetic variants that correlate with different environmental factors could be identified. However, it is likely that these effects of specific genetic variants on behaviour that mediates rGE will be small (Jaffee & Price, 2008).

Genetic factors can also influence *perception* of environment through some characteristics of personality. Questionnaires that are used to measure the features of environment may reflect individual personalities of respondents, since human perception of the world is filtered through experience, personality traits and cognitive characteristics (Plomin & Bergeman, 1991). There is evidence for the mediating role of personality for genetic and environmental influences on such phenotypes as marital relationships (Spotts et al., 2005), perception of life events (Saudino, Pedersen, Lichtenstein, McClearn, & Plomin, 1997) and recollections about family relationships (Krueger et al., 2003). Therefore, the mediating role of personality for genetic influences on perception of environment may be important in gene–environment correlations. In order to identify environmental factors free from gene–environment correlations, the information from different sources or from different types of informants need to be used.

The concept of gene–environmental correlation is often misinterpreted in psychological research: if genetic influence on an "environmental" phenotype is detected, this association is considered as pure "genetic". Therefore, the actual causes of individual differences that could have environmental origins (e.g., the availability of drugs) are confounded with genetically influenced risks (e.g., the susceptibility to drug addiction) that can lead to selection and creation of environments (Rutter et al., 2001; Rutter, 2006). The term "phenotype-environment correlation" has been proposed to reflect the complex nature of such gene–environment co-actions (Turkheimer & Waldron, 2000). Other researchers have suggested that what we think of as environmental factors are better described as "external factors" that might be partly under genetic control (Vinkhuyzen et al., 2010). The reviewed findings that environmental factors are partly under genetic control have implications on studies of gene–environment interactions (Moffitt, Caspi, & Rutter, 2005).

Gene–environment interaction (GxE)

Gene–environment interaction (GxE) refers to the mutual interdependence of genetic and environmental factors. It characterizes situations when genetic effects on a trait depend on environment, or when environmental influences depend on genetic factors. There are different types of gene–environment interactions, as explained below.

Firstly, genetic factors can influence one's response to stressful environmental conditions. This type of GxE refers to a genetic susceptibility to stressful experiences. Genetic susceptibility serves as a risk factor, determining whether some stressful experience (e.g., financial difficulties, family crises) will lead to negative outcomes (e.g., psychopathology). This type of GxE has been widely studied. For example, one study of adopted children with and without genetic susceptibility to antisocial behaviour (as indexed with antisocial behaviour of their biological parents) has demonstrated that negative parenting style has an effect on children's antisocial behaviour. This effect was only present in children with genetic susceptibility; the same parenting style exerts minimal influence on children without genetic susceptibility to antisocial behaviour (Cadoret, Yates, Woodworth, & Stewart, 1995).

A molecular genetic study, using data from the Dunedin Multidisciplinary Health and Development Study, demonstrated that the effect of childhood maltreatment on antisocial behaviour depends on the monoamine oxidase A (*MAOA*) gene: children who experienced maltreatment were at a higher risk of developing antisocial problems if they were carriers of the low-active form of the *MAOA* gene as compared to carriers of the high-active form of the *MAOA* gene (Caspi et al., 2002). Several studies have attempted to replicate this finding producing inconsistent results (Foley et al., 2004; Haberstick et al., 2014; Nilsson et al., 2006; Fergusson et al., 2011; Young et al., 2006).

Secondly, genetic factors can amplify an individual's sensitivity to both positive and negative environments. It is known as a "differential susceptibility" hypothesis (Belsky & Pluess, 2009) and is based on genetic plasticity: specific genetic variants allow the carriers to benefit most in positive environments, but put them at the highest risk in negative environments. For example, in a study of the dopamine D4 receptor (*DRD4*) gene, maternal sensitivity and behavioural problems in children, it was shown that low maternal sensitivity was associated with behavioural problems, but only in children with the 7R allele of the *DRD4* gene. Children with the same gene variant and mother with high sensitivity displayed the lowest level of behavioural problems. For children without the 7R allele of the *DRD4* gene, differences in maternal sensitivity had no effect on behavioural problems (Van Ijzendoorn & Bakermans-Kranenburg, 2006).

Thirdly, specific genetic factors can suit some environments better than others. A vivid example demonstrating this type of GxE is provided by studies of the alcohol dehydrogenase 1B (*ADH1B*) gene. This gene encodes the β subunit of alcohol dehydrogenase that is involved in the oxidative metabolism of ethanol. A polymorphic variant (Arg48His, rs1229984) of the *ADH1B* gene dramatically affects enzyme activity; the enzyme with His48 oxidizes ethanol approximately 70- to 80-fold faster than that with Arg48, and lowers the risk for alcohol dependence and abuse (Edenberg, 2007). It has been shown that the high activity allele (His48) is rare in European populations (with traditions of higher alcohol consumption) and common in Asian populations (with culturally limited alcohol consumption; Borinskaya et al., 2009; Li et al., 2007; Li et al., 2011). Such a geographically specific distribution of the *ADH1B* gene variants may be explained by natural selection in human populations that has resulted in the fixation of the gene variant (in a population gene pool) that is most appropriate for the specific environmental conditions.

Fourthly, the inherited characteristics can be revealed only in specific environmental conditions. Musical giftedness can be looked at as an example of this type of GxE. Behavioural genetic research of musical pitch recognition has demonstrated that this trait is heritable (up to 80% of its variance is accounted for by genetic factors; Drayna, Manichaikul, de Lange, Snieder, & Spector, 2001). But musical talent can only be developed in the presence of specific environmental conditions (e.g., the availability of a musical instrument and/or music lessons).

Usually behavioural genetic studies focus on investigating either rGE or GxE; however they are not necessarily independent. Including rGE in molecular genetic studies of GxE is particularly important because failing to do so can lead to biased results obtained in these studies (Jaffee & Price, 2007).

Conclusion

In conclusion, behavioural genetic approaches – both quantitative and molecular – are valuable instruments for studying the role of specific environmental factors in psychological traits across development. Cross-cultural and cross-population studies are particularly valuable in behavioural genetic research, as they allow the inclusion of a wider range of genetic and environmental contexts (see Chapter 1). Genetically informative experimental studies, where environmental factors can be directly controlled and constructed, are also of great value (Malykh, 2004).

The links between genotype and phenotype cannot be properly understood without considering the role of environmental factors, whereas environmental explanations of behaviour need to consider the role of genetic factors. The value of genetic informative research designs is in their potential to reveal mechanisms and processes that underlie the links between genes, environments and behaviours.

In our opinion, the most promising directions of future behavioural genetic research include: conducting studies that allow identification of the processes of gene–environment interplay that are unfolding over the lifespan; and developing theoretical models of different types of gene–environment interplay processes. For this type of research, longitudinal genetically informative designs are particularly important. As mentioned before, genetic factors can influence perception of trait-relevant environment; therefore future studies need to account for these influences. Although methodologically demanding, these studies would not only reveal specific genetic and/or environmental influences on a trait, but would improve the existing psychological theories of development that rarely address the role of genetic factors in psychological processes.

Another promising direction of behavioural genetic research is behavioural epigenetics. Recent findings that environmental factors (e.g., nutrition, stress and parenting) can affect gene expression and, subsequently, development, open a whole new research area that brings together genetics, neuroscience, psychology and other disciplines (Meaney, 2010; Mill & Petronis, 2008; Rutten & Mill, 2009; Unternaehrer et al., 2012).

There are numerous possibilities for practical implications of behavioural genetic research of gene–environment interplay. The development of more focused (family or person-oriented) prevention programmes for decreasing the risks of problem behaviour or psychopathology on the basis of individual genetic (or gene–environment) screening is a reality of the near future. Genetic factors that contribute to the risks or protect from the risks in specific adverse environments, once identified, will be used to design evidence-based psychological services and individualized educational programmes.

References

Abrahamson, A. C., Baker, L. A., & Caspi, A. (2002). Rebellious teens? Genetic and environmental influences on the social attitudes of adolescents. *Journal of Personality and Social Psychology, 83*(6), 1392–1408.

Asbury, K., Dunn, J. F., Pike, A., & Plomin, R. (2003). Nonshared environmental influences on individual differences in early behavioral development: A monozygotic twin differences study. *Child Development, 74*(3), 933–943.

Beaver, K. M., Shutt, J. E., Vaughn, M. G., DeLisi, M., & Wright, J. P. (2012). Genetic Influences on Measures of Parental Negativity and Childhood Maltreatment: An Exploratory Study Testing for Gene × Environment Correlations. *Journal of Contemporary Criminal Justice, 28*(3), 273–292.

Belsky, J., & Pluess, M. (2009). Beyond diathesis stress: differential susceptibility to environmental influences. *Psychological Bulletin, 135*(6), 885–908.

Billig, J. P., Hershberger, S. L., Iacono, W. G., & McGue, M. (1996). Life events and personality in late adolescence: genetic and environmental relations. *Behavior Genetics, 26*(6), 543–554.

Bokhorst, C. L., Bakermans-kranenburg, M. J., Fonagy, P., & Schuengel, C. (2003). The importance of shared environment in mother–infant attachment security: A behavioral genetic study. *Child Development, 74*(6), 1769–1782.

Borinskaya, S., Kal'ina, N., Marusin, A., Faskhutdinova, G., Morozova, I., Kutuev, I., ... & Rogaev, E. (2009). Distribution of the alcohol dehydrogenase ADH1B*47His allele in Eurasia. *American Journal of Human Genetics, 84*(1), 89–92;

Borkenau, P., Riemann, R., Angleitner, A., & Spinath, F. M. (2001). Genetic and environmental influences on observed personality: Evidence from the German Observational Study of Adult Twins. *Journal of Personality and Social Psychology, 80*(4), 655–668.

Borkenau, P., Riemann, R., Spinath, F., & Angleitner, A. (2000). Behavior genetics of personality: The case of observational studies. *Advances in Personality Psychology, 1*, 107–137.

Burt, S. A. (2008). Genes and Popularity Evidence of an Evocative Gene–Environment Correlation. *Psychological Science, 19*(2), 112–113.

Burt, S. A. (2009). Rethinking environmental contributions to child and adolescent psychopathology: a meta-analysis of shared environmental influences. *Psychological Bulletin, 135*(4), 608–637.

Burt, S. A. (2014). Research Review: The shared environment as a key source of variability in child and adolescent psychopathology. *Journal of Child Psychology and Psychiatry, 55*(4), 304–312.

Cadoret, R. J., Yates, W. R., Woodworth, G., & Stewart, M. A. (1995).Genetic-environmental interaction in the genesis of aggressivity and conduct disorders. *Archives of General Psychiatry, 52*(11), 916–924.

Caspi, A., McClay, J., Moffitt, T. E., Mill, J., Martin, J., Craig, I. W., ... & Poulton, R. (2002). Role of genotype in the cycle of violence in maltreated children. *Science, 297*(5582), 851–854.

Copp, E. F. (1916). Musical ability. Bases of it inherited by nearly everyone—difference in ability due more to training than to heredity—means for bringing latent ability into expression. *Journal of Heredity*, 7(7), 297–305.

D'Onofrio, B. M., Turkheimer, E., Emery, R. E., Maes, H. H., Silberg, J., & Eaves, L. J. (2007). A children of twins study of parental divorce and offspring psychopathology. *Journal of Child Psychology and Psychiatry, 48*(7), 667–675.

Deater-Deckard, K., Pike, A., Petrill, S. A., Cutting, A. L., Hughes, C., & O'Connor, T. G. (2001). Nonshared environmental processes in social-emotional development: An observational study of identical twin differences in the preschool period. *Developmental Science, 4*(2), F1-F6.

Dick, D. M., Agrawal, A., Schuckit, M. A., Bierut, L., Hinrichs, A., Fox, L., ... & Nurnberger, J. I. (2006). Marital status, alcohol dependence, and GABRA2: evidence for gene–environment correlation and interaction. *Journal of Studies on Alcohol, 67*(2), 185–194.

Drayna, D., Manichaikul, A., de Lange, M., Snieder, H., & Spector, T. (2001). Genetic correlates of musical pitch recognition in humans. *Science, 291*(5510), 1969–1972.

Dunn, J., & Plomin, R. (1986).Determinants of maternal behaviour towards 3-year-old siblings. *British Journal of Developmental Psychology, 4*(2), 127–137.

Edenberg, H. J. (2007). The genetics of alcohol metabolism: role of alcohol dehydrogenase and aldehyde dehydrogenase variants. *Alcohol Research & Health, 30*(1), 5–13.

Elam, K. K., Harold, G. T., Neiderhiser, J. M., Reiss, D., Shaw, D. S., Natsuaki, M. N., Gaysina, D., Barrett, D., & Leve, L. D. (2014). Adoptive parent hostility and children's peer behavior problems: Examining the role of genetically informed child attributes on adoptive parent behavior. *Developmental Psychology, 50*(5), 1543–1552.

Fergusson, D. M., Boden, J. M., Horwood, L. J., Miller, A. L., & Kennedy, M. A. (2011). MAOA, abuse exposure and antisocial behaviour: 30-year longitudinal study. *British Journal of Psychiatry, 198*(6), 457–463.

Foley, D. L., Eaves, L. J., Wormley, B., Silberg, J. L., Maes, H. H., Kuhn, J., & Riley, B. (2004). Childhood adversity, monoamine oxidase a genotype, and risk for conduct disorder. *Archives of General Psychiatry, 61*(7), 738–744.

Ge, X., Conger, R. D., Cadoret, R. J., Neiderhiser, J. M., Yates, W., Troughton, E., & Stewart, M. A. (1996). The developmental interface between nature and nurture: a mutual influence model of child antisocial behavior and parent behaviors. *Developmental Psychology, 32*(4), 574–589.

Goldsmith, H. (1993). Nature–nurture issues in the behavioral genetics context: Overcoming barriers to communication. In R. Plomin and G. E. McClearn (Eds), *Nature, Nurture & Psychology* (pp. 325–340). Washington, DC: American Psychological Association.

Haberstick, B. C., Lessem, J. M., Hewitt, J. K., Smolen, A., Hopfer, C. J., Halpern, C. T., ... & Harris, K. M. (2014). MAOA genotype, childhood maltreatment, and their interaction in the etiology of adult antisocial behaviors. *Biological Psychiatry, 75*(1), 25–30.

Hanscombe, K. B., Haworth, C. M., Davis, O. S., Jaffee, S. R., & Plomin, R. (2011). Chaotic homes and school achievement: a twin study. *Journal of Child Psychology and Psychiatry, 52*(11), 1212–1220.

Hanscombe, K. B., Trzaskowski, M., Haworth, C. M., Davis, O. S., Dale, P. S., & Plomin, R. (2012). Socioeconomic status (SES) and children's intelligence (IQ): In a UK-representative sample SES moderates the environmental, not genetic, effect on IQ. *PLoS One, 7*(2), e30320.

Harold, G. T., Aitken, J. J. and Shelton, K. H. (2007), Inter-parental conflict and children's academic attainment: a longitudinal analysis. *Journal of Child Psychology and Psychiatry*, 48 (12), 1223–1232.

Harold, G. T., Leve, L. D., Barrett, D., Elam, K., Neiderhiser, J. M., Natsuaki, M. N., ... & Thapar, A. (2013). Biological and rearing mother influences on child ADHD symptoms: revisiting the developmental interface between nature and nurture. *Journal of Child Psychology and Psychiatry, 54*(10), 1038–1046.

Haworth, C. M. A., Wright, M. J., Luciano, M., Martin, N. G., De Geus, E. J. C., Van Beijsterveldt, C. E. M., ... & Plomin, R. (2010). The heritability of general cognitive ability increases linearly from childhood to young adulthood. *Molecular Psychiatry, 15*(11), 1112–1120.

Hayden, E. P., Hanna, B., Sheikh, H. I., Laptook, R. S., Kim, J., Singh, S. M., & Klein, D. N. (2013). Child dopamine active transporter 1 genotype and parenting: Evidence for evocative gene–environment correlations. *Development and Psychopathology, 25*(01), 163–173.

Horwitz, A. V., Videon, T. M., Schmitz, M. F., & Davis, D. (2003).Rethinking twins and environments: possible social sources for assumed genetic influences in twin research. *Journal of Health and Social Behavior,* 111–129.

Jaffee, S. R., & Price, T. S. (2008). Genotype–environment correlations: implications for determining the relationship between environmental exposures and psychiatric illness. *Psychiatry, 7*(12), 496–499.

Jaffee, S. R., Strait, L. B., & Odgers, C. L. (2012). From correlates to causes: can quasi-experimental studies and statistical innovations bring us closer to identifying the causes of antisocial behavior? *Psychological Bulletin, 138*(2), 272–295.

Kendler, K. S., & Baker, J. H. (2007). Genetic influences on measures of the environment: a systematic review. *Psychological medicine, 37*(05), 615–626.

Kendler, K. S., Neale, M., Kessler, R., Heath, A., & Eaves, L. (1993). A twin study of recent life events and difficulties. *Archives of General Psychiatry, 50*(10), 789–796.

Krueger, R. F., Markon, K. E., & Bouchard Jr, T. J. (2003). The extended genotype: The heritability of personality accounts for the heritability of recalled family environments in twins reared apart. *Journal of Personality, 71*(5), 809–833.

Leve, L. D., Neiderhiser, J. M., Scaramella, L. V., & Reiss, D. (2010). The Early Growth and Development Study: Using the prospective adoption design to examine genotype–environment interplay. *Behavior Genetics, 40*(3), 306–314.

Li, H., Mukherjee, N., Soundararajan, U., Tárnok, Z., Barta, C., Khaliq, S., ... & Kidd, K. K. (2007). Geographically separate increases in the frequency of the derived ADH1B* 47His allele in eastern and western Asia. *American Journal of Human Genetics, 81*(4), 842–846.

Li, D., Zhao, H., & Gelernter, J. (2011).Strong association of the alcohol dehydrogenase 1B gene (ADH1B) with alcohol dependence and alcohol-induced medical diseases. *Biological Psychiatry, 70*(6), 504–512.

Liang, H., & Eley, T. C. (2005). A monozygotic twin differences study of nonshared environmental influence on adolescent depressive symptoms. *Child Development, 76*(6), 1247–1260.

Lick, D. W. (1969). The Remarkable Bernoulli Family. *The Mathematics Teacher*, 401–409.

Loehlin, J. C., Neiderhiser, J. M., & Reiss, D. (2003). The behavior genetics of personality and the NEAD study. *Journal of Research in Personality, 37*(5), 373–387.

Lucht, M., Barnow, S., Schroeder, W., Grabe, H. J., Finckh, U., John, U., ... & Herrmann, F. H. (2006). Negative perceived paternal parenting is associated with dopamine D2 receptor exon 8 and GABA (A) alpha 6 receptor variants: An explorative study. *American Journal of Medical Genetics Part B: Neuropsychiatric Genetics, 141*(2), 167–172.

Malykh, S. B. (2004). *Psikhogenetika: teoriya, metodologiya, experiment [Psychogenetics: theory, methodology, experiment]*. Moscow: Epidavr.

Marceau, K., Humbad, M. N., Burt, S. A., Klump, K. L., Leve, L. D., & Neiderhiser, J. M. (2012). Observed externalizing behavior: a developmental comparison of genetic and environmental influences across three samples. *Behavior Genetics, 42*(1), 30–39.

McGuire, S. (2003). Nonshared environment research. *Marriage & Family Review, 33*(1), 31–56.

Meaney, M. J. (2010). Epigenetics and the biological definition of gene x environment interactions. *Child Development, 81*(1), 41–79.

Meyer, J., Rutter, M., Silberg, J., Maes, H., Simonoff, E., Shillady, L., ... & Eaves, L. (2000). Familial aggregation for conduct disorder symptomatology: the role of genes, marital discord and family adaptability. *Psychological Medicine, 30*(4), 759–774.

Miles, D. R., & Carey, G. (1997). Genetic and environmental architecture on human aggression. *Journal of Personality and Social Psychology, 72*(1), 207–217.

Mill, J., & Petronis, A. (2008). Pre- and peri-natal environmental risks for attention-deficit hyperactivity disorder (ADHD): the potential role of epigenetic processes in mediating susceptibility. *Journal of Child Psychology and Psychiatry, 49*(10), 1020–1030.

Mills-Koonce, W. R., Propper, C. B., Gariepy, J. L., Blair, C., Garrett-Peters, P., & Cox, M. J. (2007). Bidirectional genetic and environmental influences on mother and child behavior: The family system as the unit of analyses. *Development and Psychopathology, 19*(04), 1073–1087.

Moffitt, T. E., Caspi, A., & Rutter, M. (2005). Strategy for investigating interactions between measured genes and measured environments. *Archives of General Psychiatry, 62*(5), 473–481.

Nilsson, K. W., Sjöberg, R. L., Damberg, M., Leppert, J., Öhrvik, J., Alm, P. O., ... & Oreland, L. (2006). Role of monoamine oxidase A genotype and psychosocial factors in male adolescent criminal activity. *Biological Psychiatry, 59*(2), 121–127.

Pener-Tessler, R., Avinun, R., Uzefovsky, F., Edelman, S., Ebstein, R. P., & Knafo, A. (2013). Boys' serotonin transporter genotype affects maternal behavior through self-control: A case of evocative gene–environment correlation. *Development and Psychopathology, 25*(01), 151–162.

Plomin, R. (2011). Commentary: Why are children in the same family so different? Non-shared environment three decades later. *International Journal of Epidemiology, 40*(3), 582–592.

Plomin, R., Asbury, K., & Dunn, J. (2001). Why are children in the same family so different? Nonshared environment a decade later. *Canadian Journal of Psychiatry, 46*, 225–233.

Plomin, R., & Bergeman, C. S. (1991). The nature of nurture: Genetic influence on "environmental" measures. *Behavioral and Brain Sciences, 14*(03), 373–386.

Plomin, R., & Caspi, A. (1999). Behavioral genetics and personality. In L. A. Pervin, & O. P. John (Eds), *Handbook of personality: Theory and research (2nd Edition).* (pp. 251–276). New York: Guildford Press.

Plomin, R., & Daniels, D. (1987). Why are children in the same family so different from one another? *Behavioral and Brain Sciences, 10*(01), 1–16.

Plomin, R., DeFries, J. C., Knopik, V. S., & Neiderhiser, J. M. (2013). *Behavioral genetics* (6th edition). New York: Worth.

Plomin, R., Lichtenstein, P., Pedersen, N. L., McClearn, G. E., & Nesselroade, J. R. (1990). Genetic influence on life events during the last half of the life span. *Psychology and Aging, 5*(1), 25–30.

Plomin, R., Reiss, D., Hetherington, E. M., & Howe, G. W. (1994). Nature and nurture: genetic contributions to measures of the family environment. *Developmental Psychology, 30*(1), 32–43.

Quinton, D., & Rutter, M. (1988). *Parenting breakdown: The making and breaking of inter-generational links*: Avebury, Aldershot, UK.

Reiss, D., Neiderhiser, J. M., Hetherington, E. M., & Plomin, R. (2000). *The relationship code: Deciphering genetic and social influences on adolescent development.* Cambridge, MA: Harvard University Press.

Rende, R. D., Slomkowski, C. L., Stocker, C., Fulker, D. W., & Plomin, R. (1992). Genetic and environmental influences on maternal and sibling interaction in middle childhood: A sibling adoption study. *Developmental Psychology, 28*(3), 484–490.

Rhee, S. H., & Waldman, I. D. (2002). Genetic and environmental influences on antisocial behavior: a meta-analysis of twin and adoption studies. *Psychological Bulletin, 128*(3), 490–529.

Rice, F., Harold, G. T., Boivin, J., Hay, D. F., van den Bree, M., & Thapar, A. (2009). Disentangling prenatal and inherited influences in humans with an experimental design. *Proceedings of the National Academy of Sciences, 106*(7), 2464–2467.

Riemann, R., Angleitner, A., & Strelau, J. (1997). Genetic and Environmental Influences on Personality: A Study of Twins Reared Together Using the Self-and Peer Report NEO-FFI Scales. *Journal of Personality, 65*(3), 449–475.

Rowe, D. C. (1981). Environmental and genetic influences on dimensions of perceived parenting: A twin study. *Developmental Psychology, 17*(2), 203–208.

Rowe, D. C., Jacobson, K. C., & Van den Oord, E. J. C. G. (1999). Genetic and environmental influences on vocabulary IQ: Parental education level as moderator. *Child Development, 70*(5), 1151–1162.

Rutten, B. P., & Mill, J. (2009). Epigenetic mediation of environmental influences in major psychotic disorders. *Schizophrenia Bulletin, 35*(6), 1045–1056.

Rutter, M. (2006). *Genes and behaviour: Nature–nurture interplay explained*. UK: Blackwell Publishing.

Rutter, M. (2007). Proceeding from observed correlation to causal inference: The use of natural experiments. *Perspectives on Psychological Science, 2*(4), 377–395.

Rutter, M., Dunn, J., Plomin, R., Simonoff, E., Pickles, A., Maughan, B., ... & Eaves, L. (1997). Integrating nature and nurture: Implications of person–environment correlations and interactions for developmental psychopathology. *Development and Psychopathology, 9*(02), 335–364.

Rutter, M., Pickles, A., Murray, R., & Eaves, L. (2001).Testing hypotheses on specific environmental causal effects on behavior. *Psychological Bulletin, 127*(3), 291–324.

Saudino, K. J., Pedersen, N. L., Lichtenstein, P., McClearn, G. E., & Plomin, R. (1997). Can personality explain genetic influences on life events? *Journal of Personality and Social Psychology, 72*(1), 196–206.

Scarr, S., & McCartney, K. (1983). How people make their own environments: A theory of genotype→environment effects. *Child Development*, 424–435.

South, S. C., & Krueger, R. F. (2008). Marital quality moderates genetic and environmental influences on the internalizing spectrum. *Journal of Abnormal Psychology, 117*(4), 826–837.

Spotts, E. L., Lichtenstein, P., Pedersen, N., Neiderhiser, J. M., Hansson, K., Cederblad, M., & Reiss, D. (2005). Personality and marital satisfaction: A behavioural genetic analysis. *European Journal of Personality, 19*(3), 205–227.

Spotts, E. L., Neiderhiser, J. M., Hetherington, E. M., & Reiss, D. (2001). The relation between observational measures of social problem solving and familial antisocial behavior: Genetic and environmental influences. *Journal of Research on Adolescence, 11*(4), 351–374.

Spotts, E. L., Neiderhiser, J. M., Towers, H., Hansson, K., Lichtenstein, P., Cederblad, M., ... & Reiss, D. (2004). Genetic and environmental influences on marital relationships. *Journal of Family Psychology, 18*(1), 107–119.

Stover, C. S., Connell, C. M., Leve, L. D., Neiderhiser, J. M., Shaw, D. S., Scaramella, L. V., ... & Reiss, D. (2012). Fathering and mothering in the family system: linking marital hostility and aggression in adopted toddlers. *Journal of Child Psychology and Psychiatry, 53*(4), 401–409.

Thapar, A., Harold, G., Rice, F., Ge, X. J., Boivin, J., Hay, D., ... & Lewis, A. (2007). Do intrauterine or genetic influences explain the foetal origins of chronic disease? A novel experimental method for disentangling effects. *BMC Medical Research Methodology, 7*(25).

Thapar, A., Rice, F., Hay, D., Boivin, J., Langley, K., van den Bree, M., ... & Harold, G. (2009). Prenatal smoking might not cause attention-deficit/hyperactivity disorder: evidence from a novel design. *Biological Psychiatry, 66*(8), 722–727.

Turkheimer, E., Haley, A., Waldron, M., D'Onofrio, B., & Gottesman, I. I. (2003). Socioeconomic status modifies heritability of IQ in young children. *Psychological Science, 14*(6), 623–628.

Turkheimer, E., & Waldron, M. (2000). Nonshared environment: A theoretical, methodological, and quantitative review. *Psychological Bulletin, 126*(1), 78–108.

Unternaehrer, E., Luers, P., Mill, J., Dempster, E., Meyer, A. H., Staehli, S., ... & Meinlschmidt, G. (2012). Dynamic changes in DNA methylation of stress-associated genes (OXTR, BDNF) after acute psychosocial stress. *Translational Psychiatry, 2*, e150.

Van Ijzendoorn, M. H., & Bakermans-Kranenburg, M. J. (2006). DRD4 7-repeat polymorphism moderates the association between maternal unresolved loss or trauma and infant disorganization. *Attachment & Human Development, 8*(4), 291–307.

Vinkhuyzen, A. A. E., Van Der Sluis, S., De Geus, E. J. C., Boomsma, D. I., & Posthuma, D. (2010). Genetic influences on 'environmental' factors. *Genes, Brain and Behavior, 9*(3), 276–287.

Weissman, M. M. (1987). Advances in psychiatric epidemiology: rates and risks for major depression. *American Journal of Public Health, 77*(4), 445–451.

Wichers, M., Kenis, G., Jacobs, N., Mengelers, R., Derom, C., Vlietinck, R., & van Os, J. (2008). The BDNF Val66Met × 5-HTTLPR × child adversity interaction and depressive symptoms: An attempt at replication. *American Journal of Medical Genetics Part B: Neuropsychiatric Genetics, 147B*(1), 120–123.

Young, S., Smolen, A., Hewitt, J., Haberstick, B., Stallings, M., Corley, R., & Crowley, T. (2006). Interaction between MAO-A genotype and maltreatment in the risk for conduct disorder: failure to confirm in adolescent patients. *American Journal of Psychiatry, 163*(6), 1019–1025.

3

Gene–Environment Correlations in the Context of Parenting and Peer Relationships

Jeffrey Henry, Michel Boivin, and Mara Brendgen

Most models of human development highlight the role of environmental experiences in explaining individual differences in socio-emotional and cognitive development (e.g., Bowlby, 1982). This emphasis finds its support in longitudinal studies underlining associations between characteristics of early caregiving environments and later child development (e.g., Bradley & Corwin, 2007; De Wolff & Van IJzendoorn, 1997). However, there is now substantial evidence to suggest that both genetic and environmental factors shape these developmental associations (Turkheimer, 2000). More importantly, these two classes of factors are likely to interact in complex ways, and reciprocally influence each other through gene–environment (G–E) interplay (Plomin, DeFries, & Loehlin, 1977; Scarr & McCartney, 1983).

Twin studies are particularly well suited for the investigation of genetic and environmental factors – and their interplay. The basic strength of this method lies in two features: (1) its power to document within-family similarities and differences through comparing twin siblings; and (2) its capacity to examine these within-family and between-family similarities/differences as a function of genetic relatedness. Twins can thus be seen as a natural experiment where biological processes generate monozygotic (MZ) twins, who share 100% of their genes, or dizygotic (DZ) twins, sharing on average 50% of their segregating genes. In the classic twin design, this naturally occurring variation in genetic relatedness within families is harnessed to statistically tease apart genetic from environmental sources of variation in a given phenotype.

Genetic, shared environmental, and non-shared environmental sources of variation in a given phenotype are therefore estimated by comparing the phenotypic similarity among MZ twins to that of DZ twins. Higher phenotypic similarity of MZ compared to DZ twins is assumed to reflect genetic contributions (i.e., heritability), whereas equivalent phenotypic similarity represents shared environmental sources of variance. Non-shared environmental sources of variance (e.g., experiences that are unique to each twin) contribute to differences between twins of the same family.

When specific environments are directly measured *at the child level* (i.e., independently in each twin), twin designs allow for testing various G–E processes, including hypotheses regarding the aetiology and direction of the association between specific environments and child outcomes. These strengths are not trivial when the time comes to decrypt complex, multi-determined developmental processes. The goal of the present chapter is to document one form of G–E interplay: gene–environment correlation (rGE), which postulates a directed association from genes to the environment.

The gene–environment correlation (rGE)

An rGE is found when genetic variation of a child is associated with variation in exposure to a specific environment (Jaffee & Price, 2007). Various forms of rGE have been suggested (Plomin, DeFries, & Loehlin, 1977; Scarr & McCartney, 1983); an rGE is said to be *passive* if, for example, parents provide their child with an environment congruent with their own genotype (and thus their child's genotype); it can be *evocative/reactive* when the individual makes people around him/her react to his/her genetic characteristics; finally, it can be *selective/reactive* when the individual chooses an environment on the basis of his/her genetic makeup.

As stated earlier, the logic of the twin analysis may be extended to measured environments, such as those experienced within or outside the family. In the context of rGE, the goal then is to estimate the extent to which these measured environments are shared or uniquely experienced by twin siblings, and whether they are associated to genetically influenced child characteristics. Finding genetic child effects on specific environments, and then documenting that these genetic contributions are mediated through some characteristics of the child would provide evidence of rGE. For example, one may want to determine the extent to which the degree of peer victimization is accounted for by genetic factors, and then to examine through bivariate analyses the extent to which a measured child characteristic (e.g., conduct problems) predicts this negative experience, and whether or not this predictive association is mediated by genetic factors. If genetic effects are found, this may indicate that an evocative rGE process, based on child heritable characteristics, is involved.

Examination of rGE through the study of twins across childhood and adolescence is increasingly prominent in the child development literature. Most studies have examined how child genetic and environmental vulnerabilities are linked within the family (e.g., Boivin et al., 2005; Jaffee, Caspi, Moffitt, & Taylor, 2004a) and peer contexts (e.g., Harden, Hill, Turkheimer, & Emery, 2008; Rose, 2002) to shape developmental outcomes. The goal of this chapter is to briefly review emerging evidence in these two research areas, and then illustrate how different forms of rGE may underlie early parenting and peer relations. We start by considering early parenting and then review similar evidence of rGE for peer victimization. We conclude with a discussion of caveats and future directions regarding this literature.

Gene–environment correlations: the case of early parenting

Traditional models of early parenting, such as attachment theory (Bowlby, 1982), posit that individual differences in socio-emotional development arise from differences in early parental behaviours. Indeed, a stable pattern of parent–infant bonding behaviour and attachment relationship is progressively established through interactions over the course of the first 18 months of life. These attachment behaviours and associated mental representations are considered harbingers of subsequent social relationships (Ainsworth, 1985). A central claim of attachment theory is that attachment patterns mainly reflect the history of the caregiver's responses to the needs of the infant. Parental sensitivity – the caregiver's ability to detect the infant's needs and respond to them appropriately – is likely to contribute to a secure parent–child attachment relationship, thus creating a positive context for the child's later socio-emotional adjustment (Bowlby, 1982; Bretherton & Waters, 1985; De Wolff & Van IJzendoorn, 1997). Conversely, insensitive parental care, marked by inconsistencies in parental responses and a tendency to adopt hostile, restrictive and punitive behaviours toward the child, could lead to insecure attachment and later adjustment problems in the child.

The assumption that attachment patterns in the child are environmentally mediated is grounded in early evidence linking observed sensitive parenting at home and secure attachment in the laboratory-based Strange Situation (Ainsworth, Blehar, Waters, & Wall, 1978). It is also supported by more general findings showing that maladaptive parenting practices (e.g., punishment, overprotection, lack of sensitivity) are associated with attachment disorganization (Bretherton & Waters, 1985; De Wolff & Van IJzendoorn, 1997), and more general child outcomes, including externalizing (Bradley & Corwin, 2007; Campbell et al., 2007) and internalizing difficulties (Becker, Ginsburg, Domingues, & Tein, 2010; Gray, Carter, & Silverman, 2011). Although many studies confirmed a significant link between early care and later attachment and behavioural patterns, this association has been shown to vary considerably in strength (De Wolff & van IJzendoorn, 1997). Moreover, longitudinal studies have suggested that early child features (i.e., temperament, such as difficulty or fearfulness) could also elicit several forms of parenting (e.g., Martini, Root, & Jenkins, 2004; Rubin, Burgess, Dwyer, & Hastings, 2003). Clearly, a comprehensive study of early parenting should consider various classes of factors related not only to the parenting, but also to the infant and the family context as they interact over time (Belsky & Jaffee, 2006). This idea implies multi-causality, including possible *child contributions* to parenting.

Associations between child and parental behavioural patterns may partly result from shared genes rather than being only a reflection of environmental causation (Collins et al., 2000). For example, short-tempered parents may tend to display insensitive, punitive parenting behaviours and, due to shared genes, to have children who show irritable and disorganized behaviours (i.e., passive rGE). In the context of parent–infant interactions, the rGE hypothesis can be verified by determining the extent to which genetic factors in the child account for: (1) parental behaviours, and then, (2) the association between putative child early characteristics and later parental behaviours. Such analyses are essential in that they prompt us to revisit and reinterpret the nature of the mechanisms underlying the predictive association between parental behaviours and developmental outcomes in the child. Over the last 30 years, a growing number of twin studies have indeed estimated the genetic and environmental contributions to the quality of early parenting and quality of child–parent relationship.

Twin studies have traditionally assessed genetic–environmental contributions to early parenting using two assessment approaches: (1) observed quality of *parent–child interactions*; and (2) observed or self-reported quality of *parental behaviours*. While extant work on parent–child interactions has been more sporadic, it has provided valuable insights into the nature of these interactions. Two combined studies have examined the aetiology of observed parent–child dyadic mutuality (i.e., shared positive affect, responsiveness and cooperation) as assessed during a 20-minute interaction (Deater-Deckard & O'Connor, 2000). The first study included 62 MZ twin pairs and 58 DZ twin pairs (mean age = 3 years) and their primary caregiver – in the majority of cases the mother. Heritability and non-shared environment each accounted for half of the variance in mother–child dyadic mutuality, with no shared environmental contribution. These findings were replicated in a second observational study of 102 pairs of adoptive and biological siblings in matched comparison families (Deater-Deckard & O'Connor, 2000). Taken together, these results point to a potential rGE underlying the quality of early parent–infant interactions.

However, these results may be specific to the quality of the infant–mother relationship assessed via free-play interactions, because different results were obtained in studies employing infant–mother interactions structured around a task. Roisman and Fraley (2006) assessed infant–mother relationship quality in 127 MZ and 333 DZ 9-month old twin pairs using the Nursing Child Assessment Teaching Scale (Summer & Spietz, 1995). The infant's primary caregiver was asked to teach the target child a task just beyond his/her capacity. A total infant–caregiver score was derived by summing all items within a single indicator characterizing the degree to which parent and child successfully construct a "teaching loop" where the caregiver appropriately sets up expectations, scaffolds the situation, and offers adequate and sensitive feedback to the infant. Here, shared and non-shared environmental contributions were found to be substantial, while genetic variation was non-significant, in accounting for the infant–caregiver relationship quality (Roisman & Fraley, 2006). DiLalla and Bishop (1996) also found similar results, this time assessing observed maternal behaviours when the mother was attempting to elicit vocalizations when her twins were 7 and 9 months of age. Thus, mothers tended to behave similarly towards their twins, regardless of zygosity, which suggests a mother-driven process and the absence of an evocative rGE (DiLalla & Bishop, 1996). These results of low genetic (and high shared environmental) variance may differ from those of the Deater-Deckard/O'Connor group for many reasons. Most notably, observed situations where the parent's behaviour is slanted toward a specific goal may heighten the programming effect of context on parental behaviour, and artificially reduce "child effects" on parenting, with the result of increased shared environmental variance to the detriment of genetic variance. Therefore, extant work on the quality of infant–parent interactions suggests the presence of rGE in more naturalistic settings (i.e., free-play), while shared environmental effects were found in the context of structured tasks. This indicates that child effects could be involved in the general quality of day-to-day interactions with the parents, while specific contexts may sometimes shape these interactions through programming of parental behaviours, hence the shared environmental variance.

In addition to investigating parent–infant relationship quality, some twin studies have investigated possible rGE by examining the genetic–environmental aetiology of various dimensions of early parenting. These reports were based on univariate genetic analyses of parenting, which provided an initial test of the rGE hypothesis. For example, in the Quebec Newborn Twin Study, parents answered a questionnaire regarding their parenting of 5-month old twins. Specifically, parenting dimensions such as parental self-efficacy, perceived parental impact, hostile-reactive parenting behaviour and parental overprotection were investigated. Shared environment mainly accounted for each of the four parenting dimensions. However, maternal hostile-reactive behaviours were also found to be moderately heritable, and in bivariate analyses, this association was found to be mainly mediated by infant difficultness (Boivin et al., 2005).

In a follow-up of this study, longitudinal analyses were conducted on 292 mothers' self-reported hostile-reactive behaviours towards each of their twins at 6, 20 and 32 months of age (Forget-Dubois et al., 2007). The heritability of maternal hostile-reactive behaviour was modest and significant at 6 and 32 months, but not at 20 months. Interestingly, the heritability estimates at 6 and 32 months were uncorrelated. Shared environment was the main source of variance at the three time points and these estimates were highly correlated through time (Forget-Dubois et al., 2007). Thus, stability in self-reported hostile-reactive parenting was essentially accounted for by shared environment, perhaps partly due to the mother rating of both of her twins. In other words, if children's heritable characteristics evoked maternal hostile-reactive parenting behaviours at specific times, they did not appear to be responsible for its stability from infancy to toddlerhood (Forget-Dubois et al., 2007).

Other studies have examined the genetic–environmental aetiology of the *covariance* between measured parenting and child behaviour. As mentioned above, this particular type of bivariate analysis is important in that it provides a more specific insight with respect to the nature of the longitudinal developmental associations between parental and child behavioural patterns. When such associations are mediated by shared environment, they point to possible environmental causation. On the other hand, associations that are genetically mediated may reflect the presence of rGE.

Protective cycles of parenting/child behaviour have been investigated within such an rGE framework. One such study assessed maternal sensitivity in the home at 9–10 months of age of a child, while infant attachment security was observed in the laboratory at 12 months (Fearon et al., 2006). Shared environment in maternal sensitivity accounted for some of the similarity in attachment security, and there was no evidence of genetic child contribution to the association (Fearon et al., 2006). In a second study, 485 24-month old same-sex twin pairs were used to test for G–E associations between observed parenting quality and infant attachment security (Roisman & Fraley, 2008). In line with Fearon et al's study (2006), both constructs, as well as their covariation, were accounted for by shared (85%) and non-shared (15%) environment (Roisman & Fraley, 2008). Therefore, shared environment (i.e., environmental factors that tend to make twins more similar, such as socio-economic status or parental personality) may account for associations between adaptive forms of early parenting and later child development. Other studies have found the shared environment component to explain associations between positive parenting and high prosocial behaviour (Knafo & Plomin, 2006), as well as low externalizing behaviour (Boeldt et al., 2012). Taken together, available reports point to a putative environmental causation of *protective cycles* linking adaptive parenting to child outcomes.

Potential rGE processes were also investigated in explaining associations between maladaptive forms of early parenting and child outcomes. A very informative example of such rGE was highlighted in a study using a sample of 1116 5–7-year old twin pairs (Jaffee et al., 2004a). This study found that child antisocial behaviour at age 5 did not account for the prospective association between mother-reported physical maltreatment at age 5 and teacher-reported child antisocial behaviour at age 7 (Jaffee et al., 2004a). These initial findings suggested that early maltreatment could play a unique role in the later development of antisocial behaviour, over and above "child effects". In a second study, the limits of such "child effects" were tested on parental behaviour that ranged from the normative (i.e., corporal punishment) to the non-normative (i.e., physical maltreatment; Jaffee et al., 2004b). Shared environment accounted for most of the variation in both corporal punishment and physical maltreatment. However, corporal punishment, but not physical maltreatment, was partly genetically mediated, and the genetic factors that accounted for corporal punishment were largely the same as those that accounted for child antisocial behaviour, suggesting "child effects" for corporal punishment (Jaffee et al., 2004b). Thus, although normative discipline in the form of corporal punishment may partly be a function of child genetic effects, risk factors for maltreatment were found unlikely to be specifically associated with the child characteristics, and more likely to reside in environmental features varying across families. In line with this, other bivariate reports have found substantial genetic covariation between harsh/negative – but still normative – forms of early parenting and child behaviour problems such as difficultness (Boivin et al., 2005), externalizing (Boeldt et al., 2012), antisocial (Larsson, Viding, Rijsdijk, & Plomin, 2008) and low prosocial behaviours (Knafo & Plomin, 2006).

Results from genetically sensitive studies that examined the possibility of rGE in early parenting reveal four things. First, most early parenting practices providing a positive context for child development are hardly accounted for by genetically related child characteristics (Fearon et al., 2006). Therefore, protective factors for child outcomes may lie in environmental features that differ between families. Second, adverse – but normative – parenting behaviours (e.g., harshness, corporal punishment) partly involve genetic "child effects" which are likely to be associated with temperamental features, but these contributions may be seen as developmental episodes (i.e., time-specific and not necessarily persistent over time; Forget-Dubois et al., 2007). Third, more extreme forms of adverse parenting do not result from child-enacted developmental events, as child evocation does not account for physical maltreatment (Jaffee et al., 2004b). Thus, there seem to be limits to child effects on maladaptive parenting, which may be driven, in its more extreme forms, by features that differ across families. Fourth, within-family differences (i.e., non-shared environment) operate independently of genetics or shared environment in explaining the links between early environment and child development. These within-family differences could be driven by factors which tend to make twins distinct from each other – for example, child-specific parenting, peer relations, general life experience.

There is no doubt that parent–child interactions in the first years of life hold a very important role in later socio-emotional development. However, nowadays, many children attend daycare or preschool, spending much more time in the company of peers than with their families. Experiences with peers are thus likely to start playing a developmental role very early in life. In this context, genetically sensitive work on peer-related experiences is very informative.

Gene–environment correlations: the case of negative peer-related experiences

Peer relationships serve important developmental functions. From early to middle childhood, they provide contexts where children can learn self-control and new social skills, including socially appropriate means to solve interpersonal conflicts (Boivin et al., 2013a). Unfortunately, despite their positive features, experiences with peers can also be maladaptive. A substantial number of children experience chronic peer relation difficulties, such as peer rejection and victimization (Boivin, Dodge, & Coie, 1995; Juvonen, Graham, & Schuster, 2003; Kochenderfer & Ladd, 1996). Children who experience these problematic peer relations are not only deprived of social support and learning, but they suffer and are at risk of a variety of future adjustment problems, including depression, academic failure, delinquency and substance abuse (Boivin, Vitaro, & Poulin, 2005; Rubin, Bukowski, & Parker, 2006).

Peer relation difficulties have typically been captured from two perspectives (Boivin, Hymel, & Hodges, 2001). One perspective centres on children who are disliked and negatively perceived by peers. This form of *peer rejection* may lead to various negative experiences, such as excessive teasing and active peer harassment, but also to more subtle forms of ostracism and exclusion. *Peer victimization*, a related but different type of negative peer experience, refers to a child being repeatedly exposed to negative treatment by one or more children. Thus, peer victimization is a class of negative actions by peers, whereas peer rejection reflects the attitude of the peer group, attitude that may induce certain classes of manifest behaviours by peers, including overt victimization (Boivin et al., 2001). Both markers of peer difficulties have been found to predict emotional problems, such as loneliness, depression, anxiety and suicidal ideation, as well as decreased physical health and increased conduct and school problems (Arseneault et al., 2008; Boivin, Hymel, & Bukowski, 1995; Boulton & Underwood, 1992; Brendgen & Vitaro, 2008; Dodge et al., 2003; Nishina, Juvonen, & Witkow, 2005; Rigby, 1999; Rubin et al., 2006). Such strong predictive associations clearly underline the importance of these two facets of peer difficulties in developmental psychopathology. Unfortunately, the understanding of these associations is limited by the same methodological problem impinging on the early parenting literature: an overreliance on correlational designs, which are uninformative regarding aetiology and direction of associations. Recently, however, behavioural-genetic studies have provided important insights into how genetic factors are linked to and interact with specific peer-related experiences.

Evidence for active rGE

In the context of peer relations, rGE processes may arise when a child's heritable traits affect the kind of peer environment he or she experiences through passive, active or evocative processes. A passive rGE would occur if the parents' heritable characteristics naturally influence the quality of their children's peer environment. For example, parents with a history of delinquent behaviour may be more likely than others to settle in disadvantaged neighbourhoods with a high crime rate, which increases their children's risk of being recruited by criminal gangs. Contrary to passive rGE, both active and evocative rGE involve a direct contribution of a child's heritable characteristics to his or her environmental experiences. Active rGE arises when an individual actively selects his or her own environment on the basis of his or her genetically influenced personal characteristics. For example, delinquent adolescents may deliberately choose to affiliate with other delinquent friends. Such an active rGE could explain the tendency of children with antisocial tendencies to affiliate with one another. This possibility has been assessed using the twin method (Brendgen, Boivin, Vitaro, Dionne, & Pérusse 2007; Van Lier et al., March 2007). Taken together, findings from these studies suggest that affiliation with antisocial friends is affected by active rGE in adolescence, but is largely unrelated to individuals' genetic disposition prior to adolescence. This pattern suggests that rGE gradually appears in early- to mid-adolescence (Brendgen & Boivin, 2015).

Evidence for evocative rGE in peer relations difficulties

An active rGE is unlikely to explain peer experiences that are not driven by active choice, such as peer rejection and peer victimization. A more likely mechanism for these experiences is evocative (or reactive) rGE, which occurs when the child's characteristics elicit specific reactions from the peer environment. For example, a child who displays impulsive and aggressive behaviour may elicit dismissive peer responses and, as a result, become rejected and/or victimized by his or her peers. In contrast to active rGE related to deviant peer affiliation, which seems to emerge only during the transition to adolescence, evocative rGE with respect to peer rejection/victimization may appear at an earlier age. Although relatively few genetically informative studies have been published in this regard, the findings converge in showing that, as soon as children enter school, genetic factors start to play a significant role in peer rejection and peer victimization.

Relevant evidence comes from a series of reports based on the Quebec Newborn Twin Study, an ongoing population-based longitudinal study of twins for whom classroom peer nominations were obtained at multiple times (Boivin et al., 2012). The classmates of each twin were asked to participate in a peer nomination task. The participating children were asked to nominate three classmates for each of a series of behavioural and social descriptors.

In an initial study, peer nominations by twins' classmates in kindergarten were used to assess peer rejection (Brendgen et al., 2009). All children in the class were asked to nominate three classmates they most liked to play with (positive nominations) and three classmates they least liked to play with (negative nominations). A total number of received positive nominations was calculated for each participant and *z*-standardized within classroom to create a total Liked-Most-score. Similarly, the total number of negative nominations was calculated and *z*-standardized to create a total Liked-Least-score. The Liked-Most-score was then subtracted from the Liked-Least-score to create a continuous scale, where low levels indicated high rejection by peers. Univariate ACE analyses revealed that 30% of the variance in peer rejection could be explained by heritable factors, another 15% was explained by shared environmental sources, and the remaining variance was due to non-shared environment.

These initial findings suggest that genetic factors are associated with peer difficulties. However, an important limitation affecting the above-mentioned studies is that they provide a limited view of developmental processes due to peer relationships being assessed only at a single point in time. There could be developmental changes in the relative strength of genetic contributions to peer difficulties. For instance, Ball et al. (2008) found a genetic contribution to peer victimization in middle childhood, but this finding was not replicated among kindergarten children (Brendgen et al., 2008). This differing pattern could be due to the use of different methods for assessing peer victimization (i.e., mother versus peer evaluations). It could also reflect age-related processes in associations between genetic factors and peer difficulties driven by the progressive establishment of an evocative rGE as described above.

Accordingly, a longitudinal approach was adopted in the QNTS to examine the aetiology of peer relationship difficulties during the early school years (Boivin et al., 2013a). To this end, the genetic–environmental contributions to peer relation difficulties were analysed using a combination of peer-assessed peer rejection and peer-, teacher- and self-assessed peer victimization in kindergarten (5 years), Grade 1 (6 years) and Grade 4 (9 years). Starting in kindergarten and then in later grades, genetic factors accounted for most of the variance in peer difficulties, as indexed by a *latent factor* combining peer-, teacher- and self-ratings. Specifically, genetic factors accounted for 73%, 74% and 94% of individual differences in peer difficulties in kindergarten, Grade 1 and Grade 4, respectively, leaving 27%, 26% and 6% to non-shared environment. Genetic factors also accounted for the observed stability in peer difficulties in the early school years ($r = .73$ between kindergarten and Grade 1; and .69 between Grade 1 and Grade 4). Yet, the patterns of familial aggregation for peer relation difficulties in kindergarten were generally low and unstable. Results for later grades were clearer, and univariate findings pointed toward the progressive establishment of rGE regarding peer difficulties in the later school grades. Indeed, genetic contributions emerged progressively, as did a growing consensus among informants with respect to those who experienced peer difficulties. Thus, genetic factors underlying peer difficulties appeared to be enduring: as the negative experiences crystallized, the same genetically vulnerable children became regularly entangled in a cycle of adverse peer relations.

These significant genetic effects on peer rejection and victimization indicate the presence of an evocative rGE, suggesting that heritable characteristics may at least partly explain why some children are subject to covert or overt disdain from peers. But what could these heritable characteristics be? Over and above social contexts and group processes (Boivin, Dodge, & Coie, 1995; Hymel, Wagner & Butler, 1990; Salmivalli & Voeten, 2004; Wright, Giammarino & Parad, 1986), and beyond atypical physical attributes such as speech problems, physical clumsiness and obesity (Boivin, Vitaro, & Poulin, 2005; Cramer & Steinwert, 1998; Rubin et al., 2006), social behaviour may be involved (Boivin et al., 2005; Rubin et al., 2006). Indeed, children who show disruptive behaviour problems (e.g., aggression, impulsivity) are more likely to experience peer relation difficulties (Coie & Kupersmidt, 1983; Dodge, 1983; Erhardt & Hinshaw, 1994; Hoza, 2007; Hoza et al., 2005). Unsurprisingly, these early disruptive behaviours are known to be influenced by genetic factors (Rhee & Waldman, 2002). Thus, part of the risk of peer difficulties could be traced back to genetic risk for these behaviours in the child. Until recently, these hypotheses had not been directly tested.

In the first published study to examine this issue, Ball et al. (2008) used data from the Environmental Risk (E-Risk) Longitudinal Twin Study, a cohort of over 1000 MZ and DZ twin pairs. Participants' aggressive behaviour and peer victimization at age 10 years were assessed by mothers and teachers. Seventy-five percent of the variance in peer victimization was accounted for by genetic factors, and 24% by non-shared environment. Moreover, a significant but modest part of the genetic variance in peer victimization was shared with aggressive behaviour, which was indicative of rGE.

Again, these initial findings were limited by the cross-sectional nature of the design. Implicit to the idea of rGE is the view that there is a chronological sequence linking a child characteristic to a social experience. Thus, QNTS data were recently used to test the presence of rGE linking disruptive behaviours (i.e., aggressive and impulsive/hyperactive behaviours) and peer relation difficulties from Kindergarten to Grade 1 (Boivin et al., 2013b). To this end, a cross-lagged method was used taking advantage of the developmental time lag and heuristics of the twin design to more precisely assess directional associations and the aetiological architecture of the associations between disruptive behaviours and peer difficulties. As predicted, disruptive behaviours were concurrently and predictively associated with peer difficulties, and genetic factors accounted for a substantial part of these associations, thus confirming rGE.

Taking advantage of the longitudinal approach, this study showed that additive genetic factors were essential in accounting for the association between disruptive behaviours and peer difficulties. Given that peer difficulties are an experience (i.e., not a child phenotype), an evocative rGE linking disruptive behaviours and peer difficulties may be established in the first year of school, if not earlier (Barker et al., 2008), and tends to persist over time. Hence, children displaying disruptive tendencies, perhaps due to self-regulation problems, may not only provoke bullying from the peer environment, but may also elicit little sympathy and support from classmates when they are rejected and victimized. Others may perceive these children as largely responsible for their own fate, thus solidifying their reputation as socially rejected and contributing to a constant cycle of peer difficulties (Boivin et al., 2001; Schuster, 2001; Teräsahjo & Salmivalli, 2003).

Conclusion

This chapter aimed to review the evidence for rGE with respect to early parenting and peer relations. Available reports suggest that genetics partly account for the adverse, but normative, early parenting as well as peer relationship difficulties in the early school years. With respect to peer relationships, some children are genetically at risk for disruptive behaviours, and this genetic vulnerability makes them likely to also experience peer rejection and victimization starting in kindergarten. The child characteristics responsible for rGE affecting early parenting are not as clear, although difficult temperament is a possible factor.

These rGE findings are important for our understanding of child development. Strong rGE processes suggest that genes and environments work together in a dynamic mode that may reinforce and exacerbate the initial underlying child vulnerability. In the case of peer relations, hyperactive-impulsive and aggressive behaviours seem likely to create a coercive peer context that could provide the training ground for increased behaviour problems and later offending. In other words, genetically liable children seem to be progressively affected by a dual risk: an initial genetic vulnerability for disruptiveness augmented by a cycle of negative peer experiences in school likely to accentuate their maladaptive pathway. In the case of early parenting, maladaptive forms of child-rearing practices were generally found more heritable than positive parenting and high parental support. This may indicate that initial parent–infant interactions are the initial scene of ongoing coercive cycles of bidirectional dynamics linking maladaptive forms of parenting and child behaviour. On the one hand, some genetically vulnerable children may – via early characteristics, such as difficult temperament or punishment insensitivity – elicit harsh/negative responses from their caregiving environments (i.e., evocative rGE). On the other hand, convergence in child (e.g., aggression, impulsivity) and parent (e.g., harshness) behavioural patterns may partly be the reflection of shared genes (i.e., passive rGE).

The developmental cascades implied by rGE have important implications for the early identification of children at risk and the planning of appropriate early preventive intervention. They indicate that some children are at risk from the early stages of development, and that this risk will be increased through a combination of gene–environment processes, including rGE. But most importantly, they should not be interpreted as a sign that the negative developmental pathways are set and irreversible. Rather, they underline the role of child factors in the emergence of child–parent and peer relations, and point to the need of intervening early to possibly alter those characteristics and prevent these negative experiences from becoming entrenched.

There are, however, persistent methodological challenges in characterizing the work of rGE in child development. As mentioned earlier, the findings for early parenting are not as conclusive as those for peer relationships. One of the central problems with many previous twin studies investigating potential rGE in parenting concerns the limitations in statistical power. Due to the cost of collecting such data, reliance on small samples is still typical of investigation using direct assessment, such as observation of infant–parent interactions. Sufficiently powered studies are warranted to examine potential sex differences in rGE linking environmental risk factors with child psychological development.

There is another challenge in documenting potential developmental changes in rGE. Two implicit hypotheses are prominent in rGE research in child development: (1) some rGE may cumulate and therefore play a more important role as the child grows up (Scarr & McCartney, 1983); (2) there could be changes in the nature of rGE over the course of development (Scarr & McCartney, 1983). For instance, passive rGE is more likely to occur early in development as it underlies parental influence on children's experiences through selecting and structuring their child's physical or social environment. However, as children grow, passive rGE should decline and selective rGE should increase as children become more active in shaping their own social environment. Unfortunately, because of a lack of genetically sensitive longitudinal data, there is very little evidence to support such hypotheses. In this regard, more systematic longitudinal research is warranted.

Moreover, on their own, twin designs do not provide conclusive demonstrations as to the specific form(s) of rGE involved in child development. Bivariate twin studies can show evidence of rGE, but cannot discriminate between active, passive and evocative rGE, although specific forms may be inferred according to theoretical models of child development, especially in peer relations. In the case of early parenting, adoption studies – through the assessment of birth parent(s)/adoptive parent(s)/adopted infant triads – may help specify the type of rGE involved, but such designs cannot tease out shared and non-shared environmental sources of variance. Combining twin and adoption designs in testing rGE is likely to fill some gaps in the knowledge on the nature of these rGEs.

References

Ainsworth, M. D. S. (1985). Attachments across the life span. *Bulletin of the New York Academy of Medicine, 61*, 792–812.

Ainsworth, M. D. S., Blehar, M. C., Waters, E., & Wall, S. (1978). *Patterns of attachment: A study of the strange situation*. Hillsdale, NJ, US: Erlbaum.

Arseneault, L., Milne, B. J., Taylor, A., Adams, F., Delgado, K., Caspi, A., & Moffitt, T. E. (2008). Being bullied as an environmentally mediated contributing factor to children's internalizing problems: a study of twins discordant for victimization. *Archives of Pediatrics and Adolescent Medicine, 162*, 145–150.

Ball, H. A., Arseneault, L., Taylor, A., Maughan, B., Caspi, A., & Moffitt, T. E. (2008). Genetic and environmental influences on victims, bullies and bully-victims in childhood. *Journal of Child Psychology and Psychiatry, 49*, 104–112.

Barker, E. D., Boivin, M., Brendgen, M., Fontaine, N., Arseneault, L., Vitaro, F., ..., & Tremblay, R. E. (2008). Predictive validity and early predictors of peer-victimization trajectories in preschool. *Archives of General Psychiatry, 65*, 1185–1192.

Becker, K. D., Ginsburg, G. S., Domingues, J., & Tein, J.-Y. (2010). Maternal control behavior and locus of control: examining mechanisms in the relation between maternal anxiety disorders and anxiety symptomatology in children. *Journal of Abnormal Child Psychology, 38*, 533–543.

Belsky, J., & Jaffee, S. R. (2006). The multiple determinants of parenting. In D. Cicchetti & D. J. Donald (Eds), *Developmental Psychopathology, Vol.3: Risk, Disorder, and Adaptation* (pp. 38–85). Hoboken, NJ: John Wiley & Sons.

Boeldt, D. L., Rhee, S. H., DiLalla, L. F., Mullineaux, P. Y., Schulz, H. R., et al. (2012). The association between positive parenting and externalizing behaviour. *Infant and Child Development, 21*, 85–106.

Boivin, M., Brendgen, M., Dionne, G., Dubois, L., Pérusse, D., Robaey, P., Tremblay, R. E., & Vitaro, F. (2012). The Quebec Newborn Twin Study into adolescence: 15 years later. *Twin Research and Human Genetics, 16*, 64–69.

Boivin, M., Brendgen, M., Vitaro, F., Dionne, G., Girard, A., Pérusse, D., & Tremblay, R. E. (2013a). Strong genetic contribution to peer relationship difficulties at school entry: findings from a longitudinal twin study. *Child Development, 84*(3), 1098–1114.

Boivin, M., Brendgen, M., Vitaro, F., Forget-Dubois, N., Feng, B., Tremblay, R. E., & Dionne, G. (2013b). Evidence of gene–environment correlation for peer difficulties: disruptive behaviors predict early school peer relation difficulties through genetic effects. *Development and Psychopathology, 25*, 79–92.

Boivin, M., Dodge, K. A., Coie, J. D. (1995). Individual–group behavioral similarity and peer status in experimental play groups of boys: The social misfit revisited. *Journal of Personality and Social Psychology, 69*, 269–279.

Boivin, M., Hymel, S., & Burkowski, W. M. (1995). The roles of social withdrawal, peer rejection, and victimization by peers in predicting loneliness and depressed mood in childhood. *Development and Psychopathology, 7*, 765–785.

Boivin, M., Hymel, S., & Hodges, E. V. E. (2001). Toward a process view of peer rejection and harassment. In J. Juvonen & S. Graham (Eds), *Peer harassment in school: The plight of the vulnerable and victimized* (pp. 265–289). New York, NY, US: Guilford Press.

Boivin, M., Pérusse, D., Dionne, G., Saysset, V., Zoccolillo, M., Tarabulsy, G., et al. (2005). Parent's perceptions and self-assessed behaviors toward their 5-month-old infants in a large twin and singleton sample. *Journal of Child Psychology and Psychiatry and Allied Disciplines, 46*, 612–630.

Boivin, M., Vitaro, F., & Poulin, F. (2005). Peer relationships and the development of aggressive behavior in early childhood. In R. E. Tremblay, W. W. Hartup & J. Archer (Eds), *Developmental origins of aggression* (pp. 376–397). New York: Guilford Press.

Boulton, M. J., & Underwood, K. (1992). Bully/victim problems among middle school children. *British Journal of Educational Psychology, 62*, 73–87.

Bowlby, J. (1982). *Attachment and loss* (vol. 1): *Attachment* (2nd ed.). London, UK: Hogarth Press.

Bradley, R. H., & Corwin, R. F. (2007). Externalizing problems in fifth grade: Relations with productive activity, maternal sensitivity, and harsh parenting from infancy through middle childhood. *Developmental Psychology, 43*, 1390–1401.

Brendgen, M., & Boivin, M. (2015). Childhood and Adolescence: Peer Rejection and Affiliation with Deviant Peers. In J. M. Neiderhiser and B. N. Horwitz (Eds), *Behavioral Genetics of Interpersonal Relationships across the Lifespan. Behavioral Genetics Book Series*. Thousand Oaks, CA: Sage Publications.

Brendgen, M., Boivin, M., Vitaro, F., Dionne, G., Girard, A., & Pérusse, D. (2008). Gene–environment interaction between peer victimization and child aggression. *Development and Psychopathology, 20*, 455–471.

Brendgen, M., Boivin, M., Vitaro, F., Dionne, G., & Pérusse, D. (March 2007). *Does Exposure to Aggressive Friends Moderate the Effect of Genes on Children's Social and Physical Aggression?* Paper presented at the Biennial Meeting of the Society for Research in Child Development, Boston, MA.

Brendgen, M., & Vitaro, F. (2008). Peer Rejection and Physical Health Problems in Early Adolescence. *Journal of Developmental and Behavioral Pediatrics, 29*, 183–190.

Brendgen, M., Vitaro, F., Boivin, M., Girard, A., Bukowski, W. M., Dionne, G.,..., & Pérusse, D. (2009). Gene–Environment Linkages Between Peer Rejection and Depressive Symptoms in Children. *Journal of Child Psychology and Psychiatry, 50*, 1009–1017.

Bretherton, I., & Waters, E. (Eds) (1985). *Growing points of attachment theory and research*. Monographs of the Society for Research in Child Development. Serial No. 209.

Campbell, S. B., Matestic, P., von Stauffenberg, C., Mohan, R., & Kirchner, T. (2007). Trajectories of maternal depressive symptoms, maternal sensitivity, and children's functioning at school entry. *Developmental Psychology*, 43, 1202–1215.

Coie, J. D., & Kupersmidt, J. B. (1983). A behavioral analysis of emerging social status in boy's groups. *Child development, 54*, 1400–1416.

Collins, W. A., Maccoby, E. E., Steinberg, L., Hetherington, E. M., & Bornstein, M. H. (2000). Contemporary research on parenting: The case for nature and nurture. *American Psychologist, 55*, 218–232.

Cramer, P., & Steinwert, T. (1998). Thin is good, fat is bad: How early does it begin? *Journal of Applied Developmental Psychology, 19*(3), 429–451.

Deater-Deckard, K., & O'Connor, T. G. (2000). Parent–child mutuality in early childhood: Two behavioral genetic studies. *Developmental Psychology, 36*, 561–570.

De Wolff, M. S., & Van IJzendoorn, M. H. (1997). Sensitivity and attachment: A meta-analysis on parental antecedents of infant attachment. *Child Development, 68,* 571–591.

DiLalla, L. F., & Bishop, E. G. (1996). Differential maternal treatment of infant twins: effects of infant behaviors. *Behavior Genetics, 26,* 535–542.

Dodge, K. A. (1983). Behavioral Antecedents of Peer Social Status. *Child Development, 54,* 1386–1399.

Dodge, K. A., Lansford, J. E., Burks, V. S., Bates, J. E., Pettit, G. S., Fontaine, R., & Price, J. M. (2003). Peer rejection and social information-processing factors in the development of aggressive behavior problems in children. *Child Development, 74,* 374–393.

Erhardt, D., & Hinshaw, S. P. (1994). Initial sociometric impressions of attention-deficit hyperactivity disorder and comparison boys: Predictions from social behaviors and from nonbehavioral variables. *Journal of Consulting and Clinical Psychology, 62,* 833–842.

Fearon, R. M., Van Ijzendoorn, M. H., Fonagy, P., Bakermans-Kranenburg, M. J., Schuengel, C., & Bokhorst, C. L. (2006). In search of shared and nonshared environmental factors in security of attachment: A behavior-genetic study of the association between sensitivity and attachment security. *Developmental Psychology, 42*(6), 1026–1040.

Forget-Dubois, N., Boivin, M., Dionne, G., Pierce, T., Tremblay, R. E., & Pérusse, D. (2007). A longitudinal twin study of the genetic and environmental etiology of maternal hostile-reactive behavior during infancy and toddlerhood. *Infant Behavior & Development, 30,* 453–465.

Gray, C., Carter, R., & Silverman, W. (2011). Anxiety Symptoms in African American Children: Relations with Ethnic Pride, Anxiety Sensitivity, and Parenting. *Journal of Child & Family Studies, 20,* 205–213.

Harden, K. P., Hill, J. E., Turkheimer, E., & Emery, R. E. (2008). Gene–environment correlation and interaction in peer effects on adolescent alcohol and tobacco use. *Behavior Genetics, 38,* 339–347.

Hoza, B. (2007). Peer functioning in children with ADHD. By special arrangement, published jointly in *Ambulatory Pediatrics, 7,* 101–106, & *Journal of Pediatric Psychology, 32,* 719–727.

Hoza, B., Mrug, S., Gerdes, A. C., Hinshaw, S. P., Bukowski, W. M., Gold, J. A., Kraemer, H. C., Pelham, W. E., Wigal, T., & Arnold, L. E. (2005). What aspects of peer relationships are impaired in children with attention-deficit/hyperactivity disorder? *Journal of Consulting and Clinical Psychology, 73,* 411–423.

Hymel, S., Wagner, E., & Butler, L. J. (1990). Reputational bias: View from the peer group. In S. R. Asher & J. D. Coie (Eds), *Peer rejection in childhood* (pp. 156–188). New York, NY, US: Cambridge University Press.

Jaffee, S. R., Caspi, A., Moffit, T. E., Polo-Tomas, M., Price, T., & Taylor, A. (2004b). The limits of child effects: Evidence for genetically mediated child effects on corporal punishment but not on physical maltreatment. *Developmental Psychology, 40,* 1047–1058.

Jaffee, S. R., Caspi, A., Moffitt, T. E., & Taylor, A. (2004a). Physical maltreatment victim to antisocial child: Evidence of an environmentally mediated process. *Journal of Abnormal Psychology, 113,* 44–55.

Jaffee, S. R., & Price, T. S. (2007). Gene–environment correlations: a review of the evidence and implications for prevention of mental illness. *Molecular Psychiatry, 12,* 432–442.

Juvonen, J., Graham, S., & Schuster, M. A. (2003). Bullying among adolescents: The strong, the weak, and the troubled. *Pediatrics, 112*, 1231–1237.

Knafo, A., & Plomin, R. (2006). Parental discipline and affection and children's prosocial behavior: Genetic and environmental links. *Journal of Personality and Social Psychology, 90*, 147–164.

Kochenderfer, B. J., & Ladd, G. W. (1996). Peer victimization: Cause or consequence of school maladjustment? *Child development, 67*, 1305–1317.

Larsson, H., Viding, E., Rijsdijk, F. V., & Plomin, R. (2008). Relationships between parental negativity and childhood antisocial behavior over time: a bidirectional effects model in a longitudinal genetically informative design. *Journal of Abnormal Child Psychology, 36*, 633–645.

Martini, T., Root, C., & Jenkins, J. (2004). Low and middle income mothers' regulation of negative emotion: effects of children's temperament and situational emotional responses. *Social Development, 13*, 515–530.

Nishina, A., Juvonen, J., & Witkow, M. R. (2005). Sticks and Stones May Break My Bones, but Names Will Make Me Feel Sick: The Psychosocial, Somatic, and Scholastic Consequences of Peer Harassment. *Journal of Clinical Child and Adolescent Psychology, 34*, 37–48.

Plomin, R., DeFries, J. C., & Loehlin, J. C. (1977). Genotype–environment interaction and correlation in the analysis of human behavior. *Psychological Bulletin, 84*, 309–322.

Rhee, S., & Waldman, I. D. (2002). Genetic and environmental influences on antisocial behavior: a meta-analysis of twin and adoption studies. *Psychological Bulletin, 29*, 490–529.

Rigby, K. (1999). Peer victimisation at school and the health of secondary school students. *The British journal of educational psychology, 69*, 95–104.

Roisman, G. I., & Fraley, R. C. (2006). The limits of genetic influence: A behavior–genetic analysis of infant–caregiver relationship quality and temperament. *Child Development, 77*, 1656–1667.

Roisman, G. I., & Fraley, R. C. (2008). A behavior-genetic study of parenting quality, infant attachment security, and their covariation in a nationally representative sample. *Developmental Psychology, 44*, 831–839.

Rose, R. J. (2002). How do adolescents select their friends? A behavior-genetic perspective. In L. Pulkkinen & A. Caspi (Eds), *Paths to successful development: Personality in the life course* (pp. 106–125). New York, NY, US: Cambridge University Press.

Rubin, K. H., Bukowski, W., & Parker, J. G. (2006). Peer interactions, relationships, and groups. In N. Eisenberg & W. Damon (Eds), *Handbook of child psychology* (6 ed., Vol. 3, pp. 571–645). Hoboken, NJ, US: John Wiley & Sons Inc.

Rubin, K. H., Burgess, K. B., Dwyer, K. M., & Hastings, P. D. (2003). Predicting preschoolers' externalizing behaviors from toddler temperament, conflict, and maternal negativity. *Developmental Psychology, 39*(1), 164–176.

Salmivalli, C., & Voeten, M. (2004). Connections between attitudes, group norms, and behaviors associated with bullying in schools. *International Journal of Behavioral Development, 28*, 246–258.

Scarr, S., & McCartney, K. (1983). How people make their own environments: A theory of genotype–environment effects. *Child Development, 54*, 424–435.

Schuster, B. (2001). Rejection and victimization by peers: Social perception and social behavior mechanisms. In J. Juvonen & S. Graham (Eds), *Peer harassment in school: The plight of the vulnerable and victimized* (pp. 290–309). New York, NY, US: Guilford Press.

Summer, G., & Spietz, A. L. (1995). *NCAST caregiver/parent–child interaction teaching manual, 2nd edition*. Seattle, WA, US: NCAST Publications.

Teräsahjo, T., & Salmivalli, C. (2003). "She is not actually bullied." The discourse of harassment in student groups. *Aggressive behavior, 29*, 134–154.

Turkheimer, E. (2000). Three laws of behavior genetics and what they mean. *Current Directions in Psychological Science, 9*, 160–164.

Van Lier, P., Boivin, M., Dionne, G., Vitaro, F., Brendgen, M., Koot, H., ..., & Pérusse, D. (2007). Kindergarten children's genetic vulnerabilities interact with friends' aggression to promote children's own aggression. *Journal of the American Academy of Child and Adolescent Psychiatry, 46*, 1080–1087.

Wright, J. C., Giammarino, M., & Parad, H. W. (1986). Social status in small groups: individual–group similarity and the social "misfit". *Journal of Personality and Social Psychology, 50, 523–536.*

4

Behavioural Genetic Studies of Reading and Mathematics Skills

Stephen A. Petrill

Reading Disability (RD) and Math Disability (MD) have been associated with negative consequences for academic achievement, job quality, social relationships and mental health (Boetsch, Green, & Pennington, 1996; White, Moffitt, & Silva, 1992). Because RD and MD co-occur in 30–70% of individuals (Badian, 1999; Kovas, Haworth et al., 2007; Landerl & Moll, 2010; White, Moffitt, & Silva, 1992), the purpose of this chapter is to examine the comorbidity and independence of reading and math skills from a behavioural genetics perspective. First, the chapter will describe the behavioural genetic approach, discussing the genetic and environmental etiology of RD and MD, as well as of wider reading and math skills. Second, the chapter will discuss multivariate genetic studies of the relationship between math and reading ability/disability. Third the chapter will examine molecular genetic studies, and the promise of neurobiological markers for understanding the genetic and environmental underpinnings of reading and math ability/disability. Finally, conclusions and implications for educational practice will be discussed.

The behavioural genetic approach

Behavioural genetics methods are based on the assumption that the genetic and environmental etiology of a trait can be estimated by comparing family members with differing levels of genetic relatedness. Using twins as an example, identical (also called monozygotic or MZ) twins, share 100% of their inherited genes. Fraternal twins (also called dizygotic or DZ) twins share 50% of their inherited genes, on average. If identical twins are more similar than fraternal twins in a trait, the behavioural genetic model assumes that this is due to genetic influences, referred to as heritability, or h^2. On the other hand, if identical twins and fraternal twins are equally similar, then shared family environmental influences (experiences or events that make family members similar irrespective of genetic relatedness) are assumed to be significant. This is also called the common environment, or c^2. Finally, identical twins are unique individuals. Thus, the extent to which twins are uncorrelated is assumed to be due to nonshared environment, or e^2. It is important to stress that h^2, c^2, and e^2 describe the etiology of the differences in a group of people. In other words, $h^2 = .80$ means that 80% of the differences we measure in a trait across individuals are due to genetic differences in those individuals being assessed. These statistics do not refer to the amount of attributable to genetics versus environment in a particular person.

Univariate genetic studies of reading and mathematics

This approach has been used extensively to study the genetic and environmental etiology of reading and mathematics. Reading, in particular, has been one of the most successful targets of behavioural genetic research. Almost immediately after the first published study of "congenital word blindness" (Morgan, 1896), clinicians began reporting that reading disability ran in families (Hinshelwood, 1911). As described more fully in Pennington (1999), these family studies were soon joined by twin and adoption studies suggesting that identical twins were more likely to be concordant for reading disability than fraternal twins, and that biologically related siblings were more likely to be concordant for reading disability than adoptive siblings. In other words, genetic influences contributed to why reading disability ran in families.

Beginning in the 1980s behavioural genetic researchers began to examine reading disability from a dimensional perspective (DeFries & Fulker, 1985, 1988). In particular, DeFries and Fulker selected twins with reading disability, then examined whether their cotwins were more likely to have attenuated reading abilities, even if they did not cross the threshold for reading disability. This logic of the DeFries–Fulker method is shown in Figure 4.1. Probands (affected twins) are selected on the basis of reading disability. The resulting *proband mean* is far below the threshold for reading disability. Next the MZ and DZ *cotwin means* are estimated. If there is no relationship between twins, then the cotwin mean should approach the population mean. However, if there are familial factors impacting disability, then the cotwin mean should approach the threshold for disability. As shown in Figure 4.1, if MZ cotwin means are closer to the threshold (to their proband siblings) than DZ twins, genetic effects are implicated. The main advantage of this approach is that, beyond merely showing that MZ twins are more likely to be concordant than DZ twins, estimates of h^2, c^2, and e^2 can be obtained for reading disability.

Using this approach, several groups have found evidence for significant heritability for reading disability (Davis et al., 2001; Pennington & Olson, 2005; Plomin, 1991; Stevenson, 1991; Stevenson, Graham, Fredman, & McLoughlin, 1987). Moreover, these studies generally suggest that the

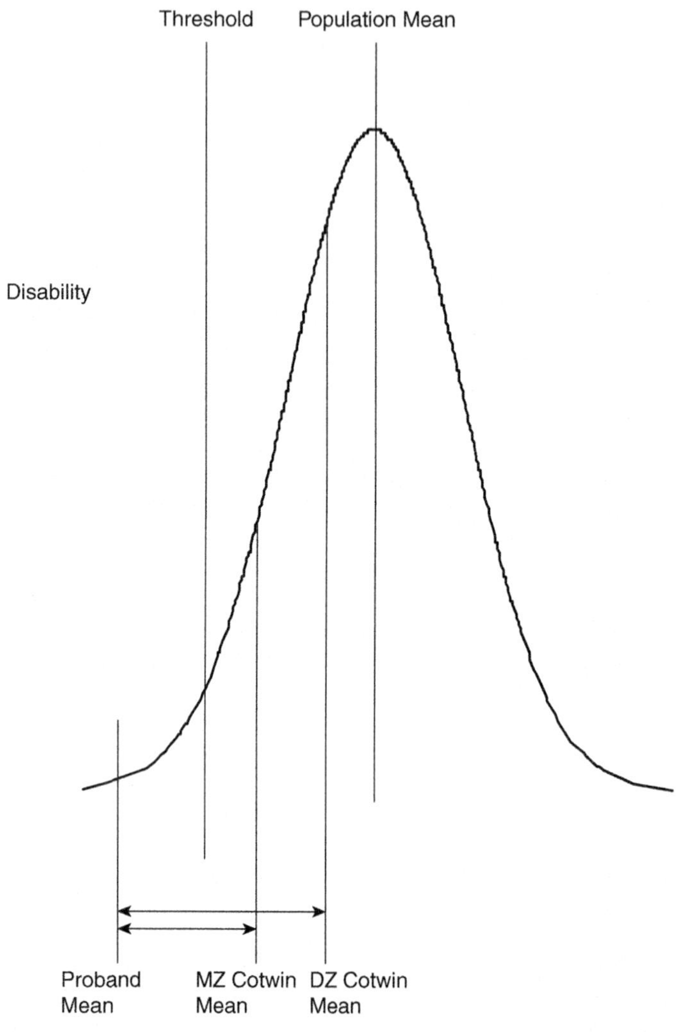

Figure 4.1 DeFries–Fulker analysis

underlying etiology of disability does not appear to be any more or less "heritable" or "environmental" than the typical range. Instead, genetic, shared environmental, and nonshared environmental factors for reading disability and math disability appear to be similar in magnitude to those found across the distribution of reading and mathematics skills. As the appreciation for the dimensionality of reading and reading disability developed, researchers also began to examine the cognitive skills associated with reading disability. The larger reading literature has consistently shown that phonological awareness, sound–letter correspondence, vocabulary, and reading fluency are important to reading (National Reading Panel, National Institute of Child & Human Development, 2000; Storch & Whitehurst, 2002). Numerous quantitative genetic studies have suggested that genetic influences are important to these components of reading processing (e.g., Olson, Gillis, Rack, DeFries, & Fulker, 1991).

In contrast to reading and reading disability, until the early 2000s, behavioural genetic studies of mathematics were confined to a handful of published papers with widely varying results (Alarcon, DeFries, Light, & Pennington, 1997; Knopik, Alarcon, & DeFries, 1997; Knopik & DeFries, 1999; Loehlin & Nichols, 1976; Thompson, Detterman, & Plomin, 1991; Vandenberg, 1962). Since that time, several research groups have conducted large, systematic behavioural genetic studies of mathematics, showing that both genetic and shared environmental influences are important to mathematics ability and disability (Hart, Petrill, Thompson, & Plomin, 2009; Haworth, Kovas, Petrill, & Plomin, 2007; Kovas, Haworth, Petrill, & Plomin, 2007; Lemelin et al., 2007; Oliver et al., 2004; Petrill & Plomin, 2007).

Multivariate behavioural genetics studies of reading and mathematics

Although the univariate genetic studies of reading and mathematics are important in that they tell us "whether and how much" genes and environments are associated with reading and mathematics skills, multivariate genetic studies have also examined the relationship between them. Considering the substantial comorbidity between reading disability and math disability, understanding the genetic and environmental etiology of this relationship is essential. An illustrative example is presented in Figure 4.2. In this model, the correlation between reading and mathematics is influenced by common genetic (A1), shared environmental (C1), and nonshared environmental (E1) factors. Additionally, the variance in mathematics not explained by reading is decomposed into unique sources of genetic (A2), shared (C2), and nonshared environmental (E2) influences. Using this model, it is possible to examine whether reading and mathematics are influenced by common genetic effects (A1 large and significant), or whether there are independent genetic influences on mathematics separate from reading (A2 is significant). The same logic applies to shared (C1, C2) and nonshared (E1, E2) environmental pathways.

Using this general approach, Knopik and DeFries (1999) found substantial genetic overlap between psychometric measures of reading and mathematics, but also found substantial genetic independence for math, separate

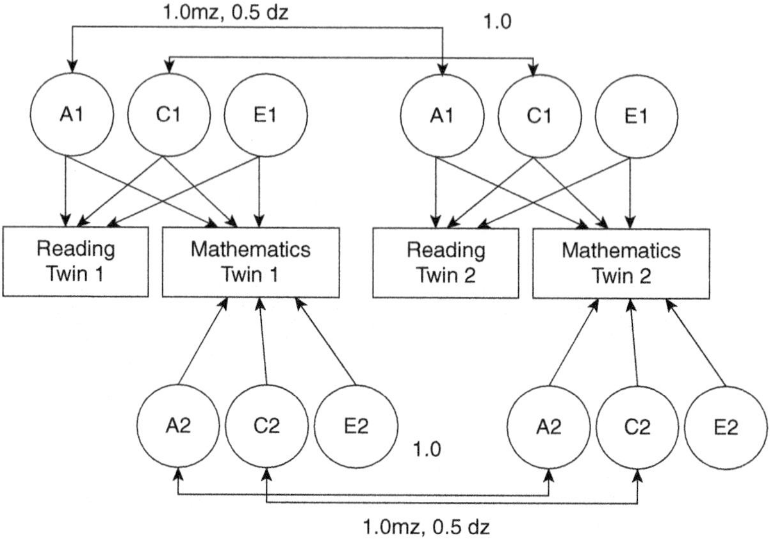

Figure 4.2 Multivariate genetic model

Note: A = Additive Genetic Effects
C = Common (Shared) Environmental Effects
E = Nonshared (Unique) Environmental Effects

from reading. More recently, the UK-based Twins Early Development (TEDS) sample indicated that although there is significant genetic overlap between math, general cognitive ability, and reading ability, roughly 1/3 of the variance in math is unique from reading and general cognitive ability (Haworth et al., 2009; Kovas, Haworth et al., 2007). Results from other studies also suggested genetic overlap and genetic independence for math, but also shared environmental overlap among mathematics, nonverbal skills, and verbal skills (e.g., Hart et al., 2009; Lemelin et al., 2007). These findings indicate that the comorbidity between reading and math skills are influenced by overlapping "domain-general" genetic factors. Moreover, these findings suggest that "domain-specific" genetic factors may also contribute to the independence of math skills from reading skills. Finally, shared environment may also be important to the overlap between reading and math skills, particularly in younger twins.

However, the majority of the quantitative genetic research on mathematics and reading involves broad behavioural traits, such as test scores or relatively broad measures of ability. Thus, it is unknown whether the domain-general overlap between reading and math is due to a small set of foundational skills, influenced by both genetics and/or shared environment, that permeate all aspects of learning disabilities, as described in the "generalist genes" hypothesis (Plomin & Kovas, 2005), or whether the overlap is due to the aggregation of a widely distributed set of neural mechanisms, of potentially different genetic/environmental etiologies that are similarly taxed by the complexities of reading or solving complex math problems. Moreover, the presence of domain-specific genetic effects for math lends credence to the discriminant validity of math disability and reading disability, but again, these are not tied to a theoretically meaningful mechanism.

As a first step, behavioural genetic studies have begun to examine measures of math processing in the context of reading and math outcomes. For example, in a series of papers, Hart et al. (2009) and Hart, Petrill, and Thompson (2010) found that timed math calculations were associated with untimed measures of mathematics as well as with timed measures of reading ability. This is interesting because the larger math literature suggest that timed measures of mathematics may be unique predictors of math ability, above and beyond untimed measures (Jordan & Hanich, 2003; Mazzocco, Devlin, & McKenney, 2008). Petrill et al. (2012) followed up on this work, examining timed and untimed measures of reading and math at one measurement occasion (Time 1), as well as timed and untimed math performance one year later (Time 2). As expected from the literature described above, reading and math outcomes were correlated through both genetic and shared environmental factors, whereas the independence of mathematics from reading was influenced primarily by genetics. Moreover, genetic and environmental factors at Time 1 completely predicted math variance at Time 2. These results suggest that the genetic and environmental influences are highly stable across time.

Interestingly, additional independent genetic effects for timed measures of reading and math were also statistically significant. The heritability for math fluency (how quickly children could solve calculation problems) at Time 1 ($h^2 = .58$) came from four sources: 1) general genetic variance shared with all reading and math measures ($h^2 = .06$, $p < .05$); 2) genetic variance for all math measures independent from reading ($h^2 = .10$, $p < .05$); 3) independent genetic effects for reading and math fluency above and beyond untimed measures of reading and math ($h^2 = .11$, $p < .05$); and 4) independent genetic variance specific to math fluency ($h^2 = .31$, $p < .05$). In contrast, the shared environment for math fluency ($c^2 = .25$) was common to all reading and math measures. This pattern was replicated at Time 2.

These results are important because the domain-general and domain-specific aspects of math and reading found in the larger learning disabilities literature appear to be reflected in overlapping and independent sources of genetic variance. We are currently collecting additional math processing measures including numeric estimation (Siegler & Opfer, 2003), and numerosity (Halberda et al., 2012; Halberda, Mazzocco, & Feigenson, 2008), as well as measures of math problem solving strategies (Geary, 2010), in order to better explain the genetic variance in mathematics independent from reading. At the same time, we are also examining skills associated with both reading and math disabilities, such as working memory (Geary, 2010).

Molecular genetics

In addition to quantitative genetic studies, there is a large molecular genetic literature on reading disability and an emerging literature on mathematics disability. Reading disability has been one of the success stories of molecular genetics. Multiple gene variants (polymorphisms) have been associated with reading disability (see Newbury et al., 2011 for an excellent review), and downstream work has begun to suggest a plausible biological mechanism for some of these gene variants. For example, a recent paper by Poelmans et al. (2011) suggests that, out of 14 of the best candidate genes that exist in the literature, ten (FMR1, DIP2A, GTF2I, DYX1C1, DCDC2, ROBO1, KIAA0319, KIAA0319L, S100B, and DOCK4) can be integrated into a plausible neurobiological network involving neuronal migration (where neurons migrate in the cerebral cortex) and neurite outgrowth (how dendrites and axons grow from the cell body). Considering the substantial influence of genetic variation on reading suggested by the quantitative genetic literature, we would expect that some of this variance may be explained by candidate genes related to how neurons involved in reading migrate and grow. We also would expect that some of these gene variants would explain a portion of the variance in mathematics.

Molecular genetic results are also beginning to emerge for mathematics. First, researchers have examined populations with rare genetically mediated conditions, including individuals with Williams syndrome, which is caused by a microdeletion on Chromosome 7 and Velocardio-facial Syndrome (VFS), which is caused by a microdeletion on Chromosome 2 (see Ansari, 2010 for a fuller discussion). In particular, individuals with William's syndrome are thought to have impaired mathematic skills related to spatial ability, but appear to have stronger verbal mathematical skills. Similarly, individuals with VFS appear to possess stronger verbal skills compared to numerosity skills.

In addition, examining the cognitive effects on mathematics arising from specific genetic conditions, Docherty et al. (2010) conducted the first study to identify gene variants related to mathematics impairment. Follow up work has suggested that these gene variants are also related to reading outcomes (Docherty, Kovas, Petrill, & Plomin, 2010) and that gene variants associated with reading are related to mathematics (Marino et al., 2011). This work is consistent with the quantitative genetic literature described above which suggests significant genetic overlap between mathematics and reading, but also some evidence for genetic specificity for mathematics.

Given these results, it is tempting to conclude that the molecular genetics results can be combined to explain the substantial heritability of reading and mathematics, as well as their substantial genetic overlap and independence. Although that is one of the ultimate goals of molecular genetics, at present, it is not possible due to the "Missing Heritability Paradox" (Maher, 2008). Until recently, it was thought that genetic influences on human traits could be best explained by two possible models at the opposite ends of Figure 4.3. The One-Gene-One-Disorder hypothesis (OGOD) holds that certain diseases/conditions are attributable to very rare genetic variants of very strong effect (Plomin, Owen, & McGuffin, 1994). Genetic variants of the OGOD type are both necessary and sufficient to cause a disease. For example, as of November 2012, Online Mendelian Inheritance in Man (OMIM) lists 533 known genetic variants associated with the term "Mental Retardation" that are best explained by an OGOD model. These variants are generally very rare, and are not found in the majority of individuals with intellectual disability (see Topper, Ober, & Das, 2011 for a review). However, these genetic variants explain almost all of the variance in the intellectual performance of individuals who possess them.

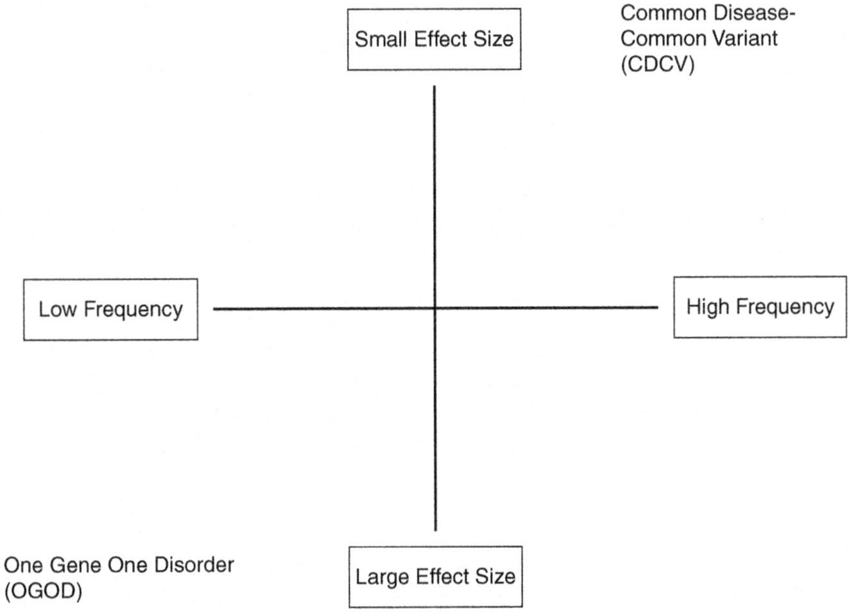

Figure 4.3 Genetic influences on behavioural traits

In contrast to the OGOD model, the Common Disease-Common Variant (CDCV) Model sits on the other side of Figure 4.3. According to this model, there is a large set of genetic variants that are high in frequency but very small in effect size. Another term for this model is the Quantitative Trait Locus (QTL) model (Plomin et al., 1994), the idea being that diseases/conditions constitute the extreme ends of normally distributed behaviour, which are influenced probabilistically by a large number of genetic variants each of very small effect size. The remaining molecular genetic results described above were hypothesized to fall into this category. The CDCV model is now testable because it is possible for researchers to efficiently examine millions of genetic differences across the human genome (see Hardy & Singleton, 2009 for a review). Currently, these approaches have not been able to identify enough gene variants to account for the substantial heritability and genetic correlations found in the quantitative genetic literature, not only for reading and mathematics, but most other complex behavioural and health related outcomes, resulting in the Missing Heritability Paradox.

One possibility (of many) is that the effect sizes and frequencies of genetic variants may operate across a continuum, between the OGOD and CDCV poles (see Figure 4.3). Consider the several hundred OGOD genetic variants of strong effect size related to mental retardation. These are clustered in the population, and defined by known syndromes (e.g., Down syndrome, Fragile X). These genes explain a portion of the overall variance of intellectual skills in the population through different biological mechanisms. Expanding this logic, there may be moderately infrequent genetic variants of small to moderate effect size that raise or lower the risk for milder forms of learning impairments (e.g., reading and math disability) through more differential, and more subtle, biological mechanisms. Put another way, understanding the multiple genetic/biological mechanisms for reading and math disability may elucidate differential pathways by which individuals may come to the same behavioural diagnosis.

This has important implications for the genetic contribution to reading and math disability. Until recently, the guiding principle was that reading and math disability represented the lower tail of a continuum of genetic, environmental, and neurobiological risk across the population, as suggested by the CDCV model. The underlying assumption of this principle was that quantitative genetic results represented the relative importance of uniform genetic variation in reading and mathematics. By considering the Missing Heritability Paradox, it is possible that a portion of the genetic variance may due to polymorphisms that are infrequent, small/moderate in effect size, and clustered in subgroups within the population. Quantitative genetics, therefore, may represent an aggregation of multiple pathways through which family members can be genetically similar or dissimilar, and thus may provide an epidemiological index of the multiple genetic pathways through which reading and math abilities and disabilities emerge (univariate genetics) and how reading and math are associated with one another (multivariate genetics).

Implications and next steps for neurocognitive studies of reading and mathematics

The possibility of aggregated genetic heterogeneity means that there may be clinically meaningful genetic subgroups of individuals that lie hidden in the population. Since the advent of functional neuroimaging, a large literature has evolved examining the neurobiological mechanisms associated with math and reading, and common themes and main theoretical perspectives have emerged for each area.

Mathematics

Over the last decade, neuroimaging work has provided important insights into the understanding of the nature of mathematical learning ability and math disability (MD). Behavioural studies of MD have identified two groups of cognitive processes that contribute to atypical math performance. The first group covers a broad range of domain-general processes fundamental for multiple areas of academic performance and not specific to mathematical learning (Geary, Bailey, & Hoard, 2009; Swanson & Jerman, 2007). These processes include phonological processing, working memory, inhibitory control, and visuospatial abilities. Note that many of these skills overlap with those found to influence reading abilities. The second group is unique to domain-specific aspects of mathematical learning and includes numerical quantity approximation and discrimination (Houde, Rossi, Lubin, & Joliot, 2010; Mazzocco, Feigenson, & Halberda, 2011). For example, with respect to numerical quantity approximation, research has supported the role of the parietal area and in particular, the domain specificity of the bilateral intraparietal sulcus to processing information about quantity (Dehaene, Piazza, Pinel, & Cohen, 2003; Schmithorst & Brown, 2004). Tasks that require an approximation of quantity, as opposed to exact calculation, have been shown repeatedly to activate the area around the intraparietal sulcus (Cantlon et al., 2009; Holloway & Ansari, 2010; Stanescu-Cosson et al., 2000).

Reading

Similar to math, there is a rich and extensive literature using functional neuroimaging in both skilled and impaired readers. Studies have explored the neural correlates of domain-general processing that have long been known to be critical for reading comprehension, such as sentence comprehension (Meyler et al., 2007; Rimrodt et al., 2009) and working memory (Beneventi, Tonnessen, & Ersland, 2009; Wolf et al., 2010). These studies also show abnormalities between typically developing readers and those with RD, although the areas of anomalous activation extend beyond those of traditional reading tasks, including prefrontal cortices. In terms of domain-specific skills, much work has been devoted to the neural correlates of word recognition/decoding. In brief, neurobiological correlates of word recognition in RD show perisylvian region dysfunction, including underactivation of left temporo-parietal and occipitotemporal regions and greater *right* temporo-parietal and occipitotemporal activation (Pugh et al., 2000; Rezaie et al., 2011; Richlan, Kronbichler, & Wimmer, 2009; Rumsey et al., 1997; Stuebing et al., 2002) as compared to typically developing readers. Typically developing readers demonstrate consistent patterns of left-hemisphere perisylvian activation, in temporo-parietal and inferior frontal regions, as well as occipitotemporal activation that are especially associated with fast and efficient reading (Fiez & Petersen, 1998; Georgiewa et al., 2002; Simos et al., 2002; Turkletaub et al., 2003).

Mechanisms for the genetic overlap and discrepancy between RD and MD

This brief review of the neuroimaging literature illustrates plausible domain-general and domain-specific mechanisms for the genetic overlap and discrepancy between RD and MD found in the behavioural genetic literature. Consider a "domain-general" skill, such as working memory, that 1) shows activation across widely distributed brain regions, 2) is highly heritable, and 3) is correlated with many complex behavioural outcomes (Koten et al., 2009). Within a twin design, one would be able to systematically examine whether structural and functional differences associated with working memory account for comorbidity between reading and math skills. Conversely, it is also possible to examine functional differences in measures related to math but not reading (e.g., numeric estimation) that demonstrate activation differences in more narrowly distributed regions (e.g., intraparietal sulcus) specific to math outcomes. When assessed in the same individuals, these relationships could be further examined by comparing patterns of genetic correlations within measures of brain function arising from the same/different tasks, occurring in the same/different neurobiological regions. Moreover, beyond genetics, results indicate that shared environmental factors for math are statistically significant (and substantial) for mathematics, but not for reading. Better understanding the structural and functional correlates of the shared environment may theoretically yield knowledge about the boundaries of genetic and environmental vulnerability at the level of the brain, in addition to the level of behaviour.

Moreover, one can also examine the question of domain-general and domain-specific factors in the context of selected groups. Over a dozen competing explanations have been proposed to account for non-artifactual comorbidity between multifactorial disorders. The extant literature suggests that a *domain-general + domain-specific* model appears most tenable (Cirino, Morris, & Morris, 2007; Moll, Fussenegger, Willburger, & Landerl, 2009; Rosselli, Matute, Pinto, & Ardila, 2006; van der Sluis, de Jong, & van der Leij, 2004); what remains unclear are the conditions under which the data is best explained by a Phenocopy form of the model (where, for example, seemingly similar behavioural math impairments in MD+RD and MD groups are influenced by deficits in different neurobiological factors) or a Correlated Liabilities form of the model (where, for example there is some overlap in neurobiological structure and function across MD, RD, and MD+RD groups, but also unique neurobiological mechanisms within each group). Moreover, examining the cotwins of the probands selected for MD, RD, and MD+RD provides a means to examine the extent that selection into groups is heritable. For example, particularly heritable aspects of the activation of the left occipitotemporal region may allow differentiation of the MD+RD group from the MD group, and may be reflected in similar patterns of functioning in the cotwins. In this way, it would be possible to systematically examine the relationships between brain structure, brain function, etiology, math outcomes, and reading outcomes with and without selection for disability.

Implications and conclusions

Although there is clearly some hard work left to understand the multiple biological, genetic, and environmental pathways through which reading and mathematics ability/disability develop, we have come a long way from asking whether genes OR environment are important. It is clear that both are important, and it is also clear that the time has come to actively consider the role of genetics and environment in educational practice. At a minimum, irrespective of the genetic and environmental origins, reading and math are highly familial. Thus, a better understanding of family history may offer an important window into the level of risk, both genetically, but also in terms of the level of environmental support outside of school. Children with reading and/or math impairments are not only likely to have other family members with learning problems, but children who come from families with a high density of learning problems may be at particular risk for intervention nonresponse.

More generally, quantitative and molecular genetic results suggest that a shift towards a biobehavioral model of education is necessary. Reading and mathematics are like most complex human traits and behaviours; involving a biologically mediated system, influenced by genetics and environment. These genetic and environmental factors may be rare but severe in their effect, may be frequent and subtle, but may also be highly heterogeneous. This is a daunting empirical challenge, but this idea is being increasingly accepted in medicine, as well as in some domains of clinical psychology. One outcome of this acceptance is that the genetic risk triggers the application of more intensive intervention strategies. The current state of affairs – removing a large portion of the variance from discussion – may limit opportunities to more quickly detect and more effectively treat learning problems through a better understanding of the many yet-to-be discovered biological processes that give rise to effective or ineffective learning.

References

Alarcon, M., DeFries, J. C., Light, J. G., & Pennington, B. F. (1997). A twin study of mathematics disability. *Journal of Learning Disabilities, 30*(6), 617–623.

Ansari, D. (2010). Neurocognitive approaches to developmental disorders of numerical and mathematical cognition: The perils of neglecting the role of development. *Learning and Individual Differences, 20*(2), 123–129.

Badian, N. (1999). Persistent arithmetic, reading, or arithmetic and reading disability. *Annals of Dyslexia, 49*(1), 43–70.

Beneventi, H., Tonnessen, F. E., & Ersland, L. (2009). Dyslexic children show short-term memory deficits in phonological storage and serial rehearsal: An fMRI study. *International Journal of Neuroscience, 119*(11), 2017–2043.

Boetsch, E. A., Green, P. A., & Pennington, B. F. (1996). Psychosocial correlates of dyslexia across the life span. *Development and Psychopathology, 8*(03), 539–562.

Cantlon, J. F., Libertus, M. E., Pinel, P., Dehaene, S., Brannon, E. M., & Pelphrey, K. A. (2009). The neural development of an abstract concept of number. *Journal of Cognitive Neuroscience, 21*(11), 2217–2229.

Cirino, P. T., Morris, M. K., & Morris, R. D. (2007). Semantic, executive, and visuospatial abilities in mathematical reasoning of referred college students. *Assessment, 14*(1), 94–104.

Davis, C., Gayan, J., Knopik, V. S., Smith, S. D., Cardon, L. R., Pennington, B. F., ..., & DeFries, J. C. (2001). Etiology of reading difficulties and rapid naming: The Colorado Twin Study of Reading Disability. *Behavior Genetics, 31*(6), 625–635.

DeFries, J. C., & Fulker, D. W. (1985). Multiple regression analysis of twin data. *Behavior Genetics, 15*(5), 467–473.

DeFries, J. C., & Fulker, D. W. (1988). Multiple regression analysis of twin data: Etiology of deviant scores versus individual differences. *Acta Geneticae Medicae et Gemellologiae: Twin Research, 37*(3–4), 205–216.

Dehaene, S., Piazza, M., Pinel, P., & Cohen, L. (2003). Three parietal circuits for number processing. *Special Issue: The Organisation of Conceptual Knowledge in the Brain: Neuropsychological and Neuroimaging Perspectives, 20*(3–6), 487–506.

Docherty, S. J., Davis, O. S. P., Kovas, Y., Meaburn, E. L., Dale, P. S., Petrill, S. A., ..., & Plomin, R. (2010). A genome-wide association study identifies multiple loci associated with mathematics ability and disability. *Genes, Brain, and Behavior, 9*(2), 234–247.

Docherty, S. J., Kovas, Y., Petrill, S. A., & Plomin, R. (2010). Generalist genes analysis of DNA markers associated with mathematical ability and disability reveals shared influence across ages and abilities. *BMC Genetics, 11*, 61–61.

Fiez, J. A., & Petersen, S. E. (1998). Neuroimaging studies of word reading. *Proceedings of the National Academy of Sciences of the United States of America, 95*(3), 914–921.

Geary, D. C. (2010). Mathematical disabilities: Reflections on cognitive, neuropsychological, and genetic components. *Learning and Individual Differences, 20*(2), 130–133.

Geary, D. C., Bailey, D. H., & Hoard, M. K. (2009). Predicting mathematical achievement and mathematical learning disability with a simple screening tool: The Number Sets Test. *Journal of Psychoeducational Assessment, 27*(3), 265.

Georgiewa, P., Rzanny, R., Gaser, C., Gerhard, U. J., Vieweg, U., Freesmeyer, D.,..., & Blanz, B. (2002). Phonological processing in dyslexic children: a study combining functional imaging and event related potentials. *Neuroscience Letters, 318*(1), 5–8.

Halberda, J., Ly, R., Wilmer, J. B., Naiman, D. Q., & Germine, L. (2012). Number sense across the lifespan as revealed by a massive Internet-based sample. *Proceedings of the National Academy of Sciences, 109*(28), 11116–11120.

Halberda, J., Mazzocco, M. M. M., & Feigenson, L. (2008). Individual differences in non-verbal number acuity correlate with maths achievement. *Nature*, 455.

Hardy, J., & Singleton, A. (2009). Genome-wide association and human disease. *New England Journal of Medicine, 360*, 1759–1768.

Hart, S. A., Petrill, S. A., & Thompson, L. A. (2010). A factorial analysis of timed and untimed measures of mathematics and reading abilities in school aged twins. *Learning and Individual Differences, 20*(2), 63–69.

Hart, S. A., Petrill, S. A., Thompson, L. A., & Plomin, R. (2009). The ABCs of math: A genetic analysis of mathematics and its links with reading ability and general cognitive ability. *Journal of Educational Psychology, 101*(2), 388–402.

Haworth, C. M. A., Kovas, Y., Harlaar, N., Hayiou-Thomas, M. E., Petrill, S. A., Dale, P. S., & Plomin, R. (2009). Generalist genes and learning disabilities: a multivariate genetic analysis of low performance in reading, mathematics, language and general cognitive ability in a sample of 8000 12-year-old twins. *Journal of Child Psychology and Psychiatry, and Allied Disciplines, 50*(10), 1318–1325.

Haworth, C. M. A., Kovas, Y., Petrill, S. A., & Plomin, R. (2007). Developmental origins of low mathematics performance and normal variation in twins from 7 to 9 years. *Twin Research and Human Genetics, 10*(1), 106–117.

Hinshelwood, J. (1911). Two Cases of Hereditary Congenital Word-Blindness. *British Medical Journal, 1*(2620), 608–609.

Holloway, I. D., & Ansari, D. (2010). Developmental specialization in the right intraparietal sulcus for the abstract representation of numerical magnitude. *Journal of Cognitive Neuroscience, 22*(11), 2627–2637.

Houde, O., Rossi, S., Lubin, A., & Joliot, M. (2010). Mapping numerical processing, reading, and executive functions in the developing brain: An fMRI meta-analysis of 52 studies including 842 children. *Developmental Science, 13*(6), 876–885.

Jordan, N. C., & Hanich, L. B. (2003). Characteristics of children with moderate mathematics deficiencies: A longitudinal perspective. *Learning Disabilities Research & Practice, 18*(4), 213–221.

Knopik, V. S., Alarcon, M., & DeFries, J. C. (1997). Comorbidity of mathematics and reading deficits: evidence for a genetic etiology. *Behavior Genetics, 27*(5), 447–453.

Knopik, V. S., & DeFries, J. C. (1999). Etiology of covariation between reading and mathematics performance: a twin study. *Twin Research and Human Genetics, 2*, 226–234.

Koten Jr, J. W., Wood, G., Hagoort, P., Goebel, R., Propping, P., Willmes, K., & Boomsma, D. I. (2009). Genetic contribution to variation in cognitive function: an fMRI study in twins. *Science, 323*(5922), 1737.

Kovas, Y., Haworth, C. M. A., Harlaar, N., Petrill, S. A., Dale, P. S., & Plomin, R. (2007). Overlap and specificity of genetic and environmental influences on mathematics and reading disability in 10-year-old twins. *Journal of Child Psychology and Psychiatry, 48*(9), 914–922.

Kovas, Y., Haworth, C. M. A., Petrill, S. A., & Plomin, R. (2007). Mathematical ability of 10-year-old boys and girls: Genetic and environmental etiology of typical and low performance. *Journal of Learning Disabilities, 40*(6), 554–567.

Landerl, K., & Moll, K. (2010). Comorbidity of learning disorders: prevalence and familial transmission. *Journal of Child Psychology and Psychiatry, 51*(3), 287–294.

Lemelin, J.-P., Boivin, M., Forget-Dubois, N., Dionne, G., Séguin, J. R., Brendgen, M., ..., & Pérusse, D. (2007). The genetic–environmental etiology of cognitive school readiness and later academic achievement in early childhood. *Child Development, 78*(6), 1855–1869.

Loehlin, J. C., & Nichols, R. C. (1976). *Heredity, Environment, & Personality: A Study of 850 Sets of Twins*. University of Texas Press.

Maher, B. (2008). Personal genomes: The case of the missing heritability. *Nature, 456*(7218), 18–21.

Marino, C., Mascheretti, S., Riva, V., Cattaneo, F., Rigoletto, C., Rusconi, M., ..., & Molteni, M. (2011). Pleiotropic effects of DCDC2 and DYX1C1 genes on language and mathematics traits in nuclear families of developmental dyslexia. *Behavior Genetics, 41*(1), 67–76.

Mazzocco, M. M. M., Devlin, K. T., & McKenney, S. J. (2008). Is it a fact? Timed arithmetic performance of children with mathematical learning disabilities (MLD) varies as a function of how MLD is defined. *Developmental Neuropsychology, 33*(3), 318–344.

Mazzocco, M. M. M., Feigenson, L., & Halberda, J. (2011). Impaired acuity of the approximate number system underlies mathematical learning disability (dyscalculia). *Child Development, 82*(4), 1224–1237.

Meyler, A., Keller, T. A., Cherkassky, V. L., Lee, D., Hoeft, F., Whitfield-Gabrieli, S., ..., & Just, M. A. (2007). Brain activation during sentence comprehension among good and poor readers. *Cerebral Cortex, 17*(12), 2780–2787.

Moll, K., Fussenegger, B., Willburger, E., & Landerl, K. (2009). RAN is not a measure of orthographic processing: Evidence from the asymmetric German orthography. *Scientific Studies of Reading, 13*(1), 1–25.

Morgan, W. P. (1896). A Case of Congenital Word Blindness. *British Medical Journal, 2*(1871), 1378.

National Reading Panel (US), National Institute of Child Health, & Human Development (US). (2000). Report of the national reading panel: Teaching children to read: An evidence-based assessment of the scientific research literature on reading and its implications for reading instruction: Reports of the subgroups. National Institute of Child Health and Human Development, National Institutes of Health.

Newbury, D., Paracchini, S., Scerri, T., Winchester, L., Addis, L., Richardson, A., ..., & Monaco, A. (2011). Investigation of dyslexia and SLI risk variants in reading- and language-impaired subjects. *Behavior Genetics, 41*(1), 90–104.

Oliver, B., Harlaar, N., Hayiou Thomas, M. E., Kovas, Y., Walker, S. O., Petrill, S. A., ..., & Plomin, R. (2004). A twin study of teacher-reported mathematics performance and low performance in 7-year-olds. *Journal of Educational Psychology, 96*(3), 504–517.

Olson, R. K., Gillis, J. J., Rack, J. P., DeFries, J. C., & Fulker, D. W. (1991). Confirmatory factor analysis of word recognition and process measures in the Colorado Reading Project. *Reading and Writing, 3*, 235–248.

Pennington, B. F. (1999). Toward an integrated understanding of dyslexia: Genetic, neurological, and cognitive mechanisms. *Development and Psychopathology, 11*(03), 629–654.

Pennington, B. F., & Olson, R. K. (2005). Genetics of Dyslexia. In M. J. Snowling & C. Hulme (Eds), *The science of reading: A handbook* (pp. 453–472).

Petrill, S., Logan, J., Hart, S., Vincent, P., Thompson, L., Kovas, Y., & Plomin, R. (2012). Math fluency is etiologically distinct from untimed math performance, decoding fluency, and untimed reading performance: evidence from a twin study. *Journal of Learning Disabilities, 45*(4), 371–381.

Petrill, S. A., & Plomin, R. (2007). Quantitative genetics and mathematical abilities/disabilities. In D. B. Berch & M. M. M. Mazzocco (Eds), *Why is math so hard for some children? The nature and origins of mathematical learning difficulties and disabilities* (pp. 307–322).

Plomin, R. (1991). Genetic risk and psychosocial disorders: Links between the normal and abnormal. In M. Rutter & P. Casaer (Eds), *Biological risk factors for psychosocial disorders* (pp. 101–138).

Plomin, R., & Kovas, Y. (2005). Generalist genes and learning disabilities. *Psychological Bulletin, 131*(4), 592–617.

Plomin, R., Owen, M. J., & McGuffin, P. (1994). The genetic basis of complex human behaviors. *Science, 264*(5166), 1733–1739.

Poelmans, G., Buitelaar, J. K., Pauls, D. L., & Franke, B. (2011). A theoretical molecular network for dyslexia: integrating available genetic findings. *Mol. Psychiatry, 16,* 365–382. 10.1038/mp.2010.105

Pugh, K. R., Mencl, W. E., Jenner, A. R., Katz, L., Frost, S. J., Lee, J. R., ..., & Shaywitz, B. A. (2000). Functional neuroimaging studies of reading and reading disability (developmental dyslexia). *Mental Retardation and Developmental Disabilities Research Reviews, 6*(3), 207–213.

Rezaie, R., Simos, P. G., Fletcher, J. M., Cirino, P. T., Vaughn, S., Papanicolaou, A. C., & others (2011). Temporo-parietal brain activity as a longitudinal predictor of response to educational interventions among middle school struggling readers. *Journal of the International Neuropsychological Society, 17*(5), 875–885.

Richlan, F., Kronbichler, M., & Wimmer, H. (2009). Functional abnormalities in the dyslexic brain: A quantitative meta-analysis of neuroimaging studies. *Human Brain Mapping, 30*(10), 3299–3308.

Rimrodt, S. L., Clements-Stephens, A. M., Pugh, K. R., Courtney, S. M., Gaur, P., Pekar, J. J., & Cutting, L. E. (2009). Functional MRI of sentence comprehension in children with dyslexia: beyond word recognition. *Cerebral Cortex, 19*(2), 402–413.

Rosselli, M., Matute, E., Pinto, N., & Ardila, A. (2006). Memory abilities in children with subtypes of dyscalculia. *Developmental Neuropsychology, 30*(3), 801–818.

Rumsey, J. M., Nace, K., Donohue, B., Wise, D., Maisog, J. M., & Andreason, P. (1997). A positron emission tomographic study of impaired word recognition and phonological processing in dyslexic men. *Archives of Neurology, 54*(5), 562–573.

Schmithorst, V. J., & Brown, R. D. (2004). Empirical validation of the triple-code model of numerical processing for complex math operations using functional MRI and group Independent Component Analysis of the mental addition and subtraction of fractions. *NeuroImage, 22*(3), 1414–1420.

Siegler, R. S., & Opfer, J. E. (2003). The development of numerical estimation: Evidence for multiple representations of numerical quantity. *Psychological Science, 14*(3), 237–243.

Simos, P. G., Papanicolaou, A. C., Castillo, E. M., Breier, J. I., Fletcher, J. M., Wheless, J. W., ..., & Constantinou, J. E. (2002). [Insights into brain function though magnetic source imaging: A review of research and clinical applications]. *Revista de Neurologia, 34*(9), 871–876.

Stanescu-Cosson, R., Pinel, P., Moortele, P.-F. v. d., Le Bihan, D., Cohen, L., & Dehaene, S. (2000). Understanding dissociations in dyscalculia. A brain imaging study of the impact of number size on the cerebral networks for exact and approximate calculation. *Brain: A Journal of Neurology, 123*(11), 2240–2255.

Stevenson, J. (1991). Which aspects of processing text mediate genetic effects? *Special Issue: Genetic and neurological influences on reading disability, 3*(3–4), 249–269.

Stevenson, J., Graham, P., Fredman, G., & McLoughlin, V. (1987). A twin study of genetic influences on reading and spelling ability and disability. *Journal of Child Psychology and Psychiatry, 28*(2), 229–247.

Storch, S. A., & Whitehurst, G. J. (2002). Oral language and code-related precursors to reading: Evidence from a longitudinal structural model. *Developmental Psychology, 38*(6), 934–947.

Stuebing, K. K., Fletcher, J. M., LeDoux, J. M., Lyon, G. R., Shaywitz, S. E., & Shaywitz, B. A. (2002). Validity of IQ-discrepancy classifications of reading disabilities: A meta-analysis. *American Educational Research Journal, 39*(2), 469–518.

Swanson, H. L., & Jerman, O. (2007). The influence of working memory on reading growth in subgroups of children with reading disabilities. *Journal of Experimental Child Psychology, 96*(4), 249–283.

Thompson, L. A., Detterman, D. K., & Plomin, R. (1991). Associations between cognitive abilities and scholastic achievement: Genetic overlap but environmental differences. *Psychological Science, 2*(3), 158–165.

Topper, S., Ober, C., & Das, S. (2011). Exome sequencing and the genetics of intellectual disability. *Clinical Genetics, 80*(2), 117–126.

Turkletaub, P. E., Gareau, L., Flowers, D. L., Zeffiro, T. A., & Eden, G. F. (2003). Development of neural mechanisms for reading. *Nature Neuroscience, 6*(6), 767–773.

Van der Sluis, S., de Jong, P. F., & van der Leij, A. (2004). Inhibition and shifting in children with learning deficits in arithmetic and reading. *Journal of Experimental Child Psychology, 87*(3), 239–266.

Vandenberg, S. G. (1962). The hereditary abilities study: hereditary components in a psychological test battery. *American Journal of Human Genetics, 14*(2), 220–237.

White, J. L., Moffitt, T. E., & Silva, P. A. (1992). Neuropsychological and socio-emotional correlates of specific-arithmetic disability. *Archives of Clinical Neuropsychology, 7*, 1–16.

Wolf, R. C., Sambataro, F., Lohr, C., Steinbrink, C., Martin, C., & Vasic, N. (2010). Functional brain network abnormalities during verbal working memory performance in adolescents and young adults with dyslexia. *Neuropsychologia, 48*(1), 309–318.

5

Studying Rare Genetic Syndromes as a Method of Investigating Aetiology of Normal Variation in Educationally Relevant Traits

Maja Rodic, Darya Gaysina, Sophia Docherty, Sergey Malykh, Kaili Rimfeld, Robert Plomin, and Yulia Kovas

Studies using genetically sensitive research designs have contributed greatly to our understanding of the aetiology of variation in educationally relevant traits such as IQ, reading, mathematics, spatial ability and, most recently, exam performance (i.e., General Certificate of Secondary Education – GCSE – in the UK (Shakeshaft et al., 2013).

Quantitative genetic research which encompasses twin, family and adoption studies suggests that all educationally relevant traits are to some extent influenced by genetic factors (Butterworth & Kovas, 2013; Kovas, Haworth, Dale, & Plomin, 2007; Kovas et al., 2007d; Plomin & Kovas, 2005). However, despite the huge advancements in molecular genetics technologies, progress has been relatively slow in identifying specific genetic variants associated with these complex traits (Plomin, 2013).

In this chapter, we review evidence from studies of educationally relevant traits that use quantitative and molecular genetic approaches. We also present a novel approach to study genetics of cognitive traits – using the perspective of rare genetic disorders. Specifically, we give a detailed description of cognitive profiles of two genetic syndromes – Williams and Prader–Willi – and discuss what can be learnt from rare genetic pathologies about normal variation in educationally relevant traits. Finally, we present results from a recent association study investigating specific genetic regions implicated in these syndromes in a normal population. We conclude by outlining future directions for behavioural genetic investigations of educationally relevant traits and potential implications of this research.

Genetics of educationally relevant traits: evidence from quantitative behaviour genetics

Heritability of educationally relevant traits

Although educational psychology has been slow in recognizing the importance of genetics, it is now widely accepted that phenotypic variation in educationally relevant traits is at least in part due to genetic factors (Plomin, DeFries, Knopik, & Neiderhiser, 2013). Within the last decade a large number of quantitative genetic studies have shown that, on average, heritability estimates for all educationally relevant traits are around 50% (Plomin et al., 2013).

Heritability refers to a proportion of phenotypic variance between individuals within a population that can be explained by their genotypic variance. Quantitative genetic research investigates the extent to which variation in specific traits or disorders is influenced by genetic and environmental factors. It is based on the notion that effects of multiple genes lead to quantitative traits. These effects are mostly of the additive type, meaning that the combined effects of *alleles* at different loci are equal to the sum of their individual effects. If a trait is influenced by genetic factors, then observed phenotypic resemblance between relatives should also increase with their genetic relatedness (Plomin et al., 2013; Trzaskowski, Davis, DeFries, Yang et al., 2013).

Furthermore, any complex trait is also influenced by multiple environments that interact with genetic effects. One of the most popular methods of studying relative contributions of genes and environments to any given trait is the twin method. This method compares resemblance between genetically identical monozygotic (MZ) twins to that of dizygotic (DZ) twins who share on average 50% of their segregating (variable) DNA. Greater observed resemblance of MZ than DZ twins indicates genetic influences on the trait being measured, and the resulting statistic is called heritability.

One of the first twin studies of mathematical abilities/disabilities using a sample of 6 to 12 -year old twins suggested a heritability estimate of 20% (Thompson, Detterman, & Plomin, 1991). However, another study reported a much higher heritability estimate for mathematics (90%) across the ability range using a sample of twins aged between 8 and 20 years (Alarcon, Knopik, & DeFries, 2000). It is possible that these differences in heritability estimates could result from differences in participants' ages in the two samples because mathematics as a trait changes throughout development. Different cognitive and motivational processes may be present at different ages which could be reflected in how mathematics is measured at different ages. It is also possible that different genes and environments may be important at different stages of development. This can lead to these differences in heritability estimates. Furthermore, both samples in the two studies were relatively small yielding insufficient power for accurate estimates. Heritability estimates are population specific and can vary in different populations.

Recent studies, using a large sample of twins from the Twin Early Development Study (TEDS), showed similar heritability estimates for mathematical abilities/disabilities across different ages. Heritability of different components of mathematics was approximately 60 to 70% at ages 7, 9, 10, 12 and 16 years (Kovas et al., 2007b; Kovas, Haworth, Dale, & Plomin, 2007a; Oliver et al., 2004; Kovas et al., 2013; Rimfeld et al., 2015). Environmental influences were also significant, with greater importance of non-shared environment (which makes two people growing up in the same family less similar on a measured trait) than of shared environment (which makes two people growing up in the same family more similar on a measured trait) (Tosto, Haworth & Kovas, 2014; Tosto et al., 2014).

Similarly to mathematics, quantitative genetic studies on reading show moderate genetic influences across populations and across different ages. Heritability estimates vary from 18% to 87% for diverse measures of reading and related phenotypes (Bates et al., 2004; Byrne et al., 2005; Harlaar, Spinath, Dale, & Plomin, 2005). Heritability of general cognitive ability (g) has been shown to vary from 40% in childhood to 80% in adulthood, increasing linearly throughout development, with the role of shared environmental factors decreasing with age (Haworth et al., 2010).

A growing body of multivariate genetic research on learning abilities/disabilities shows that average genetic correlation (the extent to which two traits are influenced by the same genetic factors) between domains of language, reading and mathematics ranges from 0.60 to 0.90. Genetic correlations between "g" and educationally relevant traits are somewhat lower ranging between 0.50 and 0.70 (Plomin, Kovas, & Haworth, 2007).

A new method that allows the use of DNA of unrelated individuals to estimate genetic influences is called Genome-wide Complex Traits Analysis (GCTA; Yang et al., 2010). This technique cannot identify specific genes associated with a trait but it allows estimation of heritability of a trait. This is achieved by correlating genetic similarity across a large number of single nucleotide polymorphisms (SNPs) with phenotypic similarity in pairs of unrelated individuals in a large sample. A population-based approach, GCTA is free from assumptions of the twin method. Results from this new approach confirm moderate heritability for educationally relevant traits, suggesting that common SNPs account for more than two thirds of heritability found in twin studies (Trzaskowski et al., 2013b; Yang et al., 2011). Bivariate GCTA analysis also confirms high genetic correlations across traits found in twin studies: for example, bivariate GCTA genetic correlations of 0.81, 0.74, and 0.89 were reported for "g" and language, "g" and mathematics, and "g" and reading, respectively (Trzaskowski et al., 2013a).

Findings of at least moderate influences of genetic factors on educationally relevant traits provide a strong basis for molecular genetic studies that aim to identify specific genetic variants involved in these traits. Moreover, because genetic correlations between educationally relevant traits are substantial, if genetic factors implicated in one trait (e.g., mathematics) are discovered, the same factors are likely to be implicated in other traits (e.g., reading).

Quantitative Trait Loci (QTLs)

It remains unclear whether aetiology of learning disabilities is similar to that of normal variation of ability. Evidence shows that phenotypic variation in all complex psychological traits can, to a certain extent, result from the effects of common DNA variants (Plomin, Haworth, & Davis, 2009). Recent quantitative genetic research suggests that largely the same genetic factors operate throughout the ability continuum. In other words, unlike some rare DNA variants (mutations) that can lead to a particular genetic disorder (e.g., phenylketonuria), many common DNA variants (polymorphisms) of a very small effect contribute to a person's position on a learning ability continuum.

If all genetic variants had large effects they would behave categorically and would work in deterministic ways. It would mean that all carriers of a particular (ability-decreasing) DNA variant would end up at the low end of the learning ability continuum. This is known as the One Gene One Disorder (OGOD) model. Indeed, many severe genetic disorders characterized by learning disability among other symptoms are caused by a single mutation. However, these abnormal DNA variants are rare, with the prevalence of the disorders in a population from 1 in 10,000 to 1 in 100,000 (Plomin et al., 2015).

On the contrary, common variation in educationally relevant traits results from many DNA variants of small effects (Plomin, Owen, & McGuffin, 1994), with the prevalence of common learning disabilities varying from 1 in 10 to 1 in 20. Therefore, common learning disabilities lie at the low end of the normal distribution of ability and stem from complex combinations of many common DNA variants and environments. This model is called Quantitative Trait Locus (QTL), and can be applied to any complex traits (Plomin & Kovas, 2005). Because each polymorphic locus contributes only very little to a trait, these loci work additively. Therefore, each ability-decreasing allele of a polymorphic locus presents only a probabilistic risk. Although the QTL model proposes small effects of multiple common genetic variants, it is possible that rare variants (unique to individuals) of small effect also exist. Furthermore, complex gene–gene and gene–environment interactions contribute to phenotypic variation. In addition, DNA polymorphisms work in pleiotropic ways, with one genetic factor contributing to many traits; and the environments can work in a similar way.

The "generalist" genes hypothesis

Recent research suggests that genetic effects of complex traits are largely general rather than specific (Plomin & Kovas, 2005). According to the "generalist" genes hypothesis, there are three types of general genetic effects. Firstly, the same genetic factors are associated with both common learning abilities (e.g., maths ability) and common learning disabilities (e.g., maths disability). Quantitative genetic research suggests that both abilities and disabilities are influenced by largely overlapping genetic and environmental factors (Kovas et al., 2007d; Petrill et al., 2009; Plomin & Kovas, 2005). Therefore, mathematical disability should not be viewed as qualitatively distinct from the continuum of mathematical ability. This means that once genes and environments that influence mathematical ability are discovered, a large proportion of these genes and environments are also implicated in mathematical disability (Plomin et al., 2009). However, some distinct rare mutations, such as those found in Downs syndrome or Williams syndrome (Bertella et al., 2005; Mazzocco, 1998; Murphy & Mazzocco, 2008), might also be involved in normal variation in mathematical abilities/disabilities. Later, in this chapter we discuss this issue.

Secondly, the same genetic factors affect different learning disabilities (e.g., maths disability and reading disability; Plomin & Kovas, 2005). A large number of studies suggest that an average genetic correlation between reading and mathematics, language and mathematics, and reading and language is at least 0.70 (Davis, Haworth, & Plomin, 2009a; Plomin & Kovas, 2005). Genetic correlations between general cognitive ability and mathematics, reading and language are also very high (up to 0.85). Many potential mechanisms through which generalist genes operate have been proposed (Plomin & Kovas, 2005).

Thirdly, the same genetic factors contribute to different aspects of ability (Kovas, Petrill, & Plomin, 2007c; Plomin & Kovas, 2005). Recent research suggests a high degree of genetic overlap among different aspects of reading (e.g., fluency and comprehension; Bishop, 2001), mathematics (e.g., computational and spatial aspects of maths; Kovas et al., 2007; Plomin & Kovas, 2005), and language (e.g., writing and speaking; Kovas et al., 2007). The ultimate support for all three aspects of the "generalist" genes hypothesis will come when specific genetic variants that contribute to variation in learning ability/disability are discovered.

Molecular genetic studies of educationally relevant traits

Almost fifty years after James Watson and Francis Crick discovered the structure of DNA, the first human genome was decoded, and all of its three billion nucleotide pairs across the 23 pairs of chromosomes were identified. Since then, our knowledge of human genetics has increased immensely;

however, the progress in identifying genetic variants responsible for the heritability of quantitative traits and common disorders has been relatively slow (Plomin, 2013; Visscher, Brow, McCherty, & Yang, 2012).

The human genome contains the complete set of genetic information stored as a deoxyribonucleic acid (DNA) sequence within the 23 chromosome pairs. DNA is a double helix made up of 3 billion pairs of smaller units – nucleotides. There are four different types of nucleotides – *adenine* (A), *thymine* (T), *cytosine* (C), and *guanine* (G). A large proportion of the genome sequence (~99%) is the same for all humans. Several types of genetic variation exist. The most common type, accounting for approximately 90% of all DNA variation, is Single Nucleotide Polymorphism (SNP), when a single *nucleotide* (*A, T, C* or *G*) differs in the population (as opposed to an invariable locus). For example, two sequenced DNA fragments from two different individuals, AGAACCTA and AGAACTTA, contain a difference in a single nucleotide. In this case there are two *alleles* of this SNP: allele C and allele T. Almost all common SNPs are bi-allelic (there is either one or the other allele in the locus, C or T in our example). Every bi-allelic polymorphism has got a major (more frequent) and a minor (less frequent) allele. Any polymorphism with minor allele frequency of 5% or more is considered to be a common polymorphism. SNPs can have functional consequences, such as amino acid changes, changes to mRNA transcript stability and changes to transcription factor binding affinity (Bush & Moore, 2012). As millions of SNPs exist in the human genome, they can be used in molecular genetic research. In the next sections, we outline two molecular genetic techniques used to identify specific genetic variants that contribute to complex traits.

Linkage studies

Linkage analysis has been used for decades as a technique for identifying genes contributing to specific traits (Visscher, Brow, McCarthy, & Yang, 2012). Linkage studies look at segments of the genome that can be associated with different traits or disorders. This method usually involves large pedigrees, where a disorder occurs in family members across several generations. The key is to identify a genetic marker that is always inherited by only those family members who are affected by the disorder. Linkage studies usually start by identifying a genetic marker on a section of a chromosome, and further narrowing the region down until the specific genetic variant involved is identified (Plomin et al., 2013).

However, identification of large families with several affected generations can be challenging. Linkage studies also lack statistical power to detect multiple genetic variants of small effects (Risch & Merikangas, 1996). This limitation can be overcome by studying a bigger number of small family groups, such as parent–offspring triads (Plomin, 2013). To date, using this method several genes have been identified for educationally relevant traits such as dyslexia (de Kovel et al., 2008; Fisher et al., 2002; König et al.,

2011; Smith, Kimberling, Pennington, & Lubs, 1983), and Attention Deficit Hyperactivity Disorder (ADHD) (Jiang et al., 2001; Saviouk et al., 2011; Vandenbergh, Persico, & Uhl, 1992).

Association studies

Genetic association studies aim to determine whether a genetic variant is associated with a disorder or a trait. When association is present, a particular allele of an SNP will be seen more often than expected by chance in individuals with the trait or the disorder. Thus, a person with one or two copies of a risk-increasing genetic variant has a higher probability of developing the associated disorder or having the associated trait. Two widely used approaches to identify genetic associations are candidate gene studies and genome-wide association studies (GWAS).

The candidate gene approach can be used for identification of genetic variants of small effects, although this approach is limited by how much we know about the biology of the trait or the disorder. This approach requires *a priori* knowledge of underlying biological mechanisms in order to choose a potential candidate gene. It begins with selection of a putative candidate gene based on its relevance for a specific trait of interest. Then a set of representative SNPs within the gene is selected based on several criteria, for example whether an SNP has a functional role (e.g., affecting gene regulation or its protein product; Kwon & Goate, 2000). Next, each of the selected SNPs is genotyped in every individual in a sample. Finally, we test for an association between each SNP and a trait by comparing the frequency of alleles in cases (people who have a trait/disorder) and in controls (people who do not have a trait/disorder) (Peters, Rodin, De Boer, & Maitland-van der Zee, 2010).

Many candidate genes studies have focused on general cognitive ability. Three candidate genes that have been studied intensively are: dopamine receptor D2 gene (*DRD2*), catechol-O-methyltransferase gene (*COMT*) and brain-derived neurotrophic factor gene (*BDNF*). Previous research suggests a strong biological basis for their involvement in cognition (Payton, 2009). Other possible candidate genes for "g" are those involved in metabolic regulation, for example the insulin signalling pathway, or regulators of fat and glucose in mammals that have been implicated in cognitive aging (Broughton & Partridge, 2009; Rodgers et al., 2005).

For reading disability, one gene that was originally proposed as a candidate gene is Dyslexia susceptibility 1 candidate gene 1 (*DYX1C1*) located on 15q21. However, a recent meta-analysis has found no association between this gene and reading disability (Tran et al., 2013). Other candidate genes – *KIAA0319*, *DCDC2* and *CMIP* – have also been used in association studies of reading ability/disability. *DCDC2* was found to be specifically associated with reading disability, whereas variants in *CMIP* and *KIAA0319* were found to be associated with reading skills across the ability range. Furthermore, the strongest associations were found with single-word reading and spelling measures (Scerri et al., 2011).

Recently, *DCDC2* and *DYX1C1* genes have also been tested for association with four mathematically related phenotypes in the sample of 203 nuclear families where at least one child suffered from dyslexia (Marino et al., 2011). *DCDC2* was found to be associated with numerical facts and *DYX1C1* with mental calculation, even after controlling for the reading skills.

Overall, there have been few candidate gene studies for other educationally relevant traits because they require knowledge of neuronal, physiological and biochemical mechanisms that are not always available for these traits. Further, candidate gene approach is not the optimal way of studying complex traits as they are influenced by many genes of very small effect. A relatively new approach for studying complex traits is the genome-wide association study (GWAS). This is a hypothesis free approach that allows for a search of associated genetic markers across the whole genome. GWAS is a powerful tool in searching for genetic variants of small effects, examining simultaneously hundreds of thousands of DNA markers for association with a trait or a disorder (Hirschhorn & Daly, 2005). This approach has been made possible due to the development of DNA microarrays ("gene chips"). Moreover, genetic variants that are not directly genotyped can be estimated (imputed) from the genetic models that consider the correlations between SNPs (i.e. linkage disequilibrium). Several reference panels such as "The International HapMap consortium" and "Thousand genomes consortium" provide highly dense maps of SNPs across the genome and serve as a comprehensive database of human genetic variation. The data from these reference panels are commonly used to impute SNPs that have not been directly genotyped in a study (Plomin & Schalkwyk, 2007).

To date, a number of GWASs for educationally relevant traits have been conducted. One of the earliest GWAS for reading ability/disability identified ten associated SNPs (Meaburn et al., 2008). However, each of the SNPs had a very small effect. Grouping all identified markers together can provide an estimate of cumulative genetic risk for reading disability, as in this case.

To date, there have been two GWASs of mathematics ability/disability (Baron-Cohen et al., 2014; Docherty et al., 2010). In the study by Docherty et al. (2010), ten SNPs were associated with mathematical variation in the sample of 10-year olds. When these ten SNPs were combined into a set, they accounted for 2.9% of the phenotypic variance in mathematics ability. This association was found across the distribution of mathematics ability, and one third of the children who carried ten or more risk alleles were nearly twice as likely to be at the lowest 15% of the distribution. These results are supportive of the QTL hypothesis, suggesting that the trait is influenced by many genetic variants of small effects that operate across the continuum. In the study by Baron-Cohen et al. (2014), five SNPs were associated with mathematical ability. However, only one of these SNPs was significant after the correction for multiple testing with the remaining four being only nominally significant. The strongest associated SNP was located in a region of chromosome 3q29 area, previously linked to learning difficulties and autism, with replication in independent studies needed.

A GWAS of educational attainment (i.e. years of education and college completion) reported three associated SNPs. All three SNPs were replicated. The effect sizes of these SNPs were ~.02% each. When all measured SNPs were combined into a linear polygenic score they collectively explained 2–3% of the variance in years of education (Rietveld et al., 2013).

A GWAS of general cognitive ability ("g") found four associated SNPs in a sample of 7-year olds. However, the effect sizes were extremely small explaining only 0.3 percent of the variance in "g" (Butcher et al., 2005) when the SNPs associated with "g" were aggregated into a composite SNP set. Furthermore, the same set of SNPs was also associated with "g" at age 2 years suggesting stability of genetic effects for "g" over time. This set was also found to be significantly associated with reading ability. These findings provide further support for the "generalist" genes hypothesis.

Although, GWAS is a powerful tool in uncovering the genetic architecture of complex traits, replication of the original findings often fails. Analysing hundreds of thousands of DNA markers may result in spurious findings (Chabris et al., 2011; Visscher, Yang, & Goddard, 2010). The gold standard for validation of any GWAS findings is a replication of original findings in a large independent sample, which is challenging not least because of the high cost of phenotyping (collecting comparable phenotypic data) and genotyping.

From genetic abnormalities to normal variation in educationally relevant traits

Recently, we have used a novel approach to explore genetic variation associated with individual differences in educationally relevant traits (Rodic et al., in press). This approach focuses on chromosomal regions known to be affected in specific genetic disorders. These disorders, caused by chromosomal mutations such as chromosomal copy or a loss of a part or a whole chromosome, can be associated with cognitive deficits and, sometimes, with advantages in certain cognitive skills. This approach is based on the assumption that cognitive deficits or superiorities in the individuals with chromosomal mutations can be related to common genetic variants within these genome regions. Therefore, common variants within these regions may be related to normal variation in cognitive ability. In the next sections, we briefly describe our research focusing on two genetic syndromes in relation to normal variation in mathematically relevant phenotypes.

Prader–Willi syndrome

Prader–Willi syndrome (PWS) is a rare genetic disorder with estimated prevalence in the population of 1 in 8,000–16,000 live births (Whittington et al., 2001). PWS is caused by the non-expression of paternally imprinted genes in the chromosomal region 15q11–13. One of the regions, commonly associated with the PWS phenotype, is HBII-85 snoRNA (Sahoo et al., 2008).

Prader–Willi syndrome is associated with a number of developmental problems, such as delayed psychomotor development, hyperphagia (obesity due to insatiable appetite), mild to moderate mental retardation, as well as obsessive-compulsive behaviour, temper tantrums and hypogonadism (reduction or absence of hormone secretion of the testes or ovaries) (Varela et al., 2005). All individuals with PWS are to some degree intellectually impaired, with most falling in the mildly mentally retarded range, with mean IQ of 60 – 70 (Cassidy, 1997). Even individuals with PWS who have IQ within the normal range may still experience learning disabilities, and their academic achievement can be poorer than expected for their mental ability (Curfs et al., 1991). Individuals with PWS can have impairments with short-term auditory memory, attention and abstract thinking. Their typical relative strengths include long-term memory, receptive language and reading abilities.

Mathematical skills and short-term memory deficits are thought to be the most profound in individuals with PWS, with all mathematical domains affected (Sulzbacher et al., 1981). For example, Bertella et al. (2015) assessed various components of mathematical ability in individuals with PWS and compared them with non-mathematical skills. Using the Wechsler Adult Intelligence Scale, they found that the lowest ranking subtests were the arithmetic and digit span tests. Compared to an IQ-matched control group with brain lesions, the only scores that differed significantly were those on the arithmetic sub-test, suggesting that low IQ could not be solely responsible for the selective impairment in mathematical abilities. This suggests that mathematical and arithmetic skills in individuals with PWS are particularly affected, even beyond the impairment one would expect from their general cognitive level.

Contrary to the findings of impaired mathematical ability of individuals with PWS their reading and visuo-spatial ability show relative strengths (Cassidy, 1997). Interestingly, individuals with PWS show strong talents for solving jigsaw puzzles, scoring twice higher than typically developing children (Dykens, 2002). Furthermore, they also score higher on spatial tasks than IQ-matched samples with unclassified mental retardation (Dykens, 2002).

One of the explanations for the jigsaw puzzle advantage is that individuals with PWS enjoy playing with them and spend more time practising. Indeed, enjoyment of puzzles has found to be one of the strongest predictors of puzzle performance (Dykens, 2002). In addition PWS individuals are often overly concerned with symmetry, exactness and the need for things to be "just right" – a part of their obsessive-compulsive tendencies (Dykens & Kasari, 1997). However, PWS individuals also outperform typical controls on the analogue number scale task, another task which relies on visuo-spatial abilities (Semenza, 2008). Their advantage in this task cannot be explained with their practice as with jigsaw puzzles. This finding suggests that the exceptional performance of individuals with PWS in visuo-spatial tasks is not solely due to practise, particular enjoyment or obsessive concern (Semenza, 2008).

Williams syndrome

Williams syndrome (WS), or Williams–Beuren syndrome (WBS), is a genetic disorder caused by micro-genetic deletion on the long arm of chromosome 7 (7q11.23), with 25–28 genes missing from this region (Ewart et al., 1993). The prevalence of WS is estimated as 1 in 20,000 live births (Morris et al., 1988).

Individuals with WS are described as cheerful, at ease with strangers and with strong language skills. They have characteristic "elfin" facial appearance, with low nasal bridge; suffer from cardiovascular problems; and suffer from noise-induced hearing loss, which is most likely caused by a malfunctioning auditory nerve. Other common phenotypic characteristics include problems with anxiety, phobias and ADHD.

Individuals with WS have a mild to moderate intellectual disability or learning problems, with average IQ ranging from 40 to 90 (Howlin, Davies, & Udwin, 1998). Their maths ability, number skills, planning and problem solving, spatial working memory, and visuo-spatial long-term memory are usually very poor, whereas their spoken language and skills in recognizing faces are well developed (Ansari et al., 2003; Bellugi, Sabo, & Vaid, 1998; Mesulam, 1982).

Mathematical and numerical ability is found to be unevenly impaired in children and adults with WS, with some aspects less impaired than others. Children with WS show impairments with processing approximate large number displays, while they show no impairments with small number displays (Paterson, Girelli, Butterworth, & Karmiloff-Smith, 2006; Van Herwegen, Ansari, Xu, & Karmiloff-Smith, 2008). This is not a surprise since the approximate number processing relies largely on visuo-spatial skills, which are found to be particularly impaired in individuals with WS (Dehaene, Kerszberg, & Changeux, 1998). On the other hand, processing small number displays relies more on the verbal skills, which are relatively strong in individuals with WS (Ansari, Donlan, & Karmiloff-Smith, 2007).

In line with this evidence, neurological research indicates decreased grey matter and hypoactivation in parietal areas in individuals WS – the areas implicated in magnitude processing, visuo-spatial abilities and visual attention. However, despite their poor mathematical skills, individuals with WS show good performance in reading numbers, and are able to use their strong verbal skills to facilitate performance on maths tasks (O'Hearn & Luna, 2009).

Based on this review of the two syndromes and their cognitive profiles we hypothesized that the genetic regions implicated in these syndromes (i.e. Williams and Prader–Willi) might contain some genetic variants that are also associated with normal variation in cognitive traits such as mathematical, numerical and spatial cognitive abilities. Further, we explored whether these genetic markers were the same across different cognitive abilities and ages. We used genotypic and phenotypic data from a sample of 3,152 12 and 16-year old children from the Twins Early Development Study (TEDS), a longitudinal study of 11,000 twins born in England and Wales in 1994, 1995 and 1996 (Oliver & Plomin, 2007). The TEDS sample is considered to be representative of the UK population in terms of demographic characteristics based on comparisons to the UK census data (Harlaar, Hayiou-Thomas, & Plomin, 2005). The sample exclusion criteria included twins who had severe medical problems or specific chromosomal abnormalities, where severe complications at birth were present. In order to decrease heterogeneity of ancestry, only families who declared themselves as white and whose first language was English were used in this study. The sample size varied from 1112 for general cognitive ability at 16 years of age to 2906 for composite mathematics score at 12 years of age dependant on the data available for the phenotype of interest.

Twelve phenotypes related to mathematical ability, numerical and spatial cognition and general intelligence were tested for association with SNPs in the two genetic regions. Three mathematically related phenotypes were used: Understanding Number, Non-numerical Processes, and Computation and Knowledge – closely linked to the curriculum requirements in the UK. Further phenotypes included three tests of spatial ability (Corsi block, Jigsaw puzzle and hidden shapes), and general cognitive ability composed of two verbal (general knowledge and vocabulary multiple choice subtest) and two non-verbal (Picture completion , and Raven's Standard Progressive Matrices) tests at 12 years of age. The same phenotypes were used for 16 years of age. The data were collected using different approaches such as teacher reports and online administration.

Affymetrix GeneChip 6.0 DNA arrays were used for genotyping nearly 700,000 SNPs. In addition imputation was carried out using the IMPUTE version 2 software (Howie, Donnelly, & Marchini, 2009) on the genotype data after application of quality control procedures, using a 1000 genomes reference panel data. From the imputed SNPs only those who had an information score ≥ 0.98 were retained. All available SNPs in the regions associated with the two genetic syndromes were used in the analyses: ~26,478 SNPs in the 7q11.22 area (Williams syndrome); and ~19,389 SNPs in the 15q11–13 area (Prader–Willi syndrome). The standard quality control criteria were applied for genotyping data: excluding those with minor allele frequency less than 1%; those with more than 2% missing data; and those deviating from Hardy–Weinberg equilibrium ($p < 10^{-3}$).

We performed association analysis for SNPs in two genetic regions using linear regression models under an additive genetic model using an SNP test software which accounts for uncertain genotype data (Marchini et al., 2007). Participants' sex and ancestry (first eight principal components of the genotype data) were included as covariates in the regression model. Phenotypic data were normalized by using the van der Waerden transformation (van der Waerden, 1952). The first step of the association analysis was conducted for three mathematically related phenotypes at age 12. Additionally, three phenotypes including the spatial ability and general cognitive ability at age 12 were tested in order to clarify whether any of our top SNP associations were specific to mathematics or general to other cognitive abilities. Further, the analyses were repeated for the three mathematically related phenotypes and three phenotypes including spatial ability and general cognitive ability at age 16 to establish whether any of the top SNP associations were age specific. The results were corrected for multiple testing using Bonferroni correction producing the corrected p-values of 1.8×10^{-7} (0.05/26,478) for Williams syndrome region, and 2.5×10^{-7} (0.05/19,389) for the Prader–Willi syndrome region.

The results showed that none of the SNP associations were able to meet the demanding significance criterion corrected for multiple testing. The attempts to repeat the top SNP associations for different phenotypes at different ages were also unsuccessful. The power calculation using the Genetic Power Calculator (Purcell & Sham, 2003) revealed 85% power to detect the effect sizes of 0.5% at $p < .05$ in our biggest sample ($N = 2906$), however, for the p-values corrected for multiple comparisons there was virtually no power to detect those effects.

The results suggest that no common genetic variants of large effects contribute to the heritability of any of the phenotypes tested in this study. In line with the previous research (Butcher et al., 2005; Docherty et al., 2010), and similar to other quantitative traits, mathematical abilities and disabilities, general cognitive ability and spatial ability are affected by many genetic variants of small effects (QTLs). However, lack of power prevented detection of these variants. Future studies with larger sample sizes are needed to address this issue.

These results are consistent with the "missing heritability" problem – discrepancy between heritability estimates, from both classical twin and GCTA studies, and effect sizes of associated genetic variants. Many potential explanations for the "missing heritability" problem have been proposed such as the presence of rare variants of small effects, gene–gene interactions and insufficient power of the present studies to detect even smaller effects (Plomin et al., 2013).

Conclusions

Significant advances in our understanding of the aetiology of complex educationally relevant traits have come from the field of behavioural genetics. It is clear that individual differences of all educationally relevant traits to some extent stem from genetic variation. Evidence from the quantitative genetic research indicates that genetic variants (QTLs) have small effects and work in generalist ways. Evidence from quantitative and molecular genetics studies suggest that learning disability represents the low end of the normal distribution of ability (Docherty et al., 2010; Kovas et al., 2007c; Plomin & Kovas, 2005).

With further advancements in molecular genetics, the sequencing of the whole genome will be commonly used leading to identifying both common and rare variants. In 2010, the Wellcome Trust in the United Kingdom launched a 10,000 genome project which allows identification of extremely rare genetic variants. Until whole-genome sequencing becomes common practice, the techniques such as the one described above – focusing on the genetic regions implicated in rare genetic disorders – could be fruitful for genetic studies of complex traits, although bigger samples are needed to maximize the probability of detection of very small effects of the genetic variants. Better understanding of genetic mechanisms underlying individual differences in educationally relevant traits can contribute to attempts to improve education and optimize learning.

References

Alarcon, M., Knopik, V. S., & DeFries, J. C. (2000). Covariation of mathematics achievement and general cognitive ability in twins. *Journal of School Psychology, 38*, 63–77.

Ansari, D., Donlan, C., & Karmiloff-Smith, A. (2007). Atypical and typical development of visual estimation abilities. *Cortex: Special Issue on Selective Developmental Disorders, 6*, 758–768.

Ansari, D., Donlan, C., Thomas, M. S., Ewing, S. A., Peen, T., & Karmiloff-Smith, A. (2003). What makes counting count? Verbal and visuo-spatial contributions to typical and atypical number development. *Journal of Experimental Child Psychology, 85*, 50–62.

Baron-Cohen, S., Murphy, L., Chakrabarti, B., Craig, I., Mallya, U., Lakatosova´, S., ... & Warrier, V. (2014). A Genome Wide Association Study of Mathematical Ability Reveals an Association at Chromosome 3q29, a Locus Associated with Autism and Learning Difficulties: A Preliminary Study. *Plos One, 9*(5), e96374.

Bates, T. C., Castles, A., Coltheart, M., Gillespie, N., Wright, M., & Martin, N. G. (2004). Behaviour genetic analyses of reading and spelling: a component processes approach. *Australian Journal of Psychology, 56*, 115–126.

Bellugi, U., Sabo, H., & Vaid, J. (1998). Spatial deficits in children with Williams syndrome. In J. Stiles-Davis, M. Kritchevsky & U. Belugi (Eds), *Spatial cognition brain bases and development* (pp. 273–297). Hillsdale, NJ: Lawrence Erlbaum.

Bertella, L., Girelli, I., Grugni, G., Marchi, S., Molinari, E., & Semenza, C. (2005). Mathematical skills in Prader–Willi syndrome. *Journal of Intellectual Disability Research, 49*, 159–169.

Bishop, D. V. M. (2001). Genetic Influences on Language Impairment and Literacy Problems in Children: Same or Different? *Journal of Child Psychology and Psychiatry, 42*(2), 189–198.

Broughton, S., & Partridge, L. (2009). Insulin/IGF-like signalling, the central nervous system and aging. *Biochemical Journal, 418*, 1–12.

Bush, W. S., & Moore, J. H. (2012). Chapter 11: Genome-Wide Association Studies. *PLoS Computational Biology, 8(12).*

Butcher, L. M., Meaburn, E., Knight, J., Sham, P., Schalkwyk, L. C., Craig, I. W., & R., P. (2005). SNPs, microarrays, and pooled DNA: identification of four loci associated with mild mental impairment in a sample of 6000 children. *Human Molecular Genetics, 14*, 1315–1325.

Butterworth, B., & Kovas, Y. (2013). Understanding Neurocognitive Developmental Disorders Can Improve Education for All. *Science, 340*(6130), 300–305.

Byrne, B., Wadsworth, S., Corley, R., Samuelsson, S., Quain, P., DeFries, J. C., ... & Olson, R. K. (2005). Longitudinal twin study of early literacy development: Preschool and kindergarten phases. *Scientific Studies of Reading, 9*, 219–235.

Cassidy, S. B. (1997). Prader–Willi Syndrome. *Journal of Medical Genetics, 34*, 917–923.

Chabris, C. F., Hebert, B. M., Benjamin, D. J., Beauchamp, J. P., Cesarini, D., van der Loos, M. J. H. M., ..., & Laibson, D. (2011). Most reported genetic associations with general intelligence are probably false positives. *Psychological Science, 23*(11), 1314–1323.

Curfs, L. G., Wiegers, A. M., Sommers, J. R., Borghgraef, M., & Fryns, J. P. (1991). Strengths and weaknesses in the cognitive profile of youngsters with Prader–Willi syndrome. *Clinical Genetics, 40*, 430–434.

Davis, O. S. P., Haworth, C. M. A., & Plomin, R. (2009a). Dramatic increase in heritability of cognitive development from early to middle childhood an 8-year longitudinal study of 8,700 pairs of twins. *Psychological Science, 20*(10), 1301–1308.

Dehaene, S., Kerszberg, M., & Changeux, J.-P. (1998). A neuronal model of a global workspace in effortful cognitive tasks. *Proceedings of the National Academy of Sciences, USA, 95*, 14529–14534.

de Kovel, C. G., Franke, B., Hol, F. A., Lebrec, J. J., Maassen, B., Brunner, H., ... & Pauls, D. (2008). Confirmation of dyslexia susceptibility loci on chromosomes 1p and 2p, but not 6p in a Dutch sib-pair collection. *American Journal of Medical Genetics Part B: Neuropsychiatric Genetics, 147*, 294–300.

Docherty, S. J., Davis, O. P. S., Kovas, Y., Meaburn, E. L., Dale, S. P., Petrill, S. A., ... & Plomin, R. (2010). A genome-wide association study identifies multiple loci associated with mathematics ability and disability *Genes, Brain and Behaviour, 9*, 234–247.

Dykens, E. M. (2002). Are jigsaw puzzles skills "spared" in persons with Prader–Willi syndrome? *Journal of Child Psychology and Psychiatry, 43*, 343–352.

Dykens, E. M., & Kasari, C. (1997). Maladaptive behavior in children with Prader–Willi syndrome, Down syndrome, and nonspecific mental retardation. *American Journal on Mental Retardation, 102*, 228–237.

Ewart, A. K., Morris, C. A., Atkinson, D., Jin, W., Sternes, K., Spallone, P., ... & Keating, M. T. (1993). Hemizygosity at the elastin locus in a developmental disorder, Williams syndrome. *Nature Genetics, 5*(11), 11–16.

Fisher, S. E., Francks, C., Marlow, A. J., Macphie, I. L., Newbury, D. F., Cardon, L. R., ... & Pennington BF, S. S., DeFries JC, Stein JF, Monaco AP. (2002). Independent genome-wide scans identify a chromosome 18 quantitative-trait locus influencing dyslexia. *Nature Genetics, 30*, 86–91.

Harlaar, N., Hayiou-Thomas, M. E., & Plomin, R. (2005). Reading and general cognitive ability: a multivariate analysis of 7-year-old twins. *Scientific Studies of Reading, 9*, 197–218.

Harlaar, N., Spinath, F. M., Dale, P. S., & Plomin, R. (2005). Genetic influences on early word recognition abilities and disabilities: a study of 7-year-old twins. *Journal of Child Psychology and Psychiatry, 46(4)*, 373–384.

Haworth, C. M. A., Wright, M. J., Luciano, M., Martin, N. G., de Geus, E. J. C., van Beijsterveldt, C. E. M., ..., & Plomin, R. (2010). The heritability of general cognitive ability increases linearly from childhood to young adulthood. *Molecular Psychiatry, 15*(11), 1112–1120.

Hirschhorn, J. N., & Daly, M. J. (2005). Genome-wide association studies for common diseases and complex traits. *Nature Reviews Genetics, 6(2)*, 95–108.

Howie, B. N., Donnelly, P., & Marchini, J. (2009). A Flexible and Accurate Genotype Imputation Method for the Next Generation of Genome-Wide Association Studies. *Plos Genetics, 5*.

Howlin, P., Davies, M., & Udwin, O. (1998). Cognitive functioning in adults with Williams syndrome. *Journal of Child Psychology and Psychiatry, 39*, 183–189.

Jiang, S., Xin, R., Lin, S., Qian, Y., Guomei, T., Wang, D., & Wu, X. (2001). Linkage studies between attention-deficit hyperactivity disorder and the monoamine oxidase genes. *American Journal of Medical Genetics, 105*, 783–788.

König, I. R., Schumacher, J., Hoffmann, P., Kleensang, A., Ludwig, K. U., & T., G. (2011). Mapping for dyslexia and related cognitive trait loci provides strong evidence for further risk genes on chromosome 6p21. *American Journal of Psychiatric Genetics Part B: Neuropsychiatric genetics, 156B*, 36–43.

Kovas, Y., Haworth, C. M. A., Dale, P. S., Petrill, S. A., & Plomin, R. (2007b). Mathematical ability of 10-year-old boys and girls: genetic and environmental etiology of typical and low performance. *Journal of Learning Disabilities, 40*, 554–567.

Kovas, Y., Haworth, C. M. A., Dale, P. S., & Plomin, R. (2007). The genetic and environmental origins of learning abilities and disabilities in early years. *Monographs of the Society for Research in Child Development, 72*, 1–158.

Kovas, Y., Haworth, C. M. A., Dale, P. S., & Plomin, R. (2007a). The genetic and environmental origins of learning abilities and disabilities in the early school years. *Monographs of the Society for research in Child Development, i-156.*

Kovas, Y., Haworth, C. M. A., Harlaar, N., Petrill, S. A., Dale, P. S., & Plomin, R. (2007d). Overlap and specificity of genetic and environmental influences on mathematics and reading disability in 10-year-old twins. *Journal of Child Psychology and Psychiatry. 48*, 914–922.

Kovas, Y., Petrill, S. A., & Plomin, R. (2007c). The origins of diverse domains of mathematics: generalist genes but specialist environments. *Journal of Educational Psychology, 99*, 128–139.

Kovas, Y., Voronin, I., Kaydalov, A., Malykh, S. B., Dale, P. S., & Plomin, R. (2013). Literacy and numeracy are more heritable than intelligence in primary school. *Psychological Science,* 24(10), 2048–2056.

Kwon, J. M., & Goate, A. M. (2000). Candidate gene approach. *Alcohol research and health, 24(3)*, 164–168.

Marchini, J., Howie, B., Myers, S., McVean, G., & Donnelly, P. (2007). A new multipoint method for genome-wide association studies by imputation of genotypes. *Nature genetics, 39*(7), 906–913.

Marino, C., Mascheretti, S., Riva, V., Cattaneo, F., Rigoletto, C., Rusconi, M., ... & Molteni, M. (2011). Pleiotropic Effects of DCDC2 and DYX1C1 Genes on Language and Mathematics Traits in Nuclear Families of Developmental Dyslexia. *Behavior Genetics, 41*(1), 67–76.

Mazzocco, M. M. M. (1998). A process approach to describing mathematics difficulties in girls with Turner syndrome. *Pediatrics, 102*(2 Pt 3), 492–496.

Meaburn, E. L., Harlaar, N., Craig, I. W., Schalkwyk, L. C., & Plomin, R. (2008). Quantitative trait locus association scan of early reading disability and ability using pooled DNA and 100K SNP microarrays in a sample of 5760 children. *Molecular Psychiatry, 13*, 729–740.

Mesulam, M. M. (1982). Slowly progressive aphasia without generalized dementia. *Annals of Neurology, 11*, 592–598.

Morris, C. A., Dempsey, S. A., Leonard, C. O., Dilts, C., & Blackburn, B. L. (1988). Natural history of Williams syndrome: Physical characteristics. *Journal of Paediatric Medicine 113*, 318–326.

Murphy, M. M., & Mazzocco, M. M. M. (2008). Rote numeric skills may mask underlying mathematical disabilities in girls with fragile X syndrome. *Developmental Neuropsychology, 33*, 345–336.

O'Hearn, K., & Luna, B. (2009). Mathematical skills in Williams syndrome: Insight into the importance of underlying representations. *Developmental Disabilities Research Reviews, 15*(1), 11–20.

Oliver, B., Harlaar, N., Hayiou-Thomas, M. E., Kovas, Y., Walker, S. O., Petrill, S. A., ... & Plomin, R. (2004). A twin study of teacher-reported mathematics performance and low performance in 7-year-olds. *Journal of Educational Psychology, 96, 504–517.*

Oliver, B., & Plomin, R. (2007). Twins' Early Development Study (TEDS): A multivariate, longitudinal genetic investigation of language, cognition and behaviour problems from childhood through adolescence. *Twin Research and Human Genetics, 10*, 96–105.

Paterson, S. J., Girelli, L., Butterworth, B., & Karmiloff-Smith, A. (2006). Are numerical impairments syndrome specific? Evidence from Williams syndrome and Down's syndrome. *Journal of Child Psychology and Psychiatry, 47*, 190–204.

Payton, A. (2009). The Impact of Genetic Research on our Understanding of Normal Cognitive Ageing: 1995 to 2009. *Neuropsychology review, 19*, 451–477.

Peters, B. J. M., Rodin, A. S., De Boer, A., & Maitland-van der Zee, A.-H. (2010). Methodological and statistical issues in pharmacogenomics. *Journal of Pharmacy and Pharmacology, 62(2)*, 161–166.

Petrill, S. A., Kovas, Y., Hart, S. A., Thompson, L. A., & Plomin, R. (2009). The genetic and environmental etiology of high math performance in 10-year-old twins. *Behavior Genetics, 39(4)*, 371–379.

Plomin, R. (2013). Child Development and Molecular Genetics: 14 Years Later. *Child Development, 84*, 104–120.

Plomin, R., DeFries, J. C., Knopik, V., & Neiderhiser. (2013). *Behavioral genetics* (6th ed.). New York: Worth.

Plomin, R., Haworth, C. M. A., & Davis, O. S. P. (2009). Common disorders are quantitative traits. *Nature Reviews Genetics, 10*, 872–878.

Plomin, R., & Kovas, Y. (2005). Generalist genes and learning disabilities. *Psychological Bulletin, 131*(4), 592–617.

Plomin, R., Kovas, Y., & Haworth, C. M. A. (2007). Generalist genes: genetic links between brain, mind and education. *Mind, Brain and Education, 1*(1), 11–19.

Plomin, R., Owen, M. J., & McGuffin, P. (1994). The genetic basis of complex human behaviors. *Science, 264*, 1733–1739.

Plomin, R., & Schalkwyk, L. C. (2007). Microarrays. *Developmental Science 10*(1), 19–23

Purcell, S. C., S. S. Sham, P. C. (2003). Genetic Power Calculator: design of linkage and association genetic mapping studies of complex traits. *Bioinformatics, 19*(1), 149–150.

Rietveld, C. A., Medland, S. E., Derringer, J., Yang, J., Esko, T., Martin, N. W., ... & Koellinger, P. D. (2013). GWAS of 126,559 Individuals Identifies Genetic Variants Associated with Educational Attainment. *Science, 340*, 1467–1471.

Rimfeld, K., Kovas, Y., Dale, P. S., & Plomin, R. (2015). Pleiotropy across academic subjects at the end of compulsory education. *Scientific Reports*, DOi: 10.1038/srep11713.

Risch, N., & Merikangas, K. (1996). The future of genetic studies of complex human diseases. *Science, 273*(5281), 1516–1517.

Rodgers, J. T., Lerin, C., Haas, W., Cygi, S. P., Spiegelman, B. M., & P., P. (2005). Nutrient control of glucose homeostasis through a complex of PGC-1α and SIRT1.*Nature, 434*, 113–118.

Rodic, M., Gaysina, D., Docherty, S., Malykh, S., Rimfeld, K., Kovas, Y., & Plomin, R. (in press). Can we use rare genetic syndromes to learn about normal variation in educationally-relevant traits?

Sahoo, T., del Gaudio, D., German, J. R., Shinawi, M., Peters, S. U., Person, R. E., ... & Beaudet, A. L. (2008). Prader–Willi phenotype caused by paternal deficiency for theHBII-85 C/Dbox small nucleolar RNA cluster. *Nature Genetics, 40*, 719–721.

Saviouk, V., Hottenga, J.-J., Slagboom, E. P., Distel, M. A., de Geus, E. C., Willemsen, G., & Boomsma, D. (2011). ADHD in Dutch adults: Heritability and linkage study. *American Journal of Medical Genetics Part B: Neuropsychiatric Genetics, 156*, 352–362.

Scerri, T. S., Morris, A. P., Buckingham, L.-L., Newbury, D. F., Miller, L. L., Monaco, A. P., ..., & Paracchini, S. (2011). DCDC2, KIAA0319 and CMIP Are Associated with Reading-Related Traits. *Biological Psychiatry, 70(30*, 237–245.

Semenza, C. (2008). Number Processing. In B. Stemmer & H. Whitaker (Eds), *Handbook of the neuroscience of language* (pp. 219–227). Oxford: Elsevier.

Shakeshaft, N. G., Trzaskowski, M., McMillan, A., Rimfeld, K., Krapohl, E., Haworth, C. M. A., ..., & Plomin, R. (2013). Strong Genetic Influence on a UK Nationwide Test of Educational Achievement at the End of Compulsory Education at Age 16. *Plos One, 8*(12).

Smith, S. D., Kimberling, W. J., Pennington, B. F., & Lubs, H. A. (1983). Specific reading disability: identification of an inherited form through linkage analysis. *Science, 219*, 1345–1347.

Sulzbacher, S., Wong, B., McKeen, J., Glock, J., & MacDonald, B. (1981). Long term therapeutic effects of a three month intensive growth group. *The Journal of Clinical Psychiatry, 42(4)*, 148–153.

Thompson, L. A., Detterman, D. K., & Plomin, R. (1991). Associations between cognitive abilities and scholastic achievement: Genetic overlap but environmental differences. *Psychological Science*, 158–165.

Tosto, M., Haworth, C. M. A., & Kovas, Y. (2014). Behavioural Genomics of Mathematics. In R. C. Kadosh & A. Dowker (Eds), *Oxford Handbook of Numerical Cognition*. Oxford Handbooks online.

Tosto, M. G., Petrill, S. A., Halberda, J., Trzaskowski, M., Tikhomirova, T. N., Bogdanova, O. E., Ly, R., Wilmer, J. B., Naiman, D. Q., Germine, L., Plomin, R., & Kovas, Y. (2014). How heritable is number sense? A genetic investigation into numerosity estimation abilities in 16 year-old students. *Intelligence, 35–46.*

Tran, C., Gagnon, F., Wigg, K. G., Feng, Y., Gomez, L., Cate-Carter, T. D.,…, & Barr, C. L. (2013). A family-based association analysis and meta-analysis of the reading disabilities candidate gene DYX1C1. *American Journal of Medical Genetics Part B, 162(2),* 146–156.

Trzaskowski, M., Davis, O. S. P., DeFries, J. P., Visscher, P. M., & Plomin, R. (2013a). DNA Evidence for strong genome-wide pleiotropy of cognitive and learning abilities. *Behavior Genetics, 43*(4), 267–273.

Trzaskowski, M., Eley, T. C., Davis, O. S. P., Docherty, S. J., Hanshombe, K. B., Meaburn, E. L.,…, & Plomin, R. (2013b). First genome-wide association study on anxiety-related behaviours in childhood. *PLoS One, 8(4).*

van der Waerden, B. L. (1952). Order tests for the two-sample problem and their power. *Indagationes Mathematicae,* 453–458.

Van Herwegen, J., Ansari, D., Xu, F., & Karmiloff-Smith, A. (2008). Small and larger number processing in infants and toddlers with Williams syndrome. *Developmental Science, 11,* 637–643.

Vandenbergh, D. J., Persico, A. M., & Uhl, G. R. (1992). A human dopamine transporter cDNA predicts reduced glycosylation, displays a novel repetitive element and provides racially-dimorphic TaqI RFLPs. *Molecular Brain Research, 15,* 161–166.

Varela, M. C., Kok, F., Setian, N., Kim, S. A., & Koiffmann, C. P. (2005). Impact of molecular mechanisms, including deletion size on Prader–Willi syndrome phenotype: study of 75 patients. *Clinical Genetics, 67, 47–52.*

Visscher, P. M., Brow, B. A., McCarthy, M. I., & Yang, J. (2012). Five tears of GWAS discovery. *American Journal of Human Genetics, 90(1),* 7–24.

Visscher, P. M., Yang, J., & Goddard, M. E. (2010). A commentary on "Common SNPs explain a large proportion of the heritability for human height" by Yang et al. (2010). *Twin research and Human Genetics, 13,* 517–524.

Whittington, J. E., Holland, A. J., Webb, T., Butler, J., Clarke, D., & Boer, H. (2001). Population prevalence and estimated birth incidence and mortality rate for people with Prader – Willi syndrome in one UK Health Region. *Journal of Medical Genetics, 38,* 792–798.

Yang, J., Benyamin, B., McEvoy, B. P., Gordon, S., Henders, A. K., Nyholt, D. R., ... et al., (2010). Common SNPs explain a large proportion of the heritability for human height. *Nature Genetics, 42,* 565–569.

Yang, J., Manolo, T. A., Pasquale, L. R., Boerwinkle, E., Caporaso, N., Cunningham, J. M., ... et al., (2011). Genome partitioning of genetic variation for complex traits using common SNPs. *Nature Genetics, 43(6),* 519–525.

6

Self in the Mirror of Behavioural Genetics: Reflections from Twin Studies on Self-Esteem and Self-Concept

Yu L. L. Luo and Huajian Cai

The pursuit of self-knowledge possesses a long history in philosophy around the world (in both East and West). But in the field of psychology, the first account of the self did not appear until 1890 when William James devoted an entire chapter to it in his classic book *Principles of Psychology* (James, 1890). Thereafter, the self has remained one of the most researched topics in psychology (Tesser, 2000), particularly since the 1960s. In this chapter, we review the cumulative literatures on self-esteem and self-concept, two core components of the self, from the perspective of behavioural genetics.

Self-concept answers the question "Who am I?" and represents the cognitive component of the self, whereas self-esteem corresponds to "How do I feel about who I am?" and reflects the affective component of the self. Correspondingly, self-concept has been defined as perception of the self (Hoelter, 1985; Shavelson, Hubner, & Stanton, 1976; Watkins & Dhawan, 1989) and self-esteem has been distinguished as affective evaluation of the self (Brown, 2007). Both self-concept and self-esteem are built upon two primary foundations: social interaction and personal merits (Cooley, 1902; James, 1890; Mead, 1913; Rosenberg, 1965). Self-concept has been assumed to be multidimensional and hierarchical, where a general self-concept may be located at the apex of hierarchy while academic and nonacademic self-concepts cluster around a middle level. Academic self-concept is further divided into diverse domain-specific self-concepts such as mathematic self-concept and language self-concept, while nonacademic self-concept includes social, emotional, and physical self-concepts (Marsh, 1990; Marsh & Shavelson, 1985; Shavelson et al., 1976). Likewise, self-esteem may be characterized as multidimensional and hierarchical, with global self-esteem at the apex of hierarchy and multiple specific self-esteem or self-evaluations at lower levels (Brown, 2007).

Obviously, significant possibilities exist for overlap between self-concept and self-esteem. Indeed, self-concept and self-esteem often have been used interchangeably, with general and specific self-concepts corresponding to global and specific self-esteems, respectively (Baumeister, Campbell, Krueger, & Vohs, 2003; Marsh, Byrne, & Shavelson, 1988; Pelham & Swann, 1989; Shavelson et al., 1976). These concepts, however, should not be construed as the same. Perceptions of the self may or may not implicate affective evaluations of the self (e.g., I am a quiet person). Hence, self-concept and self-esteem may be viewed as similar, but at once, remain dissimilar. Previous studies on self-concept usually have focused on a specific domain, whereas those on self-esteem often target a global level. In line with this approach, in this chapter, we refer to global self-evaluation or -perception as self-esteem, while specific self-perception or -evaluation is identified as self-concept. Moreover, we treat both self-esteem and self-concept as trait-like constructs, reflecting stable individual differences in the self.

Research on self-esteem and self-concept has thrived for over half a century. We have charted more than 40,000 hits through June 2012 when searching either term in abstracts appearing in the PsycINFO database, which covers a range of fields including personality psychology, social psychology, clinical psychology, and cognitive neuroscience. To date, research has shown that self-esteem and self-concept are predictive of qualities of close relationships, academic performance, and psychological well-being (Diener & Diener, 1995; Larson, Anderson, Holman, & Niemann, 1998; Pottebaum, Keith, & Ehly, 1986; Orth, Robins, & Widaman, 2012). In the last two decades, researchers have started to explore genetic and environmental origins of individual differences in self-esteem and self-concept, given their importance and trait-like natures (Greenier, Kernis, & Waschull, 1995).

In this chapter, we examine the body of extant behavioural genetic studies on self-esteem and self-concept and provide an overview of the factors responsible for individual differences in each of them. We will first review the literature on self-esteem and follow with a discussion of self-concept. For each construct, evidence of heritability will be addressed first, the aetiology of its stability thereafter, and finally, the genetic basis for its associations with various constructs. In the conclusion to this chapter, we will summarize the main findings and consider the limitations of current research as well as directions for future studies.

Self-esteem

The heritability of self-esteem

With the help of behavioural genetics, psychologists have examined the genetic and environmental origins of variation in self-esteem. By comparing the resemblance of monozygotic (MZ) and dizygotic (DZ) twin pairs on observed traits, behavioural genetic studies can decompose variance within a trait and covariance between traits into additive genetic (A), non-additive genetic (D)/shared environmental (C), and non-shared environmental (E, E also includes measurement error) effects (Plomin, DeFries, Knopik, & Neiderhiser, 2012). MZ twins are 100% genetically identical, whereas DZ twins are on average 50% identical for additive genetic effects and 25% identical for non-additive genetic effects. In the case of twins reared together, higher resemblance between MZ twins than that between DZ twins is attributed to the greater genetic similarity of MZ twins. The observed trait, in other words, is heritable. A non-additive genetic effect is suggested when the DZ twin correlation is less than half of the MZ twin correlation, whereas a shared environmental effect is indicated when the DZ twin correlation is more than half of the MZ twin correlation. The term "heritability" denotes the proportion of the variance of a trait (or the covariance between traits) explained by the genetic effects. While a shared environment contributes to the similarity of twins growing up in the same family, a non-shared environment contributes to the dissimilarity.

Earlier studies have identified moderate genetic effects on self-esteem (Table 6.1). McGuire and his colleagues examined the global self-worth of 720 same-sex siblings (including twins, full siblings, half siblings, and unrelated siblings), whose ages ranged from 9 to 18 years (McGuire, Neiderhiser, Reiss, Hetherington, & Plomin, 1994). The MZ twin correlation ($r = .32$) was notably higher than that of other sibling groups (e.g., $r_{DZ} = .03$, $r_{fullsib} = -.02$). The heritability of global self-worth was estimated at 29%, although this was statistically insignificant. Environmental influences on self-worth were mainly contributed by non-shared environmental factors. Another study by Neiderhiser and McGuire (1994) found that the heritability of self-esteem in children was 49% at age 9, though non-significant, but dropped to 0% at age 10. The lack of statistical significance and consistency over time may be due to the limited sample size, which decreased from 109 pairs at 7 years to 58 pairs at 10 years. A longitudinal study of female twins aged between 17 to 55, which involved 601 pairs in the first wave and 738 in the second wave found significant genetic influences on self-esteem (Roy, Neale, & Kendler, 1995). Heritability was estimated at 40% during the first wave and 36% during the second wave (16 months later on average). The environmental influences predominantly derived from non-shared environments. Kendler, Gardner, and Prescott (1998) have offered the most definitive evidence by investigating 3,793 twin pairs between 18 and 60 years. They found that individual differences in self-esteem were due to genetic factors and non-shared environmental factors. Moreover, they concluded that genetic influences on males and females were identical with only a slight difference in magnitude (female = 32%, male = 29%). The moderate heritability of self-esteem was also observed in other cultures. A study of 81 Japanese adolescent twins (MZ: 50, DZ: 31) found that genetic influences accounted for 49%

of individual differences in self-esteem, with 51% of the variance attributed to non-shared environments (Kamakura, Fujimoto, Itoi, Suzuki, & Kobori, 2001).

More recent studies have yielded similar results (Table 6.1). Using the Rosenberg Self-esteem Scale, Kamakura, Ando, and Ono (2007) found moderate heritability of self-esteem in 100 pairs of Japanese adolescent twins at two waves of assessment, with an interval of 1.3 years (time 1: 31%, time 2: 49%). Non-shared environments accounted for the remainder of individual differences. Another longitudinal study targeting Finnish twins estimated the heritability of self-esteem to be 62% for boys and 40% for girls at age 14, whereas at age 17, the figures dropped to 48% for boys and 29% for girls (Raevuori et al., 2007). By tracking 741 twin pairs for over 7 years, Jonassaint (2010) established that self-esteem was moderately heritable during adolescence (42.5% and 45% at 14.9 and 16.5 years average age, correspondingly), but the heritability dropped to 13% in young adulthood (21.8 years average age).

In sum, the heritability of self-esteem is moderate, with a mean of 38%, deriving from the average score of all the genetic effects listed in Table 6.1. Non-shared environmental influences on self-esteem are substantial (mean = 57%). However, in the majority of cases shared environments impart no significant influence on the variance of self-esteem. The same genetic factors contribute to the self-esteem of females and males, though to different extents. Finally, the heritability of self-esteem varies across ages.

The aetiology of the stability of self-esteem

As a personality trait, the self-esteem of most individuals remains relatively stable across their lifespan (Robins & Trzesniewski, 2005), although feelings of self-regard can fluctuate across life events and situations (Burke, 1980; Campbell & Tesser, 1985; Wells & Marwell, 1976). With the help of multivariate genetic analysis, researchers have examined the genetic and environmental influences on the stability of self-esteem. For two or more traits (or the same trait measured several times), cross-trait cross-twin correlation, such as the correlation between one twin's trait 1 (or time 1) and the co-twin's trait 2 (or time 2), is used to assess the strength of association between twin members across multiple observed variables. By examining the cross-trait cross-twin correlations, the extent to which the phenotypic covariance between two traits (or times) is attributed to common genetic factors, can be estimated in terms of bivariate/multivariate heritability. Bivariate/multivariate shared environmental and non-shared environmental effects can also be estimated. Another set of estimates, such as genetic, shared environmental, and non-shared environmental correlations, may also be used to measure the extent to which two traits are affected by the same genes, shared environments, and non-shared environments, respectively.

Table 6.1 Heritability of self-esteem

Study	Sample	Age	Measure	a^2	c^2	e^2
McGuire et al., 1994	720 siblings	9–18	Harter's SPPA: global self-worth	29%	2%	**69%**[a]
Neiderhiser and McGuire, 1994	Between 109 and 58[c] 58	9 10	Harter's SPPC: self-worth	49%[b] 0%	9% 11%	42% 89%
Roy et al., 1995	Wave 1[d]: 363 MZ, 238 DZ Wave 2[d]: 430 MZ, 308 DZ	17–55	Rosenberg (7 items)	**40%** **36%**	– –	**60%** **64%**
Kendler et al., 1998	1359 MZ, 2434 DZ	18–60	Rosenberg (9 items)	Female: **32%** Male: **29%**	– –	**66%** **72%**
Kamakura et al., 2001	50 MZ, 31 DZ	14–28	Rosenberg (9 items)	**49%**	–	**51%**
Neiss et al., 2005	344 MZ, 534 DZ	44.82 (SD = 12.07)	Ryff's Personal Acceptance Subscale (3 items) self-satisfaction (1 item)	**45%**	–	**55%**
Neiss et al., 2006	78 MZ, 105 DZ	10–19	Rosenberg (10 items)	Phase 1: **47%** Phase 2: 19%	2% 30%	**51%** **52%**
Kamakura et al., 2007	68 MZ, 32 DZ	Phase 1: 19.8 (SD = 4.4) Phase 2: 21.1 (SD = 4.4)	Rosenberg (9 items)	**31%** **49%**	– –	**69%** **51%**
Raevuori et al., 2007	Phase 1: 663 MZ, 1361 DZ Phase 2: 636 MZ, 1280 DZ	14.1 (SD = 0.08) 17.6 (Male SD = 0.24, Female SD = 0.27)	Rosenberg (10 items)	Female: **40%** Male: **62%** Female: **29%** Male: **48%**	**31%** – **34%** –	**30%** **38%** **37%** **52%**
Neiss et al., 2009	431 MZ, 275 DZ	14.57 (SD = 0.57)	Piers-Harris Children's Self-Concept Scale	59%	–	41%
Jonassaint, 2010	289 MZ, 452 DZ	Time 1: 14.9 (SD = 1.6) Time 2: 16.5 (SD = 1.7) Time 3: 21.8 (SD = 1.7)	Rosenberg (4 items)	**42.5%** **45.0%** **13.0%**	– – –	**57.5%** **54.9%** **87.0%**

Note: MZ: monozygotic twin pairs; DZ: dizygotic twin pairs; SD: standard deviation; -: estimate at 0 or negligible. a) Significant estimates are shown in boldface type when statistical significance is reported in the original article. b) Estimates from Neiderhiser and McGuire (1994) are approximate values based on figures in the original article. c) Numbers are for twin pairs, otherwise is specified. d) Participants are all females.

As stated earlier, Roy et al. (1995) assessed self-esteem at two points in time. Their analysis yielded a heritability estimate for self-esteem stability of 53%, which was higher than the heritability estimate for self-esteem level at either single point in time (i.e., average heritability of 38%). A longitudinal twin study reported that genetic influences accounted for nearly all of the phenotypic stability in adolescents' global self-worth across a period of approximately 2.6 years, while non-shared environment explained changes in global self-esteem (McGuire et al., 1999). Similar results were found for Japanese twins in adolescence, where genetic factors accounted for 75% of the stability (Kamakura et al., 2007). In addition, Jonassaint (2010) ascertained that self-esteem, as a latent construct underlying multiple measures, was stable across all three points in time and far more heritable (75%) than any single measurement.

A previously mentioned study also examined sex differences in the aetiology of the stability of self-esteem (Raevuori et al., 2007). The phenotypic correlation (r = .44) between ages 14 and 17 was primarily due to genetic effects in males (82%), while the correlation (r = .46) was moderately influenced by genetic factors in females (31%). The environmental contributions also differed between genders: shared environments accounted for a large portion (61%) of the stability in females, whereas environmental influences on males were modest (18%) and non-shared.

To sum up, stability of self-esteem is attributed largely to genetics. Environmental influences mostly result from factors unique to each individual. Still, there is evidence showing that shared environments contribute considerably to the stability of self-esteem in females.

Self-esteem and emotion

As the affective component of the self, self-esteem is closely related to emotion. Low self-esteem has been linked to negative emotions, such as depression, negative affectivity, neuroticism, anxiety, sadness, anger, and shame (Leary & MacDonald, 2005; Smart & Walsh, 1993; Watson & Clark, 1984). In contrast, high self-esteem has been linked to great happiness, life satisfaction, and subjective well-being (DeNeve & Cooper, 1998; Diener & Diener, 1995; Furnham & Cheng, 2000). Recently, researchers have started to examine how the relationship between self-esteem and emotion is shaped.

Neiss et al. (2005) assessed self-esteem and negative affectivity in 878 pairs of twins, with an average age of 44.82 (27 ~ 74) years. Forty-one percent of the genetic variance in negative affectivity was accounted for by genetic factors common to negative affectivity and self-esteem. Shared environmental effects did not explain covariance between them. Although non-shared environmental effects on self-esteem and negative affectivity were mostly unique to each of them, there was modest overlap due to common factors underlying them.

In a more recent study, Neiss and her colleagues (2009) investigated the genetic and environmental relationship among self-esteem, negative emotionality, and depression. Results showed that genetic factors accounted for the majority of the overlap among the three constructs (69%), while non-shared environmental factors accounted for the rest (30%). Most of the genetic variation of self-esteem (76%) and negative emotionality (86%) is due to shared genetic influences. The specific genetic effects on both of them were modest and not statistically significant. However, genetic influences on depression were largely specific (83%), with modest overlap of those influences on self-esteem and negative emotion. Non-shared environmental influences on self-esteem and negative emotionality showed overlap to a substantial extent, while those influences on depression were generally unique.

In another study concerning 428 pairs of Italian twins between 23 and 24 years of age, Capara et al. (2009) assessed the genetic and environmental overlap underlying the participants' self-esteem, life satisfaction, and optimism. The genetic correlation between any two of them was substantial (r_A: .80 ~ .87), whereas the non-shared environmental correlation was moderate (r_E: .18 ~ .31). Moreover, the same genetic factors accounted for more than 75% of the genetic influences on self-esteem, life satisfaction, and optimism. But the non-shared environmental influences common to the three constructs only made up 19% of the environmental influences on self-esteem and life satisfaction, while contributing more to optimism (56%).

Although the quantity of genetic research on self-esteem and emotion is limited, the findings are rather consistent, at least for the surveyed emotions. It is chiefly for genetic reasons that self-esteem is associated with emotion. Environmental influences are modest and function solely via non-shared environmental factors. Nevertheless, more empiric evidence using diverse samples is needed.

Self-concept

The heritability of self-concept

Since the emergence of the first evidence on the heritability of self-perception (McGue, Hirsch, & Lykken, 1993), we have seen an outgrowth of behavioural genetic studies on self-concept. Up to now, genetic research on self-concept has dealt with some crucial domains, including academic performance, athletic capacity, physical appearance, and social competence (Figure 6.1). The estimate of the heritability of academic self-concept ranges from 0% to 71%, with an average of 40%; the estimate of shared environmental effects is much lower (0% ~ 40%); and the estimate for non-shared environment is comparable to the heritability (21% ~ 68%) (Greven, Harlaar, Kovas, Chamorro-Premuzic, & Plomin, 2009; Hur, 2005; Hur, McGue, & Iacono, 1998; Luo, Haworth, & Plomin, 2010; Luo, Kovas, Haworth, &

Plomin, 2011; McGue et al., 1993; McGuire et al., 1994; McGuire et al., 1999; Neiderhiser & McGuire, 1994; Spinath, Spinath, & Plomin, 2008). Spinath's team (2008) examined potential sex differences in genetic and environmental influences on self-perceived abilities in English language and mathematics, but no disparities were identified.

For athletic self-concept, the heritability demonstrates a wide range from 0% to 61% (average = 45%). The contribution of shared environment is minimal (0% ~ 9%), while that of non-shared environment is substantial (30% ~ 81%) (McGue et al., 1993; McGuire et al., 1994; McGuire et al., 1999; Neiderhiser & McGuire, 1994). Genetic influences on self-concept of physical appearance are significant (average = 59%, 37% ~ 87%), whereas environmental influences are largely from factors unique to siblings (C: 0% ~ 16%, E: 0% ~ 63%) (Hur, 2005; Hur et al., 1998; McGuire et al., 1994; McGuire et al., 1999; Neiderhiser & McGuire, 1994). These studies also identified modest heritability of self-concept for social competence (average = 30%, 0% ~ 50%) and large non-shared environmental effects (40% ~ 93%), with 0% to 15% of the individual differences accounted for by shared environmental effects.

Genetic and environmental influences on other aspects of self-concept have also been examined (Figure 6.1). McGuire and his colleagues found that self-concept for friendship was modestly heritable (10% ~ 40%) as was self-concept for morality (18% ~ 24%) (McGuire et al., 1999; McGuire et al., 1994). Meanwhile, the environmental effects were primarily due to non-shared environments. Three separate studies showed that individual differences in self-perception for behaviour problems largely originated in non-shared environments, with modest genetic (0% ~ 46%) and

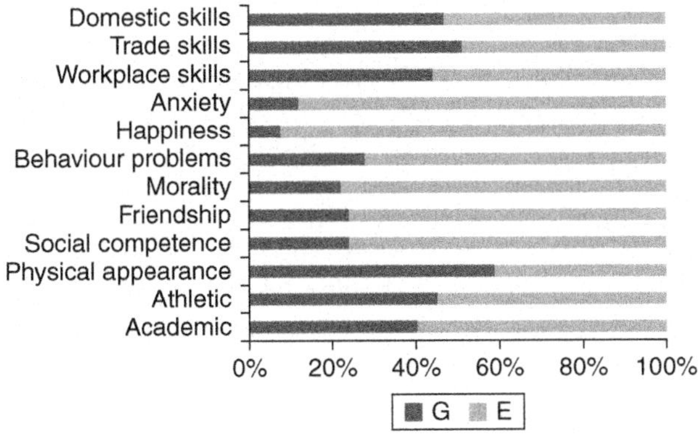

Figure 6.1 Genetic and environmental contributions to self-concept for various domains.

Note: Means of genetic (G) and environmental (E) effects are presented if there are more than one study available. For those domains with only one study, the original estimate is used.

shared environmental (0% ~ 13%) influences (Hur, 2005; Hur et al., 1998; Neiderhiser & McGuire, 1994). Hur's team (1998) further studied the heritability of self-perception of happiness and anxiety, and obtained estimates at 23% and 36%, correspondingly. In addition, they found that the environmental contributions to the variation of self-reported happiness and anxiety manifested similar patterns of other self-concepts; that is, non-shared environments explained a sizeable portion of individual differences (happiness: 71%, anxiety: 55%), whereas shared environment accounted for a narrow proportion (happiness: 6%, anxiety: 9%). Hur's later study (2005), however, failed to show any significant genetic effect on self-reported happiness or anxiety, but the results of environmental effects were comparable to the earlier study. Finally, McGue and his colleagues (1993) assessed self-concepts for workplace, trade, and domestic skills of adult twin pairs and discovered that nearly half of the variation of those self-concepts was caused by genetic factors and the other half by non-shared environmental factors.

To summarize, self-concepts for various domains are heritable to varying extents, with the heritability ranging from 20% to 50% for most domains. The environmental contributions above all stem from non-shared environments. The negligible influences of shared environment seem to be consistent across domains. There is considerable variation among these genetic and environmental estimates, which may arise from differences in self-concept measures.

The aetiology of the stability of self-concept

As the cognitive component of the self, self-perception of one's attributes generally remains stable, despite the fact that the relative accessibility and salience of certain self-perception are context-dependent (Markus & Wurf, 1987; Schlenker, 1985). To date, empirical studies on genetic and environmental influences vis-à-vis the stability of self-concept are relatively rare.

Neiderhiser and McGuire (1994) investigated the extent to which genetic and environmental factors influenced the stability of self-concept. Genetic effects accounted for the majority of stability in twins' ratings of their own physical appearance, athletic competence, and scholastic competence. Stability in behaviour conduct showed a slight genetic influence. Shared environments contributed modestly to stability in children's self-perception of their physical appearance, behaviour conduct, scholastic competence, social acceptance. Non-shared environmental influences were large for stability in behaviour conduct and social acceptance self-concepts, modest for stability in physical appearance and athletic competence self-concepts, but absent for scholastic competence self-concept. Another longitudinal study revealed a somewhat different picture (McGuire et al., 1999): first, the stability of self-perceived athletic competence was entirely due to genetic reasons; second, genetic contributions to the stability of social acceptance self-concept were noted; third, non-shared environment influenced the stability of scholastic competence self-concept. Results on the stability of self-concept of physical appearance, however, were similar across these studies. In a more recent study by Luo and colleagues (2010), the stability of self-perceived academic ability (i.e., how good people think they are at academic activities) from ages 9 to 12 ($r = .44$) was modestly accounted for by genetic (30%) and shared environmental (11%, non-significant) factors, with the majority attributed to non-shared environmental (59%) factors. Specifically for mathematics self-concept, the longitudinal correlation ($r = .47$) from self-perception at 9 years to that at 12 years was primarily due to genetic contributions (61%), together with moderate non-shared environmental contributions (39%) (Luo et al., 2011). (Though beyond the focus of these two studies, the genetic and environmental estimates of stability were not reported in the papers, but can be calculated from the model estimates presented.)

As we have seen, genetic influences are important for the stability of self-concept among most domains. Although the magnitude of environmental influences varies across different domains, the influences are largely from non-shared environments. Nevertheless, given the dearth of research in this area, it is too early to draw further conclusions regarding the stability of domain-specific self-concept.

Self-concept and cognition

Self-concept is a multidimensional construct that includes academic and non-academic facets (Marsh & Shavelson, 1985). Academic self-concept has consistently shown moderate association with cognition. In one study, Ackerman and Wolman (2007) revealed small to large correlations between self-concept for verbal, math, and spatial abilities and parallel objective tests. A recent meta-analysis on a total of 154 effect sizes reported in 41 published studies indicated a moderate correlation of 0.33 between self-estimated and psychometrically measured cognitive abilities (Freund & Kasten, 2012).

The first study on the aetiology of the association between self-concept and cognition examined self-perceived ability, IQ, and academic achievement in a UK sample of 3,785 pairs of twins at ages 7, 9, and 10 (Greven et al., 2009). Results showed considerable genetic correlations between self-perceived ability and IQ ($r_A = .53$) and between self-perceived ability and achievement ($r_A = .65$) at 9 years, suggesting that the genetic basis underlying self-perceived ability and cognition had significant overlap. Results also showed moderate non-shared environmental correlation between self-perceived ability and IQ ($r_E = .38$) and between self-perceived ability and achievement ($r_E = .30$). But both shared environmental correlations were not significant. In addition, the authors analysed IQ at age 7, self-perceived ability at 9, and achievement at 10. The results were similar to those of the same age, that is, high genetic and moderate non-shared environmental correlations were found between self-perceived ability at 9 years old and IQ at 7 years old, but no significant shared environmental correlations were found. Furthermore, the analysis identified significant genetic effects on the association between self-perceived ability at 9 years and achievement at 10 years independent of IQ at 7 but trivial for environmental effects, suggesting that prior self-perceived ability predicted later achievement mainly for genetic reasons.

Using the same sample as Greven et al. (2009) but with assessments at different ages, we extended this investigation by adopting a cross-lagged genetic model (Luo et al., 2010). The cross-lagged association between self-perceived ability at age 9 and school achievement in general at age 12 (r = .11) was explained by genetic influences (28%) as well as shared (55%) and non-shared (16%) environments. The reverse cross-lagged association from school achievement at 9 to self-perceived ability at 12 (r =.24) was primarily genetically mediated (A = 73%, C = 20%, E = 5%). We specifically examined the cross-lagged associations between mathematical self-concept (i.e., self-perception of mathematical ability and interest) and mathematical achievement, adjusted for IQ (Luo et al., 2011). Genetic factors accounted for 98% of the modest cross-lag from self-concept at age 9 to achievement at 12 (r = .11), with non-shared environment accounting for the rest. The "reverse" cross-lag from achievement at 9 to self-concept at 12 (r = .05) was also primarily due to genetic influences (A = 62%, E = 38%).

Clearly, genetic influences exist in the association between academic self-concept and cognition. Possibly, particular genes are present that influence both self-concept and cognition. More important, these influences manifest themselves not only simultaneously but also prospectively. Shared environmental influences, however, are minimal in most cases. Non-shared environment moderately contributes to the concurrent connection between self-concept and cognition, but its contribution to over-time connection is rather limited.

Summary and future directions

Based on the above discussion, a general impression emerges whereby genetic contributions account for a moderate proportion of the individual differences in self-esteem and self-concept, as well as in the stability of self-concept and self-esteem. Meanwhile, environmental contributions are primarily attributed to non-shared environments rather than shared environments. In other words, individuals possess different levels of self-regard due to factors inherited from their parents and their unique circumstances and perceptions. We should be aware, however, that estimates from quantitative genetic studies may shift with changes in the genetic pool or the overall environment of the population. Furthermore, high heritability is not equal to immutability. Acknowledging the genetic influences on self-esteem and self-concept, therefore, is not to presume that genes are the only determinant.

The findings indicating few shared environmental influences seem contradictory to some widely held beliefs, which emphasize the importance of family and school circumstances in shaping children's self-esteem and self-concept. This is not to say that family, schooling, and additional experiences shared by siblings are trivial. It is likely that such conditions as parenting and teaching per se do not make a difference, but what may matter more is precisely how children perceive and internalize these experiences. None of the extant studies, moreover, have considered gene-environment interplay. Such interactions have been documented in intelligence, reading ability, and depressive symptoms, where shared environments were important within sub-groups of the population (Lau & Eley, 2008; Rowe, Jacobson, & Van den Oord, 1999; Taylor, Roehrig, Soden Hensler, Connor, & Schatschneider, 2010). Future studies may examine shared environmental influences on self-esteem and self-concept of different segments of the population.

Another salient conclusion is that genetic factors account for the majority of the associations between self-concept and cognition as well as between self-esteem and emotion. In other words, it is likely that a group of genes influence the self, cognition, and emotion to a substantial extent. In a typical multivariate genetic analysis, genetic overlap would imply pleiotropy – the same genes influencing both traits. A set of genes that make people smart, for instance, may also make people feel good about themselves. The understanding of the mechanism of such pleiotropic associations is quite limited. Nevertheless, current evidence helps to clarify at least one important issue: no environmentally determined link exists between self-perception and cognition or emotion. Given the lack of empirical evidence, whether the major role of genetic effects holds true for the relationship between self-esteem/self-concept and other unexamined (or even those examined) traits, is still unclear. This lingering question invites more genetically-sensitive research on self-esteem/self-concept and cognition, emotion, and beyond. This area can be much enriched by studies on links between non-academic self-concept and cognition, self-esteem and positive emotion, and self-esteem and well-being.

In spite of the merits mentioned above, the limitations of extant research are also worth noting. First, most of the studies are conducted among western populations. As shown by cultural psychologists in the past two decades, people from different cultures differ in the ways they embody self-esteem and self-concept, even if the need for feeling good about oneself is universal (Brown, 2010; Cai, Brown, Deng, & Oakes, 2007; Cai, Sedikedis, Gaertner, Wang, Carvallo, Xu, Mara, & Jackson, 2011; Cai, Wu, & Brown, 2009; Heine, Lehman, Markus, & Kitayama, 1999). To date, there are only three reports in circulation regarding the heritability of self-esteem and self-concept specifically in Asia (Hur, 2005; Kamakura et al., 2007; Kamakura et al., 2001). The significance of those studies was hampered, moreover, due to the small samples (less than 120 twin pairs in each study).

Second, most extant studies use samples consisting of participants who are essentially normal people. It is possible that genetic and environmental factors at the extremes of self-esteem/self-concept may differ from those of individual differences across a normal range. For instance, in a study of language development in young children, heritability is much higher and the shared environmental effect is lower for the lowest 5% of the participants (Dale et al., 1998). It would be intriguing to determine whether extremely high or low self-esteem/self-concept is more affected by genetic factors than that of what is considered to be normal self-esteem/self-concept.

Third, all available studies have focused on explicit self-cognition, but have ignored implicit self-esteem and self-concept. So far research has established that self-esteem and self-concept could manifest themselves at both conscious and unconscious levels (Greenwald & Banaji, 1995; Devos & Banaji, 2003). Notably, implicit self-esteem, known as the introspectively unidentified (or inaccurately identified) effect of the self-attitude on evaluation of self-associated and self-dissociated objects, is largely independent of explicit self-esteem (Cai, 2003; Greenwald & Banaji, 1995; Yamaguchi et al., 2007). The genetic and environmental origins of implicit self-esteem/self-concept are possibly different from those of explicit self-esteem/self-concept. Hence, future attempts to comprehend the heritability of implicit self-esteem should shed light on this issue.

Lastly, though often treated as a stable trait in behavioural genetic research, self-esteem manifests short-term fluctuations, and the pattern varies across individuals. Research has established the importance of the short-term stability of self-esteem, independent of self-esteem level, in psychological functioning (Kernis, 1993). For instance, compared to people with stable self-esteem, people with unstable self-esteem report a higher frequency of depressive symptoms when facing daily challenges, possibly because they tend to over-generalize the negative implications of domain-specific failure (Kernis et al., 1998). Neiss, Sedikides, and Stevenson (2006) have examined the perceived self-esteem stability and found substantial heritability. However, we still do not know how genetic and environmental factors would influence the real short-term stability of self-esteem. Future genetic research on the fluctuations of self-esteem is definitely needed.

Conclusions

In the last two decades, a series of quantitative genetic studies has identified moderate genetic and non-shared environmental influences on the individual differences in self-esteem and self-concept. More recent multivariate genetic studies of self-cognition have revealed that self-esteem/self-concept is associated with certain cognitions or emotions primarily because of a common genetic basis. Nevertheless, we are far from a thorough understanding of the genetic and environmental mechanisms underlying self-esteem and self-concept. More genetic research on both self-esteem and self-concept is required, particularly research with a multivariate and longitudinal design that focuses on various aspects of each of them.

Acknowledgement

The writing of this chapter was supported by National Natural Science Foundation of China [31300871]; the Knowledge Innovation Program [KSCX2-EW-J-8], and the Hundred Talents Program [Y1CX153003] of Chinese Academy of Sciences.

References

Ackerman, P. L., & Wolman, S. D. (2007). Determinants and validity of self-estimates of abilities and self-concept measures. *Journal of Experimental Psychology: Applied, 13*(2), 57–78.

Baumeister, R. F., Campbell, J. D., Krueger, J. I., & Vohs, K. D. (2003). Does high self-esteem cause better performance, interpersonal success, happiness, or healthier lifestyles? *Psychological Science*, 1–44.

Brown, J. D. (2007). *The Self.* New York: Psychology Press.

Brown, J. D. (2010). Across the (not so) great divide: Cultural similarities in self-evaluative processes. *Social and Personality Psychology Compass, 4*, 318–330.

Burke, P. J. (1980). The self: Measurement implications from a symbolic interactionist perspective. *Social Psychology Quarterly, 43*, 18–29.

Cai, H. J. (2003). The effect of implicit self-esteem and the relationship of explicit self-esteem and implicit self-esteem. *Acta Psychologica Sinica, 35*(6), 796–801.

Cai, H., Brown, J. D., Deng, C., & Oakes, M., (2007). Self-esteem and culture: Differences in cognitive self-evaluations or affective self-regard? *Asian Journal of Social Psychology, 10*, 162–170.

Cai, H. J., Wu, Q. P., & Brown, J. D. (2009). Is self-esteem a universal need? Evidence from the People's Republic of China. *Asian Journal of Social Psychology, 12*(2), 104–120.

Cai, H., Sedikedis, C., Gaertner, L., Wang, C., Carvallo, M., Xu, Y., Mara, E., & Jackson, L. E. (2011). Tactical Self-Enhancement in China: Is Modesty at the Service of Self-Enhancement in East Asian Culture. *Social Psychological and Personality Science.* 2(1), 59–64.

Campbell, J., & Tesser, A. (1985). Self-evaluation maintenance processes in relationships. In S. Duck & D. Perlman (Eds), *Understanding personal relationships: An interdisciplinary approach* (Vol. 1, pp. 107–135). London: Sage Publications.

Capara, G. V., Fagnani, C., Alessandri, G., Steca, P., Gigantesco, A., Cavalli Sforza, L. L., & Stazi, M. A. (2009). Human optimal functioning: the genetics of positive orientation towards self, life, and the future. *Behavior Genetics, 39*(3), 277–284.

Cooley, C. H. (1902). *Human nature and the social order.* New York: C. Scribner's sons.

Dale, P. S., Simonoff, E., Bishop, D. V., Eley, T. C., Oliver, B., Price, T. S., ..., & Plomin, R. (1998). Genetic influence on language delay in two-year-old children. *Nature Neuroscience, 1*(4), 324–328.

DeNeve, K. M., & Cooper, H. (1998). The happy personality: A meta-analysis of 137 personality traits and subjective well-being. *Psychological Bulletin, 124*(2), 197–229.

Devos, T., & Banaji, M. R. (2003). Implicit self and identity. In M. Leary & J. Tangney (Eds), *Handbook of Self and Identity* (pp. 153–175). New York: The Guilford Press.

Diener, E., & Diener, M. (1995). Cross-cultural correlates of life satisfaction and self-esteem. *Journal of Personality and Social Psychology, 68*, 653–663.

Freund, P. A., & Kasten, N. (2012). How smart do you think you are? A meta-analysis on the validity of self-estimates of cognitive ability. *Psychological Bulletin, 138*(2), 296–321.

Furnham, A., & Cheng, H. (2000). Perceived parental etiologyself-esteem and happiness. *Social Psychiatry and Psychiatric Epidemiology, 35*(10), 463–470.

Greenier, K. D., Kernis, M. H., & Waschull, S. B. (1995). Not all high or low self-esteem people are the same: Theory and research on the stability of self-esteem. In M. H. Kernis (Ed.), *Efficacy, agency, and self-esteem* (pp. 51–68). New York: Plenum.

Greenwald, A. G., & Banaji, M. R. (1995). Implicit Social Cognition – Attitudes, Self-Esteem, and Stereotypes. *Psychological Review, 102*(1), 4–27.

Greenwald, A. G., & Farnham, S. D. (2000). Using the implicit association test to measure self-esteem and self-concept. *Journal of Personality and Social Psychology, 79*(6), 1022–1038.

Greven, C. U., Harlaar, N., Kovas, Y., Chamorro-Premuzic, T., & Plomin, R. (2009). More than just IQ: School achievement is predicted by self-perceived abilities-but for genetic rather than environmental reasons. *Psychological Science, 20*(6), 753–762.

Heine, S. J., Lehman, D. R., Markus, H. R., & Kitayama, S. (1999). Is there a universal need for positive self-regard? *Psychological Review, 106*(4), 766–794.

Hoelter, J. W. (1985). The structure of self-conception: conceptualization and measurement. *Journal of Personality and Social Psychology, 47*, 213–231.

Hur, Y. M. (2005). Genetic and environmental influences on self-concept in female preadolescent twins: Comparison of Minnesota and Seoul data. *Twin Research and Human Genetics, 8*(4), 291–299.

Hur, Y. M., McGue, M., & Iacono, W. G. (1998). The structure of self-concept in female preadolescent twins: a behavioral genetic approach. *Journal of Personality and Social Psychology, 74*(4), 1069–1077.

James, W. (1890). *Principles of Psychology* (1983 Paperback ed.). Harvard University Press.

Jonassaint, C. R. (2010). Heritability of self-esteem from adolescence to young adulthood. *The New School Psychology Bulletin, 7*(1), 3–15.

Kamakura, T., Ando, J., & Ono, Y. (2007). Genetic and environmental effects of stability and change in self-esteem during adolescence. *Personality and Individual Differences, 42*, 181–190.

Kamakura, T., Fujimoto, M., Itoi, H., Suzuki, K., & Kobori, Y. (2001). Genetic and environmental influences on self-esteem in a sample of middle-aged Japanese twins. *Behavior Genetics, 31*(5), 457–457.

Kendler, K. S., Gardner, C. O., & Prescott, C. A. (1998). A population-based twin study of self-esteem and gender. *Psychological Medicine, 28*, 1403–1409.

Kernis, M. H. (1993). The roles of stability and level of self-esteem in psychological functioning. In R. F. Baumeister (Ed.), *Self-esteem: The puzzle of low self-regard* (pp. 167–182). New York: Plenum Press.

Kernis, M. H., Whisenhunt, C. R., Waschull, S. B., Greenier, K. D., Berry, A. J., Herlocker, C. E., & Anderson, C. A. (1998). Multiple facets of self-esteem and their relations to depressive symptoms. *Personality and Social Psychology Bulletin, 24*(6), 657–668.

Larson, J. H., Anderson, S. M., Holman, T. B., & Niemann, B. K. (1998). A longitudinal study of the effects of premarital communication, relationship stability, and self-esteem on sexual satisfaction in the first year of marriage. *Journal of Sex and Marital Therapy, 24*, 193–206.

Lau, J. Y., & Eley, T. C. (2008). Disentangling gene-environment correlations and interactions on adolescent depressive symptoms. *Journal of Child Psychology and Psychiatry, 49*(2), 142–150.

Leary, M. R., & MacDonald, G. (2005). Individual differences in self-esteem: A review and theoretical integration. In M. Leary & J. P. Tangney (Eds), *Handbook of self and identity* (pp. 401–418). New York: Guilford Press.

Luo, Y. L. L., Haworth, C. M. A., & Plomin, R. (2010). A novel approach to genetic and environmental analysis of cross-lagged associations over time: The cross-lagged relationship between self-perceived abilities and school achievement is mediated by genes as well as the environment. *Twin Research and Human Genetics, 13*(5), 426–436.

Luo, Y. L. L., Kovas, Y., Haworth, C. M. A., & Plomin, R. (2011). The etiology of mathematical self-evaluation and mathematics achievement: Understanding the relationship using a cross-lagged twin study from ages 9 to 12. *Learning and Individual Differences, 21*(6), 710–718.

Markus, H., & Wurf, E. (1987). The dynamic self-concept: A social psychological perspective. *Annual Review of Psychology, 38*, 299–337.

Marsh, H. W. (1990). The structure of academic self-concept: The Marsh/Shavelson model. *Journal of Educational Psychology, 82*, 623–636.

Marsh, H. W., Byrne, B. M., & Shavelson, R. J. (1988). A multifaceted academic self-concept: Its hierarchical structure and its relation to academic achievement. *Journal of Educational Psychology Today, 80*, 366–380.

Marsh, H. W., & Shavelson, R. J. (1985). Multidimensional adolescent self-concepts: Their relationship to age, sex and academic measures. *American Educational Research Council, 22*, 422–444.

McGue, M., Hirsch, B., & Lykken, D. T. (1993). Age and the self-perception of ability: a twin study analysis. *Psychological Aging, 8*(1), 72–80.

McGuire, S., Manke, B., Saudino, K. J., Reiss, D., Hetherington, E. M., & Plomin, R. (1999). Perceived competence and self-worth during adolescence: A longitudinal behavioral genetic study. *Child Development, 70*(6), 1283–1296.

McGuire, S., Neiderhiser, J. M., Reiss, D., Hetherington, E. M., & Plomin, R. (1994). Genetic and environmental influences on perceptions of self-worth and competence in adolescence: a study of twins, full siblings, and step-siblings. *Child Development, 65*(3), 785–799.

Mead, G. H. (1913). The Social Self. *Journal of Philosophy, Psychology and Scientific Methods, 10*, 374–380.

Neiderhiser, J. M., & McGuire, S. (1994). Competence during middle childhood. In J. C. DeFries, R. Plomin & D. W. Fulker (Eds), *Nature and nurture during middle childhood* (pp. 141–151). Malden, MA: Blackwell.

Neiss, M. B., Sedikides, C., & Stevenson, J. (2006). Genetic Influences on Level and Stability of Self-Esteem. *Self and Identity, 5*, 246–266.

Neiss, M. B., Stevenson, J., Legrand, L. N., Iacono, W. G., & Sedikides, C. (2009). Self-Esteem, Negative Emotionality, and Depression as a Common Temperamental Core: A Study of Mid-Adolescent Twin Girls. *Journal of Personality, 77*(2), 327–346.

Neiss, M. B., Stevenson, J., Sedikides, C., Kumashiro, M., Finkel, E. J., & Rusbult, C. E. (2005). Executive self, self-esteem, and negative affectivity: Relations at the phenotypic and genotypic level. *Journal of Personality Social Psychology, 89*(4), 593–606.

Orth, U., Robins, R., & Windaman, K. (2012). Life-Span Development of Self-Esteem and Its Effects on Important Life Outcomes. *Journal of Personality and Social Psychology, 102*(6), 1271–1288.

Pelham, B. W., & Swann, W. B., Jr. (1989). From self-conceptions to self-worth: on the sources and structure of global self-esteem. *Journal of Personality Social Psychology, 57*(4), 672–680.

Plomin, R., DeFries, J. C., Knopik, V. S., & Neiderhiser, J. M. (2012). *Behavioral genetics* (6th ed.). New York, NY: Worth Publishers.

Pottebaum, S. M., Keith, T. Z., & Ehly, S. W. (1986). Is there a causal relation between self-concept and academic achievement? *Journal of Educational Research, 79,* 140–144.

Raevuori, A., Dick, D. M., Keski-Rahkonen, A., Pulkkinen, L., Rose, R. J., Rissanen, A.,..., & Silventoinen, K. (2007). Genetic and environmental factors affecting self-esteem from age 14 to 17: a longitudinal study of Finnish twins. *Psychological Medicine, 37*(11), 1625–1633.

Robins, R. W., & Trzesniewski, K. H. (2005). Self-esteem development across the lifespan. *Current Directions in Psychological Science,* 14(3), 158–162.

Rosenberg, M. (1965). *Society and the adolescent self-image.* Princeton, N.J.: Princeton University Press.

Rowe, D. C., Jacobson, K. C., & Van den Oord, E. J. (1999). Genetic and environmental influences on vocabulary IQ: parental education level as moderator. *Child Development, 70*(5), 1151–1162.

Roy, M. A., Neale, M. C., & Kendler, K. S. (1995). The Genetic Epidemiology of Self-esteem. *British Journal of Psychiatry, 166,* 813–820.

Schlenker, B. R. (1985). Identity and self-identification. In B. R. Schlenker (Ed.), *The self and social life* (pp. 65–99). New York: McGraw-Hill.

Shavelson, R. J., Hubner, J. J., & Stanton, G. C. (1976). Self-concept: Validation of construct interpretation. *Review of Educational Research, 46,* 407–441.

Smart, R. G., & Walsh, G. W. (1993). Predictors of depression in street youth. *Adolescence, 28*(109), 41–53.

Spinath, F. M., Spinath, B., & Plomin, R. (2008). The nature and nurture of intelligence and motivation in the origins of sex differences in elementary school achievement. *European Journal of Personality, 22*(3), 211–229.

Taylor, J., Roehrig, A. D., Soden Hensler, B., Connor, C. M., & Schatschneider, C. (2010). Teacher quality moderates the genetic effects on early reading. *Science, 328*(5977), 512–514.

Tesser, A. (2000). On the confluence of self-esteem maintenance mechanisms. *Personality and Social Psychology Review, 4,* 290–299.

Watkins, D., & Dhawan, N. (1989). Do we need to distinguish the constructs of self-concept and self-esteem? *Journal of Social Behavior and Personality, 4,* 555–562.

Watson, D., & Clark, L. A. (1984). Negative affectivity: the disposition to experience aversive emotional states. *Psychological Bulletin, 96*(3), 465–490.

Wells, L. E., & Marwell, G. (1976). *Self-esteem: its conceptualization and measurement.* Beverly Hills, Calif.: Sage Publications.

Yamaguchi, S., Greenwald, A. G., Banaji, M. R., Murakami, F., Chen, D., Shiomura, K., Kobayashi, C., Cai, H., Krendl, A. (2007). Apparent universality of positive implicit self-esteem. *Psychological Science, 18*(6), 498–500.

7
The Nature and Nurture of Wellbeing

Claire M. A. Haworth

Psychological research has contributed to considerable progress in understanding and treating mental illness, however, we know much less about mental health and wellbeing, and how we can promote human flourishing (Seligman & Csikszentmihalyi, 2000). Studying what goes well in life, and how to improve wellbeing and foster human potential are central aims in positive psychology. Professor Martin Seligman highlighted the need to move beyond the disease model in his inaugural speech as President of the American Psychological Association in 1998 and, in doing so, galvanized the field of positive psychology. Since 1998, positive psychology has blossomed, and has contributed to some important insights into what it means to be mentally healthy, and how wellbeing is linked to successful outcomes, including physical health and longevity (e.g., Fredrickson, 1998; Howell, Kern, & Lyubomirsky, 2007; Lyubomirsky, King, & Diener, 2005).

A key question is why some individuals are happier than others. And beyond happiness, what causes individuals differences in wellbeing, including life satisfaction and feelings of meaning and purpose in life. Behavioural genetic methods are well suited for the investigation of the underlying causes of individual differences in complex traits, like wellbeing. There is growing interest in applying behavioural genetic methods to study wellbeing (Nes, 2010; Sprangers et al., 2010), especially as many large twin and family studies already have some brief measures of wellbeing in their previous assessment batteries. These twin and family studies are beginning to uncover the roles of genes and environments in creating individual differences in wellbeing, as well as contributing to our understanding of why wellbeing is related to other important life outcomes. And as the field grows, research is being directed at understanding which specific genes and environments are involved, how they might interact, and how they contribute to stability and change across development. This chapter provides an overview of the current findings regarding the nature and nurture of wellbeing, as well as highlighting areas in need of further research.

How heritable is wellbeing?

A growing number of studies are addressing the role of genes and environments in explaining individual differences in wellbeing; however, it is difficult to compare directly between these studies because of the wide range of measures used. The majority of studies focus on subjective wellbeing, a couple of studies have considered psychological wellbeing, and one study included character strengths. Even within the studies on subjective wellbeing there are few examples of the same measure being used across studies. Subjective wellbeing is sometimes indexed by just a measure of life satisfaction, or a combination of life satisfaction and measures of positive and negative affect.

The first study of subjective wellbeing was based on the Minnesota Twin Registry (Tellegen et al., 1988), including both reared together and reared apart twins. The results of this study were augmented with more samples in a later study that focused specifically on wellbeing (Lykken & Tellegen, 1996).Wellbeing was measured when the twins were middle-aged using a subscale of the Multidimensional Personality Questionnaire in a sample of 1380 pairs of reared together twins and 111 pairs of reared apart twins. The correlation between identical (monozygotic; MZ) twins reared together for wellbeing was 0.44, and for non-identical (dizygotic, DZ) twins 0.08, indicating moderate heritability around 44%, no shared environmental influences and moderate non-shared environmental influences (around 56%). Results for the sub-sample of twins reared apart were strikingly similar (rMZ = 0.52, rDZ = 0.02), confirming the absence of shared environmental influences and the importance of genetic factors in explaining individual differences in wellbeing. The twin correlations were also consistent with both additive and non-additive genetic effects because the DZ twin correlation was less than half the MZ twin correlation.

Another early study on subjective wellbeing considered only life satisfaction and capitalized on the Swedish Adoption/Twin Study of Ageing (Bergeman, Plomin, Pedersen, & McClearn, 1991). Life satisfaction data were available for 424 twin pairs aged 50 years and above, including a subsample of reared apart twins. Life satisfaction was assessed using a 13-item self-report questionnaire (Wood, Wylie, & Sheafor, 1969). Genetic influences explained 25% of the variance in life satisfaction, with the remaining variance accounted for by non-shared environmental factors. Further larger studies on life satisfaction followed, including two (partly overlapping) Dutch studies of more than 2000 twin pairs (Bartels & Boomsma, 2009; Stubbe, Posthuma, Boomsma, & De Geus, 2005), which used the 5-item Satisfaction with Life Scale (Diener, Emmons, Larsen, & Griffin, 1985), and a study of 1000 pairs of adolescent twins, which used a single item "How satisfied are you with your life as a whole?" to assess life satisfaction (De Neve, Christakis, Fowler, & Frey, 2010). Heritability of life satisfaction was estimated at 47% (Bartels & Boomsma, 2009), 38% (Stubbe, Posthuma, Boomsma, & De Geus, 2005), and 33% (De Neve, Christakis, Fowler, & Frey, 2010), with the remaining variance explained by non-shared environmental factors. In agreement with the first study on wellbeing, the genetic influences in two of these studies (Bartels & Boomsma, 2009; Stubbe, Posthuma, Boomsma, & De Geus, 2005) were non-additive in nature, and none of these studies showed evidence of shared environmental influences on wellbeing.

Positive affect is another important component of subjective wellbeing. Baker and colleagues (Baker, Cesa, Gatz, & Mellins, 1992) used the Bradburn Affect Balance Scale (Bradburn, 1969) in a sample of 105 adult twin pairs to estimate the role of genetic and environmental factors in both positive and negative affect. This study included both a twin sample and a more complex intergeneration family study. The twin correlations for positive affect were 0.34 for MZ twins and 0.23 for DZ twins, indicating modest heritability (22%) and shared environment (12%) and moderate non-shared environmental factors (66%). Of note, the more complex analyses including the intergenerational family members estimated no genetic influence and a stronger shared environmental component, as well as evidence for assortative mating for positive affect. These inconclusive results require follow-up and replication in larger samples. Bartels and Boomsma (2009) have investigated a measure of positive affect in a larger sample of 2000 pairs of twins. Using data from the Netherlands Twin Registry, and the 4-item subjective happiness scale (Lyubomirsky & Lepper, 1999), they found heritability estimates of 40%, with both additive and non-additive genetic factors involved. Bartels and colleagues were able to extend this initial study by adding additional older participants with subjective happiness data (Bartels et al., 2010). In this larger study of about 4000 pairs of twins, they found no significant non-additive genetic effects, and the additive genetic effects were significantly different for males and females. For males, subjective happiness was 22% heritable, compared to 41% for females; non-shared environmental influences were 78% and 59%, respectively. There is a need for future studies to replicate these modest sex differences for happiness.

Finally for subjective wellbeing, a handful of studies have reported results from an overall composite of subjective wellbeing that combines indices of life satisfaction, and affect, in line with the way subjective wellbeing is currently conceptualized in the literature (Diener, Suh, & Lucas, 1999). Røysamb and colleagues (Røysamb et al., 2002) used a measure of subjective wellbeing that included four items: "When you think about your life at present, would you say you are mostly satisfied with your life, or mostly dissatisfied?"; "Are you usually happy or dejected?"; "Do you mostly feel strong and fit or tired and worn out?", and "Over the last month, have you suffered from nervousness (felt irritable, anxious, tense, or restless)?" Item responses were coded so that higher values reflected greater wellbeing, and a mean of the items was taken to represent overall subjective wellbeing.

The sample included 2570 complete twin pairs who were aged between 18 and 25 years. Results indicated small (but significant) sex differences in the aetiology of subjective wellbeing. The heritability of subjective wellbeing was higher for females (54%) than for males (46%). The best-fitting model included only additive genetic influences and non-shared environmental influences, there were no significant non-additive or shared environmental influences in this sample. Using the same 4-item measure of subjective wellbeing, Nes and colleagues (Nes, Czajkowski, & Tambs, 2010a) used an extended twin and sibling design to bring more power to bear on the question of non-additive versus additive genetic influences and the presence of sex differences for subjective wellbeing. The sample included 3310 complete twin pairs aged 18–31 years and more than 50,000 individuals from nuclear families (mother–offspring, father–offspring and spouse pairs). In this larger sample, spanning a wider age range, both additive and non-additive genetic factors were important for subjective wellbeing, with the total heritability estimated as 33% for females and 36% for males. These estimates are lower than those previously reported for this same measure of wellbeing (Røysamb et al., 2002). The differences are likely to be due to a combination of age differences, as well as biases included in the estimates when only twins are used, as opposed to the extended twin family design that can model more parameters together. A further study has investigated subjective wellbeing using a different measure, which combines a 6-item scale of positive affect and a single item assessing life satisfaction (Keyes, Myers, & Kendler, 2010). The authors refer to this measure as emotional wellbeing. Using data from 670 pairs of middle-aged twins, heritability was estimated as 50% and non-shared environment as 50%, with no evidence for sex differences in these estimates.

In summary, for measures of subjective wellbeing, there is evidence for significant genetic effects and an absence of shared environmental influences. Genetic influences ranged between 22% and 55%, and the results were inconsistent concerning the presence of non-additive genetic effects and sex differences. The harmonization of measures across large twin and family studies should help to resolve these issues, and the identification of the specific genes and environments involved in subjective wellbeing will provide stronger tests of the mode of genetic influence and its impact on males and females across the life span.

Two studies to date have assessed psychological wellbeing (Gigantesco et al., 2011; Keyes, Myers, & Kendler, 2010), which refers to measures that go beyond the more hedonic aspects of wellbeing that are captured by measures of subjective wellbeing, and instead tap into the eudaimonic aspects of wellbeing like meaning and purpose in life and self-realization (Ryff & Singer, 2008). Both studies included six scales, each with three items, which indexed self-acceptance, positive relations with others, personal growth, purpose in life, environmental mastery, and autonomy (Ryff & Keyes,

1995). The first study combined these six scales into an overall composite of psychological wellbeing (Keyes, Myers, & Kendler, 2010), and in a sample of 670 middle-aged twin pairs, heritability was 52%. The second study analysed the subscales separately in their sample of 284 young adult twin pairs (Gigantesco et al., 2011), with heritability estimates ranging from 37% for personal growth to 64% for positive relations with others. Neither study found evidence for shared environmental influences on psychological wellbeing, and little evidence for non-additive genetic effects or sex differences, although these studies were too small to effectively test for any subtle differences by sex or mode of genetic influence.

Beyond subjective and psychological wellbeing, one study to date has used the twin method to investigate the role of genes and environments on character strengths (Steger et al., 2007). Ninety-one complete pairs of twins provided data on the values in action inventory of strengths, which includes 24 character strengths, such as curiosity, integrity, kindness, fairness, and forgiveness (Peterson & Seligman, 2004). Significant genetic influences were found for 21 of the 24 strengths; heritability estimates ranged from 26% to 59%, with the remaining variance accounted for by non-shared environmental factors.

Twin and family studies investigating the nature and nurture of wellbeing are in their infancy, and many questions remain about the relative influence of genes and environments, how these might change across the life span, or may be different across the sexes. Nevertheless, the studies to date have produced two consistent findings: genetic influences matter, and shared environmental influences do not. The pattern of significant genetic influences and minimal shared environmental factors fits with the pattern seen in many other traits in behavioural genetics (Plomin, DeFries, McClearn, & McGuffin, 2008). Given the similar pattern of results within wellbeing, and beyond wellbeing, the next question is whether the same genes and environments impact different measures of wellbeing, and mediate the relationships between wellbeing and other behavioural and health outcomes. The next section describes how multivariate twin analyses have been used to address the degree of genetic and environmental overlap between measures.

Multivariate twin analyses of wellbeing

Multivariate twin analyses decompose both the variance of each trait and the covariance between different traits into genetic and environmental sources (Plomin, DeFries, McClearn, & McGuffin, 2008). They allow an estimation of the genetic and environmental overlap between traits, and a test of different theoretical models that underlie the association between traits. At the simplest level, multivariate twin analyses can show whether the genetic influences on trait A also influence trait B and trait C, but increasingly complex models can be used, for example, to assess whether the genetic influences on traits A, B and C can be indexed by one over-arching latent factor. More information about the types of multivariate twin analyses available can be found elsewhere (Neale & Maes, 2001). To date, only two studies have used multivariate twin analyses to understand why different measures of wellbeing are related. The first study focused on subjective wellbeing (Bartels & Boomsma, 2009) in 2000 pairs of twins aged between 13 and 28 years. Bartels and Boomsma (2009) included four measures of wellbeing: quality of life in general, life satisfaction, quality of life at present, and subjective happiness. These measures were moderately correlated at the phenotypic level, with an average correlation across all of the measures of 0.54, so the study set out to examine the extent to which shared genes and/or shared environmental influences could explain these correlations. Results from the multivariate twin analyses supported an independent pathway model for the additive and non-additive genetic factors, that is, there were common genetic influences between the different measures, but these common genetic influences could vary in their impact on each trait. There were also significant genetic influences that were specific to quality of life in general and subjective happiness. For non-shared environmental influences there were significant common and specific influences for each trait. On average, almost 60% of the correlation between these traits is due to shared genetic influences. These results suggest strong genetic overlap between these different measures of subjective wellbeing.

The second study that has used multivariate twin analyses to understand the relationship between different measures of wellbeing included composites of subjective wellbeing (called emotional wellbeing in this study), social wellbeing, and psychological wellbeing (Keyes, Myers, & Kendler, 2010). Using data from 670 pairs of middle-aged twins, the authors investigated the extent to which the same genes (and environments) are involved in these different areas of wellbeing. In this study a common pathway model was the best-fitting model, that is, the three indices of wellbeing loaded onto an overall latent factor of wellbeing (similar to a factor analysis), which was influenced by additive genetic and non-shared environmental factors. There were also significant genetic and environmental influences that were specific to each of the domains. For example, for subjective wellbeing the overall genetic influence explained 50% of the variance, about 65% of this genetic influence comes from the common factor, and the remaining 35% of the genetic influence is specific to subjective wellbeing. For psychological wellbeing genetic influences were almost entirely driven by the common pathway, with negligible genetic influences specific to psychological wellbeing. Non-shared environmental influences were more likely to be specific to each area of wellbeing. These results indicate strong genetic overlap between these different areas of wellbeing. However, as in the other multivariate example, there were some specific genetic influences, and non-shared environmental influences were more likely to be specific to each domain.

It is difficult to compare the results of these two studies directly because the estimates are based on different models, however, the basic implication of these results is that if genes are identified for one aspect of wellbeing, there is a high chance that those genes will also influence other aspects of wellbeing. Nevertheless, these measures are not exactly the same; although shared genes can account for a lot of the overlap, these traits are not 100% correlated at the phenotypic level. Further work is needed to understand the genetic and environmental overlap between other measures of wellbeing. Understanding how these measures relate to each other at the aetiological level may help clarify the conceptualization of these measures at the phenotypic level, allowing the field to move towards more unified measures of wellbeing across different studies.

Multivariate twin analyses can also be used to understand why measures of wellbeing are related to other outcomes such as physical health, personality, and mental illness. For physical health, the assumption is that the link with wellbeing is due to environmental links – for example, because a happy person may take better care of their health. However, the evidence to date from cross-sectional studies, suggests that there is genetic overlap between wellbeing and health (Røysamb et al., 2003), particularly for self-perceived health and muscular-skeletal pain, but there is no genetic (or phenotypic) overlap with allergies. A smaller study in older adults has suggested developmental differences in the link between health and wellbeing (Harris et al., 1992), with the genetic link between life satisfaction and wellbeing appearing only once the participants were over the age of 65 years. It will be important to understand the dynamic interplay between health and wellbeing across the lifespan to understand how one may lead to the other, and in doing so, suggest how interventions could be used to improve health and wellbeing across the lifespan, and particularly in old age. As discussed below, there is a study on wellbeing and longevity (Sadler, Miller, Christensen, & McGue, 2011) that has provided some clues to the developmental pattern. A handful of studies have considered more specific health-related behaviours, demonstrating, for example, genetic overlap between wellbeing and exercise behaviour (Stubbe, de Moor, Boomsma, & De Geus, 2007), and between wellbeing and sleep problems (Nes et al., 2005). Even from the limited literature in this area, it is clear that genetic factors are important in explaining why wellbeing is related to other important life outcomes, like health.

An important question in the literature is whether wellbeing can be explained via personality characteristics. At the phenotypic level there is a strong, but not complete, overlap between wellbeing and personality (DeNeve & Cooper, 1998). Behavioural genetic analyses can contribute to this discussion by investigating the genetic and environmental overlap between wellbeing and personality. A small study of 91 twin pairs found genetic overlap between personality and character strengths (Steger et al., 2007), with shared genes accounting for most of the genetic influence on character strengths. In a larger study of about 1000 pairs of middle-aged twins from the United States, Weiss and colleagues (Weiss, Bates, & Luciano, 2008) investigated the genetic and environmental overlap between the five-factor model of personality and a three item measure of life satisfaction. The phenotypic correlations between the five personality domains and life satisfaction ranged from .14 to .35 in magnitude, with the expected direction of effects (i.e. positive relations with extraversion, openness, agreeableness and conscientiousness, and a negative correlation with neuroticism). The twin analyses indicated that there were no specific genetic influences for life satisfaction once the shared genetic influences with personality were taken into account. In contrast non-shared environmental influences were almost entirely specific to life satisfaction. The shared genetic aetiology suggests a higher-order factor, which the authors suggest may reflect "life history strategy". Another possibility is that this genetic overlap is explained by a tendency towards positive self-presentation. Further studies using multiple methods to assess personality and wellbeing will help to clarify the relationship.

There is growing interest in the relationship between mental health and illness. Is happiness the other end of the distribution to depression? If these indices of mental health and illness represent the same distribution, then we would expect high phenotypic correlations between mental health and illness, as well as strong aetiological overlap. An early study demonstrated that the degree of genetic and environmental influence was not the same for positive and negative affect (Baker, Cesa, Gatz, & Mellins, 1992), suggesting at least partially distinct constructs. However, Baker and colleagues did not conduct bivariate analyses to assess the degree to which the genetic and environmental influences on positive affect were shared with those on negative affect. Kendler and colleagues have recently considered the genetic and environmental overlap between wellbeing and internalizing psychopathology (Kendler, Myers, Maes, & Keyes, 2011b), and between wellbeing and externalizing disorders (Kendler, Myers, & Keyes, 2011a). About half of the genetic influences on wellbeing were accounted for by genetic influences on internalizing disorder, showing that although there is some genetic overlap, there are also genetic influences specific to wellbeing that are independent of those on mental illness. For externalizing disorder only 7% of the genetic influence on wellbeing was accounted for by genetic factors that influence externalizing behaviours. When both internalizing and externalizing disorders are included in the model there are still significant unique genetic influences on wellbeing (Kendler, Myers, & Keyes, 2011a). These results suggest that wellbeing is more than the absence of mental illness, and that the dimensions of mental health and illness are partially distinct.

How stable are genetic influences on wellbeing?

The importance of genetic and environmental influences can change with development (Bergen, Gardner, & Kendler, 2007). For example, for general cognitive ability, heritability rises linearly from around 41% in childhood, to 55% in adolescence and 66% in young adulthood (Haworth et al., 2010). Similar increases have been found for body mass index, with heritability increasing from 48% in early childhood to 78% in early adolescence; this finding has also been replicated at the DNA level, with the importance of the FTO gene also increasing with age (Haworth et al., 2008). Heritability can also decrease, and shared environmental influences can increase in importance, as seen for school science performance in childhood (Haworth, Dale, & Plomin, 2009). What about wellbeing? Does the relative importance of nature and nurture remain constant across the lifespan, and to what extent do genetic factors contribute to the stability of wellbeing as environmental influences accumulate with age?

One of the earliest investigations of genetic influences on wellbeing also assessed the degree to which genes were important for the stability in wellbeing across time (Lykken & Tellegen, 1996). Using a subsample of 127 pairs of twins assessed at age 20 and then again at age 30, the authors showed that wellbeing was relatively stable across time ($r = 0.50$). Stable genetic influences explained 80% of this cross-age correlation of 0.50, indicating that the reason why levels of wellbeing are similar at different ages is because of genes that have a constant impact on wellbeing. These results have been replicated in a larger sample of more than 2000 pairs of twins assessed six years apart (Nes et al., 2006). Nes and colleagues (Nes et al., 2006) also found evidence for both genetic and environmental influences that were specific to each assessment, with non-shared environmental influences being particularly important for changes in wellbeing across time.

The continuity of genetic influences on wellbeing provides some support for the idea of a "genetic set-point" for wellbeing (Fujita & Diener, 2005), however, it is important to remember that finding genetic influence does not mean that wellbeing cannot be improved. Genetic and environmental estimates are entirely context- and sample-specific, indicating what currently explains the variance in wellbeing in this population at this time. If the environmental context changes, then this could alter the balance of nature and nurture (Haworth & Davis, 2014; Plomin & Haworth, 2010), and the influences on variation in a trait need not be related to the factors that impact on the mean value for that trait (Plomin, DeFries, McClearn, & McGuffin, 2008). For example, average body mass index is increasing in most populations, however, the degree to which genes influence whether individuals are heavier or lighter, has remained consistently high (Wardle, Carnell, Haworth, & Plomin, 2008).

Twin and family studies have provided an informative background about the relative influences of genes and environments on wellbeing. However, these study designs cannot tell us which specific DNA variants are involved, nor which specific environments can account for the estimates of environmental influence (Haworth & Plomin, 2010). Molecular genetic analyses, which use DNA to investigate genetic influence, and environmental studies that measure specific aspects of the environment are needed to pinpoint the specific pathways involved in the aetiology of wellbeing. The next sections include an overview of the molecular genetic studies for wellbeing, and then a discussion of research that has used complex study designs to understand the interplay between genes and environments.

Molecular genetic investigations of wellbeing

DNA can be collected easily from study participants, usually via blood or saliva samples (e.g., Freeman et al., 1997). This DNA can then be used in molecular genetic studies that aim to identify which genes, or genetic variants, are involved in the aetiology of wellbeing. There are very few examples of molecular genetic analyses for wellbeing – there is one genome-wide linkage study of subjective happiness (Bartels et al., 2010), and a candidate gene study of life satisfaction (De Neve, 2011).

Bartels and colleagues (Bartels et al., 2010) had genome-wide linkage data available for 1157 offspring from 441 families and phenotypic data on subjective happiness. Although no significant genomic loci were found, there was suggestive linkage to regions on chromosomes 1 and 19, which should be followed up in larger samples. In the candidate gene analysis, the serotonin transporter gene (5-HTTLPR) was associated with life satisfaction in a sample of 2574 adolescents (De Neve, 2011). However, replication of this effect is sorely needed, especially as non-replication is common for candidate gene studies (e.g., Ioannidis, Ntzani, Trikalinos, & Contopoulos-Ioannidis, 2001; Tabor, Risch, & Myers, 2002). In addition, as with most genetic associations with complex traits, this gene accounts for only a small proportion of the variance so there will be many genetic variants involved that together explain the heritability of wellbeing. The next step for molecular genetic analyses in wellbeing is to conduct a genome-wide association study (GWAS), where millions of DNA variants spread across the genome are analysed (McCarthy et al., 2008), in contrast to a single candidate gene, or the sparse coverage of the genome provided by linkage studies. Very large samples are needed to provide adequate power to deal with both the small effects we expect to find and the GWAS multiple testing burden of millions of tests. This has led to a field change in genetics: now, rather than each lab conducting a small study, large groups of labs are coming together to combine their samples in international consortia. For wellbeing, a new group has been created as part of the Social Science Genetic Association Consortium (www.ssgac.org; Benjamin et al., 2012). A genome-wide association analysis is currently underway for subjective wellbeing, and with a sample size of 100,000 individuals, this project offers hope of achieving the aim of identifying some of the genetic variants that contribute to individual differences in wellbeing.

Environmental investigations of wellbeing

At least half of the variance in wellbeing is explained by environmental factors, but the environments we assume matter for wellbeing, like education, income, marital status and socio-economic status, explain only a small proportion of the variance in wellbeing (Myers & Diener, 1995). For example, in a sample of 2000 adults, none of these characteristics explained more than 3% of the variance in wellbeing (Lykken & Tellegen, 1996). Just as we are finding it difficult to identify which genes are involved in wellbeing, it is proving difficult to explain the environmental component of wellbeing too. Results from twin and family studies can help to guide us: we know that the important environments tend to be unique to individuals (i.e. non-shared environments), and that in most cases these environmental influences are specific to developmental stages, that is, they contribute to changes in wellbeing across time (Nes et al., 2006). Genetically sensitive study designs, like twin and family studies, can help in another way too, by allowing the investigation of environmental influences in the context of genetic factors (Plomin, DeFries, McClearn, & McGuffin, 2008). For example, twin studies can be used to control for genetic influence and focus on environmental effects, they also allow us to consider how genetic propensities lead us to seek out environments that lead to wellbeing, and they can be used to understand how our genes and environments may interact. Very few studies like this exist for wellbeing; for example, there is one study on whether wellbeing is causally related to longevity, and a gene–environment interaction study on the effects of marriage on the aetiology of wellbeing.

Subjective wellbeing has been consistently linked to increased longevity (e.g., Chida & Steptoe, 2008; Lyubomirsky, Sheldon, & Schkade, 2005) but is increased wellbeing the cause of longevity and health? There is a chance that wellbeing and longevity are related because of shared aetiology: genes and environments that influence both outcomes. If genetic confounding is present, there will be a higher correlation between wellbeing and longevity in identical twins compared to non-identical twins. Using this co-twin control method, Sadler and colleagues (Sadler, Miller, Christensen, & McGue, 2011) have been able to demonstrate that the link between wellbeing and longevity is not due to genetic (or shared environmental) confounding, that is, results are consistent with a causal influence of wellbeing on longevity.

There is some evidence for gene–environment interaction in wellbeing, particularly in the case of marital relationships. For example, the heritability of subjective wellbeing is higher in unmarried compared to married individuals, indicating that specific environmental experiences can affect the expression of genetic influence on wellbeing (Nes et al., 2010b). There is also evidence for an interaction between genes and environments for adolescent quality of life (van der Aa et al., 2010), and for the role of genes in explaining the link between marital quality and women's wellbeing (Spotts et al., 2005).

For complex traits, such as wellbeing, there is likely to be some interplay between genetic and environmental factors. Using genetically sensitive study designs, like twin studies, to investigate this interplay, and to study environments un-confounded by genetic effects, is going to be crucial if we are to fully understand the genetic and environmental aetiology behind wellbeing.

Future research directions for positive mental health

It is clearly an exciting time in the study of the nature and nurture of wellbeing. There is growing evidence for the role of genes and non-shared environments in creating individual differences in positive mental health, and many studies trying to uncover the specific genes and environments involved are now underway. Given the growing interest in this area, the time seems right to reflect upon areas of research that should attract more attention in the future, and how we can design future studies to address the questions that really matter for positive psychology.

This brief review aimed to highlight some of the current findings from behavioural genetics, but it is by no means a complete survey of the literature available. One crucial difficulty has been the heterogeneity in the way in which wellbeing is measured and assessed. Rarely do two studies use the same measures, meaning it is difficult to compare across studies, as well as know which studies to include in this review. The most important future direction for the entire field of wellbeing is to build a robust taxonomy of wellbeing, and a collection of gold-standard measures to assess positive mental health. The field is young, and there is much work to be done to delineate what it means to be a mentally healthy and flourishing human being. Behavioural genetics can help inform this phenotypic discussion by outlining the underlying genetic and environmental aetiologies, and understanding which wellbeing measures have similar biological (and environmental) pathways. Including a broad range of wellbeing indicators collected in one study will be most effective for understanding the links between different measures. This broad battery should go beyond subjective wellbeing to assess other under-researched areas of wellbeing, including eudaimonic aspects of wellbeing like meaning and purpose in life and self-realization. It may also help to include longer measures of wellbeing, rather than relying on brief or even single-item measures, doing so should increase the reliability of the measures, which is important because heritability is capped by the reliability of the measure. It is possible that part of the large non-shared environmental component for wellbeing is due to measurement error, so it is crucial for future studies to minimize the amount of measurement error.

Other gaps in our knowledge relate to the development of wellbeing across the lifespan, and why wellbeing is linked to other outcomes. For example, there is a notable absence of behavioural genetic investigations of wellbeing in childhood samples, and very few studies have repeated measures across the lifespan so that longitudinal analyses can be conducted. More studies are also needed that address the links between mental health and illness, and especially those that can inform the debate about whether positive emotions are the other end of the distribution of negative emotions. Understanding the aetiological links between positive and negative emotions should greatly benefit the design of interventions to reduce negative mood and promote mental health.

Finally, there is much interest in interventions designed to increase wellbeing, and accumulating evidence that simple activities can lead to sustainable improvements (Lyubomirsky, Sheldon, & Schkade, 2005; Sin & Lyubomirsky, 2009). It is important to remember that finding significant genetic influences on individual differences in wellbeing does not limit the ability to change wellbeing (Haworth & Davis, 2014; Plomin & Haworth, 2010). An exciting new direction for wellbeing is to combine genetically sensitive approaches with intervention science to understand what happens to the balance of nature and nurture when you perform environmental interventions to improve wellbeing (Haworth & Davis, 2014).

As positive psychology matures as a research field, behavioural genetics continues to make significant contributions to our understanding of why some individuals are happier and more fulfilled than others in life. The time is now ripe to conduct more detailed analyses of the nature and nurture of wellbeing that will lead to important implications in the way we treat mental illness and promote mental health across the lifespan.

Acknowledgement

The writing of this chapter was supported by a research fellowship from the British Academy.

References

Baker, L., Cesa, I., Gatz, M., & Mellins, C. (1992). Genetic and environmental influences on positive and negative affect: Support for a two-factor theory. *Psychology and Aging, 7*(1), 158–163.

Bartels, M., & Boomsma, D. I. (2009). Born to be Happy? The Etiology of Subjective Well-Being. *Behavior genetics, 39*(6), 605–615.

Bartels, M., Saviouk, V., de Moor, M. H. M., Willemsen, G., van Beijsterveldt, T. C. E. M., Hottenga, J.-J., de Geus, E. J. C., et al. (2010). Heritability and Genome-Wide Linkage Scan of Subjective Happiness. *Twin Research and Human Genetics, 13*(2), 135–142.

Benjamin, D. J., Cesarini, D., Chabris, C. F., Glaeser, E. L., Laibson, D. I., Guonason, V., Harris, T. B., et al. (2012). The Promise and Pitfalls of Genoeconomics. *Annual Review of Economics, 4*, 627–662.

Bergeman, C., Plomin, R., Pedersen, N. L., & McClearn, G. (1991). Genetic mediation of the relationship between social support and psychological well-being. *Psychology and Aging, 6*(4), 640–646.

Bergen, S. E., Gardner, C. O., & Kendler, K. S. (2007). Age-related changes in heritability of behavioral phenotypes over adolescence and young adulthood: a meta-analysis. *Twin Research and Human Genetics, 10*(3), 423–433.

Bradburn, N. M. (1969). *The Structure of Psychological Well-Being.* Chicago: Aldine.

Chida, Y., & Steptoe, A. (2008). Positive psychological well-being and mortality: a quantitative review of prospective observational studies. *Psychosomatic Medicine, 70*, 741–756.

De Neve, J.-E. (2011). Functional polymorphism (5-HTTLPR) in the serotonin transporter gene is associated with subjective well-being: evidence from a US nationally representative sample. *Journal of Human Genetics, 56*(6), 456–459.

De Neve, J.-E., Christakis, N., Fowler, J. H., & Frey, B. S. (2010). Genes, economics, and happiness. *CESifo working paper no. 2946.*

DeNeve, K. M., & Cooper, H. (1998). The happy personality: a meta-analysis of 137 personality traits and subjective well-being. *Psychological Bulletin, 124*(2), 197–229.

Diener, E., Emmons, R. A., Larsen, R. J., & Griffin, S. (1985). The satisfaction with life scale. *Journal of personality assessment, 49*(1), 71–75.

Diener, E., Suh, E. M., & Lucas, R. E. (1999). Subjective well-being: Three decades of progress. *Psychological Bulletin, 125*(2), 276–302.

Fredrickson, B. L. (1998). What good are positive emotions? *Review of General Psychology, 2*(3), 300–319.

Freeman, B., Powell, J., Ball, D., Hill, L., Craig, I., & Plomin, R. (1997). DNA by mail: an inexpensive and noninvasive method for collecting DNA samples from widely dispersed populations. *Behavior Genetics, 27*(3), 251–257.

Fujita, F., & Diener, E. (2005). Life Satisfaction Set Point: Stability and Change. *Journal of Personality and Social Psychology, 88*(1), 158–164.

Gigantesco, A., Stazi, M. A., Alessandri, G., Medda, E., Tarolla, E., & Fagnani, C. (2011). Psychological well-being (PWB): a natural life outlook? An Italian twin study on heritability of PWB in young adults. *Psychological Medicine, 41*(12), 2637–2649.

Harris, J., Pedersen, N. L., Stacey, C., McClearn, G. E., & Nesselroade, J. R. (1992). Age differences in the etiology of the relationship between life satisfaction and self-rated health. *Journal of Aging and Health, 4*(3), 349–368.

Haworth, C. M. A., & Davis, O. S. P. (2014). From observational to dynamic genetics. *Frontiers in Genetics, 5*(6).

Haworth, C. M. A., & Plomin, R. (2010). Quantitative Genetics in the Era of Molecular Genetics: Learning Abilities and Disabilities as an Example. *Journal of the American Academy of Child & Adolescent Psychiatry, 49*(8), 783–793.

Haworth, C. M. A., Carnell, S., Meaburn, E. L., Davis, O. S. P., Plomin, R., & Wardle, J. (2008). Increasing Heritability of BMI and Stronger Associations With the FTO Gene Over Childhood. *Obesity, 16*, 2663–2668.

Haworth, C. M. A., Dale, P., & Plomin, R. (2009). The etiology of science performance: Decreasing heritability and increasing importance of the shared environment from 9 to 12 years of age. *Child development, 80*(3), 662–673.

Haworth, C. M. A., Wright, M. J., Luciano, M., Martin, N. G., De Geus, E. J. C., van Beijsterveldt, C. E. M., Bartels, M., et al. (2010). The heritability of general cognitive ability increases linearly from childhood to young adulthood. *Molecular Psychiatry, 15*(11), 1112–1120.

Howell, R. T., Kern, M. L., & Lyubomirsky, S. (2007). Health benefits: Meta-analytically determining the impact of well-being on objective health outcomes. *Health Psychology Review, 1*(1), 83–136.

Ioannidis, J. P. A., Ntzani, E. E., Trikalinos, T. A., & Contopoulos-Ioannidis, D. G. (2001). Replication validity of genetic association studies. *Nature Genetics, 29*(3), 306–309.

Kendler, K. S., Myers, J. M., & Keyes, C. L. M. (2011a). The Relationship Between the Genetic and Environmental Influences on Common Externalizing Psychopathology and Mental Wellbeing. *Twin Research and Human Genetics, 14*(6), 516–523.

Kendler, K. S., Myers, J. M., Maes, H. H., & Keyes, C. L. M. (2011b). The Relationship Between the Genetic and Environmental Influences on Common Internalizing Psychiatric Disorders and Mental Well-Being. *Behavior Genetics, 41*(5), 641–650.

Keyes, C. L. M., Myers, J. M., & Kendler, K. S. (2010). The structure of the genetic and environmental influences on mental well-being. *American Journal of Public Health, 100*(12), 2379–2384.

Lykken, D., & Tellegen, A. (1996). Happiness is a stochastic phenomenon. *Psychological Science, 7*(3), 186–189.

Lyubomirsky, S., & Lepper, H. S. (1999). A measure of subjective happiness: Preliminary reliability and construct validation. *Social Indicators Research, 46*, 137–155.

Lyubomirsky, S., King, L., & Diener, E. (2005). The Benefits of Frequent Positive Affect: Does Happiness Lead to Success? *Psychological Bulletin, 131*(6), 803–855.

Lyubomirsky, S., Sheldon, K. M., & Schkade, D. (2005). Pursuing Happiness: The Architecture of Sustainable Change. *Review of General Psychology, 9*(2), 111–131.

McCarthy, M. I., Abecasis, G. R., Cardon, L. R., Goldstein, D. B., Little, J., Ioannidis, J. P. A., & Hirschhorn, J. N. (2008). Genome-wide association studies for complex traits: consensus, uncertainty and challenges. *Nature Reviews Genetics, 9*(5), 356–369.

Myers, D. G., & Diener, E. (1995). Who is happy? *Psychological Science, 6*(1), 10–19.

Neale, M. C., & Maes, H. M. (2001). *Methodology for Genetic Studies of Twins and Families*. Dordrecht: Kluwer Academic Publishers.

Nes, R. (2010). Happiness in behaviour genetics: Findings and implications. *Journal of Happiness Studies, 11*, 369–381.

Nes, R. B., Czajkowski, N., & Tambs, K. (2010a). Family Matters: Happiness in Nuclear Families and Twins. *Behavior Genetics, 40*(5), 577–590.

Nes, R., Røysamb, E., Harris, J., Czajkowski, N., & Tambs, K. (2010b). Mates and Marriage Matter: Genetic and Environmental Influences on Subjective Wellbeing Across Marital Status. *Twin Research and Human Genetics, 13*(4), 312–321.

Nes, R., Røysamb, E., Reichborn-Kjennerud, T., Tambs, K., & Harris, J. (2005). Subjective wellbeing and sleep problems: a bivariate twin study. *Twin Research and Human Genetics, 8*(5), 440–449.

Nes, R., Røysamb, E., Tambs, K., Harris, J., & Reichborn-Kjennerud, T. (2006). Subjective well-being: genetic and environmental contributions to stability and change. *Psychological Medicine, 36*(07), 1033–1042.

Peterson, C., & Seligman, M. (2004). *Character Strengths and Virtues: A Handbook and Classification* (1st ed.). Oxford University Press, USA.

Plomin, R., & Haworth, C. M. A. (2010). Genetics and Intervention Research. *Perspectives on Psychological Science*, 5(5), 557–563.

Plomin, R., DeFries, J. C., McClearn, G. E., & McGuffin, P. (2008). *Behavioral Genetics* (5th ed.). New York: Worth Publishers.

Ryff, C. D., & Keyes, C. L. M. (1995). The structure of psychological well-being revisited. *Journal of Personality and Social Psychology, 69*, 719–727.

Ryff, C. D., & Singer, B. H. (2008). Know thyself and become what you are: A eudaimonic approach to psychological well-being. *Journal of Happiness Studies, 9*, 13–39.

Røysamb, E., Harris, J., Magnus, P., Vittersø, J., & Tambs, K. (2002). Subjective well-being. Sex-specific effects of genetic and environmental factors. *Personality and Individual Differences, 32*, 211–223.

Røysamb, E., Tambs, K., Reichborn-Kjennerud, T., Neale, M. C., & Harris, J. (2003). Happiness and health: environmental and genetic contributions to the relationship between subjective well-being, perceived health, and somatic illness. *Journal of Personality and Social Psychology, 85*(6), 1136–1146.

Sadler, M., Miller, C., Christensen, K., & McGue, M. (2011). Subjective Wellbeing and Longevity: A Co-Twin Control Study. *Twin Research and Human Genetics, 14*(3), 249–256.

Seligman, M. E. P., & Csikszentmihalyi, M. (2000). Positive psychology: An introduction. *American Psychologist, 55*(1), 5–14.

Sin, N. L., & Lyubomirsky, S. (2009). Enhancing well-being and alleviating depressive symptoms with positive psychology interventions: a practice-friendly meta-analysis. *Journal of Clinical Psychology, 65*(5), 467–487.

Spotts, E. L., Pederson, N. L., Neiderhiser, J. M., Reiss, D., Lichtenstein, P., Hansson, K., & Cederblad, M. (2005). Genetic effects on women's positive mental health: Do marital relationships and social support matter? *Journal of Family Psychology, 19*(3), 339–349.

Sprangers, M. A. G., Bartels, M., Veenhoven, R., Baas, F., Martin, N. G., Mosing, M., Movsas, B., et al. (2010). Which patient will feel down, which will be happy? The need to study the genetic disposition of emotional states. *Quality of Life Research, 19*(10), 1429–1437.

Steger, M. F., Hicks, B. M., Kashdan, T. B., Krueger, R. F., & Bouchard, T. J., Jr. (2007). Genetic and environmental influences on the positive traits of the values in action classification, and biometric covariance with normal personality. *Journal of Research in Personality, 41*(3), 524–539.

Stubbe, J., de Moor, M. H. M., Boomsma, D. I., & De Geus, E. J. C. (2007). The association between exercise participation and well-being: A co-twin study. *Preventive Medicine, 44*(2), 148–152.

Stubbe, J., Posthuma, D., Boomsma, D. I., & De Geus, E. J. C. (2005). Heritability of life satisfaction in adults: a twin-family study. *Psychological Medicine, 35*, 1581–1588.

Tabor, H. K., Risch, N. J., & Myers, R. M. (2002). Candidate-gene approaches for studying complex genetic traits: practical considerations. *Nature Reviews Genetics, 3*(5), 391–396.

Tellegen, A., Lykken, D., Bouchard, T., Wilcox, K. J., Segal, N. L., & Rich, S. (1988). Personality similarity in twins reared apart and together. *Journal of Personality and Social Psychology, 54*(6), 1031–1039.

van der Aa, N., Boomsma, D. I., Rebollo-Mesa, I., Hudziak, J. J., & Bartels, M. (2010). Moderation of Genetic Factors by Parental Divorce in Adolescents' Evaluations of Family Functioning and Subjective Wellbeing. *Twin Research and Human Genetics, 13*(2), 143–162.

Wardle, J., Carnell, S., Haworth, C. M. A., & Plomin, R. (2008). Evidence for a strong genetic influence on childhood adiposity despite the force of the obesogenic environment. *American Journal of Clincal Nutrition, 87,* 398–404.

Weiss, A., Bates, T., & Luciano, M. (2008). Happiness is a personal(ity) thing: The genetics of personality and wellbeing in a representative sample. *Psychological Science, 19,* 205–210.

Wood, V., Wylie, M. L., & Sheafor, B. (1969). An analysis of a short self-report measure of life satisfaction: Correlation with rater judgments. *Journal of Gerontology, 24*(4), 465–469.

8

Molecular Genetic Investigations of Personality: From Candidate Genes to Genome-wide Associations

Anastasiya Kazantseva, Sergey Malykh, and Elza Khusnutdinova

Temperament is traditionally defined as a dynamic feature of human mental activity such as its rates, rhythm and intensity. Psychologists differentiate temperament and character traits and consider temperament traits as heritable ones manifesting early in ontogenesis, while character traits represent values and goals formed during an individual's development and socio-cultural learning (Van Gestel & Van Broeckhoven, 2003). The study of personality traits as dynamic characteristics of human behaviour originates from Hippocrates (V–IV cent. BC) and Galen (II cent. BC), who characterized the main types of the human behaviour. To date, multiple theories and concepts of personality exist reflecting hypothesized biological correlates including central nervous system functioning.

According to behavioural genetic studies, 40–60% of individual differences in personality are attributed to genetic factors. Research also suggests that different aspects of personality are affected by genetic and environmental factors to a different extent. Genetic effects are complex and include additive and epistatic processes. Moreover, manifestation of personality traits has been shown to result from multiple genes with small effects, environmental factors and gene–environment interactions (Benjamin et al., 2008; Reif & Lesch, 2003; Van Gestel & Van Broeckhoven, 2003).

Models of personality

Nowadays, the following two models are used in behavioural genetic studies of personality most frequently: 1) the psychobiological model of personality (Cloninger, Svrakic, & Przybeck, 1993); and 2) the five-factor model of personality (Costa & McCrae, 1992). The first model is based on the biological mechanisms that control the level of neurotransmitters – proteins regulating various processes in the brain. The second model represents a psychometric model based on factor analysis of self-reported data, expert assessments data and behavioural data.

The psychobiological model of personality was developed based on the results of longitudinal twin and family studies, as well as neuropharmacological and neurophysiological research (Cloninger, Svrakic, & Przybeck, 1993). This model differentiates between temperament and character. The Temperament and Character Inventory (TCI) is an inventory for personality traits based on the psychobiological model. The TCI operates with seven dimensions of personality traits: four so-called temperaments and three so-called characters. These temperament scales are: 1) Novelty seeking; 2) Harm avoidance; 3) Reward dependence; and 4) Persistence. The TCI model also includes three character scales: 1) Self-directedness; 2) Cooperation; and 3) Self-transcendence – a sense of being inalienable from the universe. Cloninger has postulated the hierarchical principle of temperament development in the course of phylogenesis: from Harm avoidance (that involves the inhibition system or complete inhibition of behaviour) observed in all species to Novelty seeking and Reward dependence appeared in high order species. According to the model, variation in temperament traits is mediated by the differences in neurotransmitter systems functioning: Harm avoidance is primarily related to serotonergic system; Novelty seeking is associated with the dopaminergic system; and Reward dependence is primarily associated with the noradrenergic system (Cloninger, Svrakic, & Przybeck, 1993).

The five-factor model of personality is based on factor and cluster analysis and assesses five main personality traits: 1) Neuroticism; 2) Extraversion; 3) Agreeableness; 4) Conscientiousness; and 5) Openness to experience. These traits are universal, since they can be used to describe an individual's personality regardless of age, gender, ethnicity, religion, socio-economic level and place of residence (Costa & McCrae, 1992).

Sex differences in personality

Numerous studies have demonstrated the presence of sex differences in personality traits. Increased levels of Neuroticism and Harm avoidance (anxiety-related traits) were observed in females compared to males (Costa, Terracciano, & McCrae, 2001; Chapman, Duberstein, Sörensen, & Lyness, 2007). The results in relation to approach-related traits are inconsistent: both increased and decreased levels of Extraversion were detected in females (Rouff, Sussman, & Strube, 2005; Else-Quest, Hyde, Goldsmith, & Van Hulle, 2006). Males tend to have higher levels of Persistence, Openness to experience, Self-confidence and Self-esteem (Feingold, 1994; Costa et al., 2001), while females demonstrate higher levels of Conscientiousness, Reward dependence, Self-transcendence, Cooperation, and lower levels of Self-directedness (Chapman et al., 2007; Vaidya, Grippo, Johnson, & Watson, 2004).

Explanations for sex differences in personality traits have been proposed based on biological (e.g., hormonal differences) and socio-psychological (e.g., differences in social roles) theories (Costa et al., 2001). In line with biological explanations, sexual dimorphism in the brain anatomy has been observed: females on average have a larger hippocampus volume (the brain region characterized by oestrogen synthesis), while males on average have a larger amygdala volume (the brain region with the highest level of androgens) (Cosgrove, Mazure, & Staley, 2007). It is known that sex hormones can affect expression of the neurotransmitter system genes. Multiple studies have provided the evidence for the increased serotonergic activity in females. For instance, an increased prolactin response to D-fenfluramine and a high level of 5-hydroxyindoleacetic acid (5-HIAA) in the cerebrospinal fluid in females as compared to males was observed (Mann et al., 2001). Diminished oestrogen levels in females were correlated with reduced serotonergic activity (Cosgrove, Mazure, & Staley, 2007). Furthermore, increased dopaminergic activity in females as a result of oestrogen action was observed (Becker, 1999; Thompson & Certain, 2005). Moreover, sexual dimorphism in anxiety-related traits with respect to the level of brain derived neurotrophic factor (BDNF) was reported: plasma levels of this protein were positively correlated with anxiety-related traits (e.g., Neuroticism) and negatively – with Conscientiousness and Extraversion in males but not in females (Terracciano et al., 2010a).

Population differences in personality

Behavioural genetic research into personality traits should take into account population differences. Each population has a long history of development and is characterized by different frequencies of genetic alleles as a result of evolutionary adaptation to socio-demographic, climatic and other factors (Khusnutdinova et al., 2008). At the same time, different populations are characterized by distinct cultural and religious backgrounds. As a result of cultural and genetic positive selection, every culture may have the average levels of personality traits that are the most adaptive for successful functioning in this particular population. This concept is known as culture–gene coevolution (Chiao & Blizinsky, 2010).

A large-scale study of healthy individuals from 51 countries has shown that geographically, ethnically and historically related countries (e.g., Germany and Austria, USA and Canada) demonstrate similar personality profiles (McCrae & Terracciano, 2005). It has been reported that individuals from Western Europe and USA have higher scores on Extraversion, Novelty seeking and Reward dependence compared to individuals from Asia and Africa, who have higher scores on Harm avoidance (Richter, Eisemann & Richter, 2000; McCrae & Terracciano, 2005; Miettunen et al., 2006). Moreover, the differences between various European countries have been reported: individuals from Southern and Eastern Europe have higher scores on Neuroticism and lower scores on Conscientiousness compared to individuals from the northern European countries (McCrae and Terracciano, 2005). The data can be explained by possible cultural influences: more collectivistic countries are characterized on average by lower scores on Extraversion, Openness to experience and Agreeableness, whereas countries with more individualistic social systems have on average higher scores on Extraversion and Openness to experience (e.g., North America and Western Europe; McCrae, 2001; McCrae & Terracciano, 2005).

Despite the presence of cross-cultural variations in personality traits, the magnitude of these differences is very small compared to the individual differences within each population. Analysis of individual differences in personality using the data obtained with the respondent method in 51 countries showed that 95% of the variance was explained by differences within each culture, and only 5% by differences across cultures (McCrae & Terracciano, 2008). In another study of personality traits, using the self-survey method, cross-cultural differences were somewhat larger (e.g., 16% for Neuroticism and 17% for Extraversion; Poortinga & van Hemert, 2001).

Genetic factors in individual differences in personality

One approach to the study of genetic factors involved in individual differences in personality is the candidate gene approach. This approach is based on the known functional significance of genes and focuses on associations between a trait and genes involved in biological pathways related to this trait. The psychobiological model suggests that personality traits are mediated by the functioning of neurotransmitter systems. Therefore, the neurotransmitter system genes are considered as functional candidate genes for personality.

Genome-wide association study (GWAS) is another, more contemporary, approach to the study of molecular genetic basis of personality traits. GWAS is based on genotyping of hundreds of thousands of single nucleotide polymorphisms (SNPs) spanning the whole genome. This approach allows testing for association between allelic variation in any genomic region (not in a specific candidate gene) and the trait of interest (Aulchenko, 2010; De, Bush & Moore, 2014). SNPs located sufficiently close to each other are often in linkage disequilibrium (i.e., are more likely to be inherited together). Usually, in GWASs, only one or a few loci from a group of linked SNPs are genotyped. Therefore, any association between some SNP and a trait may indicate the involvement of another SNP located in the same linkage disequilibrium group rather than the involvement of this particular SNP. To date, the use of these two approaches – candidate gene and GWAS – has led to the identification of a number of genetic factors involved in personality.

Serotonergic system genes

The central serotonergic brain system is involved in behavioural inhibition and regulation of mood, aggression and motor activity (Hamon & Blier, 2013). Serotonin transporter gene (*5-HTT, SLC6A4*) encoding the protein responsible for neurotransmitter reuptake from the synaptic cleft is one of the most widely studied candidate genes of personality. Transcriptional activity of the *5-HTT* gene is affected by insertion–deletion polymorphism in a regulatory region (*5-HTTLPR*) representing the presence (*L*-allele) or absence (*S*-allele) of the fragment 44 bp in length. According to the results of a study using lymphoblast cell culture, exogenic *5-HTT* mRNA level was increased threefold in the presence of the *L*-allele (Nakamura, Ueno, Sano, & Tanabe, 2000). The first molecular genetic study aimed at identifying an association between this gene and personality revealed that the *5-HTTLPR* explained 3–4% of variance in Neuroticism (Lesch et al., 1996). However, numerous attempts to confirm this finding resulted in inconsistent results (Joo et al., 2007; Lang et al., 2004; Munafò et al., 2003; Vormfelde et al., 2006). Recent meta-analyses based on 60 studies (Munafò et al., 2009b; Schinka, Busch, & Robichaux-Keene, 2004; Sen, Burmeister & Ghosh, 2004), as well as the findings from three large-scale studies (N > 4000 in each; Munafò et al., 2009b; Terracciano et al., 2009; Willis-Owen et al., 2005) failed to confirm the association of *5-HTTLPR* and anxiety-related traits. At the same time, there is evidence for the significant role of another *5-HTT* gene polymorphism representing a variable number of tandem repeats in intron 2 (*STin2*) in anxiety-related traits (i.e., Neuroticism, Harm avoidance). For example, it was demonstrated that female carriers of *STin2 10*-allele or *S-10*-haplotype were characterized by lower levels of anxiety-related traits as compared to female carriers of *STin2 12/12*-genotype and *S-12* haplotype, respectively (Kazantseva et al., 2008).

The family of 14 serotonin receptors detected in the human brain belongs to the group of G-coupled proteins, as distinct from the type 3 serotonin receptors belonging to ionic channels receptors. Serotonin 2A receptor (HTR2A) is involved in both peripheral systems and the CNS functioning, including platelet aggregation, mild muscle contraction, control over sexual activity, and regulation of sleep and motion. A significant correlation was found between Harm avoidance, hopelessness and HTR2A binding index in the sample of patients that attempted suicide (van Heeringen et al., 2003). A number of the *HTR2A* polymorphisms, such as *–1438A>G* (*rs6311*) and *102T>C* (*rs6313*), were associated with anxiety-related traits (Nomura & Nomura, 2006), Novelty seeking (Serretti et al., 2007), and communicative traits (Rybakowski et al., 2006). A study of 200 SNPs in 17 serotonergic system genes revealed the involvement of the *HTR2A rs2770296* and *rs927544* in Novelty seeking variation, and these findings were replicated in an independent sample (Heck et al., 2009).

Tryptophan hydroxylase (TPH) is an enzyme which limits the rate of serotonin biosynthesis and is involved in tryptophan conversion to 5-hydroxytryptophan followed by decarboxylation into serotonin. There are two tryptophan hydroxylase forms encoded by *TPH1* and *TPH2* genes with a high degree of homology in coding regions and expressing in the frontal cortex, thalamus, hippocampus, hypothalamus and amygdala (Walther & Bader, 2003). Two polymorphisms in intron 7 of the *TPH1* gene – *218A>C* and *779A>C* – are in complete linkage disequilibrium in Caucasians (Nielsen et al., 1997). The *A*- and *C*-alleles of these two polymorphisms are designated as *U*-allele (upper, associated with high gene expression) and *L*-allele (lower, resulting in decreased gene expression), respectively. A number of molecular genetic studies have reported association of the *TPH1* polymorphisms with anxiety-related traits (Anghelescu et al., 2005), communicative traits (Nielsen et al., 1997) and creativity (Reuter, Roth, Holve, & Hennig, 2006).

Tryptophan hydroxylase 2 isoform gene (*TPH2*) plays the fundamental role in a reaction of serotonin synthesis in the CNS and is expressed in the brain specifically. Several studies demonstrated association of enhanced expression of the *TPH2* gene in distinct brain regions with lower anxiety in rats (Hiroi, McDevitt, & Neumaier, 2006). A study of *TPH2* gene polymorphisms in human personality suggested association of this gene with anxiety-related personality traits (Yang et al., 2010).

Dopaminergic system genes

The dopaminergic system is known to be responsible for motor functions, mood, cognitive abilities and sensitivity to reward (Vriend et al., 2014). One of the genes of the dopaminergic system, dopamine 4 receptor gene (*DRD4*), has been studied extensively with respect to Novelty seeking (Benjamin et al., 2008). A variable number of tandem repeats (*VNTR)* in exon 3 represents a highly polymorphic region of 48 bp in length consisting of 2–11 repeats and encoding the third cytoplasmic loop of a receptor (region important for binding with G-proteins) with 32–176 amino acids, depending on the number of repeats (Asghari et al., 1995). A study of the *DRD4* gene structure has revealed that the allele containing four repeats (4R) is an ancestor allele in relation to the alleles with two (2R) and six (6R) repeats. The allele with seven repeats (7R) occurred as a result of a rare mutation 30,000–50,000 years ago; and this allele has spread in the human population due to the positive selection in the Upper Palaeolithic period, one of extensive expansion from Africa (Ding et al., 2002; Wang et al., 2004). It is possible that personality traits such as Novelty seeking and Persistence were increased in migrants (Wang et al., 2004). It has been shown that the *7R*-allele is rare in Asian populations; however, the *2R*-allele substitutes the *7R*-allele in these populations. Some, but not all, studies reported the association of the *7R*-allele with higher scores on

Novelty seeking and Extraversion (Ebstein & Belmaker, 1997; Becker et al., 2005; Nederhof et al., 2011).

The dopamine D2 receptor (DRD2) is involved in signal pathways inhibiting cAMP synthesis while binding with G1-proteins (Kaiser et al., 2001). The most studied *DRD2* gene polymorphism in relation to personality traits is *Taq1A* (*rs1800497* or *32806C>T*) located in the 3'-untranslated gene region (UTR). However, this locus also results in amino acid substitution (Glu713Lys) in another gene (i.e., *ANKK1*; Neville et al., 2004). Carriers of the *DRD2 A1*-allele demonstrated dopamine deficit (Ritchie & Noble, 2003), as well as an increased liability to addiction (Noble, 2003). Moreover, *DRD2 A1*-allele carriers are characterized by higher scores on Novelty seeking (Lin et al., 2007), Reward dependence (Lee et al., 2003), and sensitivity to reward (Lee et al., 2007). However, higher scores on Novelty seeking in individuals with the *DRD2 A2*-allele have also been reported (Noble et al., 1998). There is some evidence for sex specific association of *Taq1A* polymorphism with Novelty seeking and Reward dependence. For example, it was demonstrated that *Taq1A A2/A2*-genotype was associated with higher scores on Novelty seeking and lower scores on Reward dependence in males only (Kazantseva et al., 2011).

Dopamine transport plays a key role in the process of regulation of dopaminergic transmission via dopamine re-uptake from the synaptic cleft and its transfer to the presynaptic terminal (Gowrishankar et al., 2014). A *VNTR* polymorphism, with a repeated region of 40 bp in length and 3–11 repeats, in 3'-UTR of the dopamine transporter gene (*SLC6A3*) has been shown to affect the transcription efficacy and mRNA stability, as well as gene regulation under binding with specific microRNAs (Presutti et al., 2006). It has been reported that the *SLC6A3 9R*-allele is associated with lower scores on Novelty seeking (Sabol et al., 1999; Van Gestel et al., 2003), Cooperation (Pełka-Wysiecka et al., 2012) and Reward dependence (Samochowiec et al., 2001). Previously, sex-specific effects of the *SLC6A3* gene on personality had been demonstrated. For example, the *SLC6A3 10R-G* haplotype (based on *VNTR* and *rs27072*, respectively) was associated with higher scores on Persistence, and this effect was only observed in women (Kazantseva et al., 2011).

Catechol-*O*-methyltransferase (*COMT*) is an enzyme catalysing the first stage of catecholamine (dopamine, epinephrine, and norepinephrine) degradation via methyl group transfer from S-adenosylmethionine to the hydroxyl group of catecholamines. The *COMT* gene contains the functional polymorphic marker *Val158Met* (*472A>G, rs4680*) known to affect temperature sensitivity of the encoded enzyme. In particular, the *Met/Met*-genotype results in the synthesis of the enzyme with 4-fold lower activity compared to the enzyme encoded with the *Val/Val*-genotype (Chen et al., 2004). Therefore, the *Met/Met*-genotype results in a delay of dopamine degradation in the prefrontal cortex. Individuals with the *Val/Met*-genotype demonstrated an intermediate level of enzymatic activity compared to homozygotes. Studies of association of this polymorphism with personality traits have demonstrated inconsistent results. For example, some studies have reported associations of *Met*-allele and/or *Met/Met* genotype with higher scores on anxiety-related traits (Hashimoto et al., 2007) and Reward dependence (Tsai et al., 2004), as well as lower scores on Novelty seeking and Extraversion (Tsai et al., 2004; Stein et al., 2005; Golimbet et al., 2007). However, other studies have reported higher Neuroticism in those with the *Val/Val*-genotype (Pełka-Wysiecka et al., 2012). Inconsistency of the results can be partly explained by the absence of analysis of environmental influences that can modify genetic associations (for more details see chapter 2). Moreover, association of the *COMT* gene with personality traits can differ between sexes: for example, the *Val/Val* genotype was associated with higher negative emotionality in men only (Chen et al., 2011).

Tyrosine hydroxylase (TH) represents an enzyme limiting the rate of reaction of catecholamine (dopamine, norepinephrine) synthesis. Several studies have reported associations of the functional polymorphism of the *TH* gene (*–824C/T,* or *rs10770141*) with personality traits in a sex-specific manner: for example, *–824C/T T*-allele (associated with increased catechoLamine levels) was related to lower scores on novelty seeking in men only (Sadahiro et al., 2010).

Noradrenergic system genes

Dysfunctions in the noradrenergic system may result in irritability, hostility, anxiety, depression and addiction (Gallagher et al., 2013; Schmidt & Weinshenker, 2014). The norepinephrine transporter (NET, or SLC6A2) plays an important role in regulation of neurotransmission and signal transduction of catecholamines (norepinephrine, dopamine) (Morón et al., 2002). Diminished NET level, resulting in enhanced levels of epinephrine, norepinephrine and its metabolite 4-metoxy-4-hydroxyphenylglicol (MHPG) were observed in individuals with increased anxiety, attention deficit and high Novelty seeking (Gerra et al., 1999) and Reward dependence (Garvey et al., 1996). Polymorphic loci *–182T>C* and *1287A>G,* located in the promoter region and exon 9 of the *NET* gene, respectively, have been extensively studied with respect to personality traits. For example, *1287A>G A*-allele was associated with higher Reward dependence (Samochowiec et al., 2002), whereas *–182T>C T*-allele was associated with higher Novelty seeking (Lee et al., 2008) and Reward dependence (Ham et al., 2005).

Alpha-2-adrenergic receptors (i.e., ADRA2A, ADRA2B, ADRA2C) belong to G-coupled proteins and act as terminal and somatodentritic autoreceptors for epinephrine and norepinephrine, as well as presynaptic heteroreceptors for serotonin. Several studies have demonstrated association of the *ADRA2A –1291C/G* polymorphism with Harm avoidance in adults (Comings et al., 2000b), and Extraversion in adolescents (Mäestu et al., 2008).

Neurotrophic factor genes

Brain derived neurotrophic factor (BDNF) is one of the most prevalent neurotrophins in the brain that plays an important role in brain development, proliferation, neuronal differentiation, synaptic plasticity and neuronal survival (Rasmusson et al., 2002; Nakazato et al., 2003). BDNF is known to mediate the effects of chronic stress on neurogenesis and neuronal survival in hippocampus. Several studies have demonstrated a decrease in BDNF concentration in individuals with high Neuroticism and depression (Hwang et al., 2006). Accordingly, the *BDNF* gene might be related to variation in anxiety-related traits. Most of the studies of the *BDNF* gene have focused on the *Val66Met* (*rs6265*) polymorphism that is functionally significant: the gene variant encoding methionine (*Met*) is related to lower level of gene expression (Egan et al., 2003). Several studies have revealed the association between this polymorphism and anxiety-related traits (Lang et al., 2005) and Conscientiousness (Hiio et al., 2011). A recent meta-analysis (N = 15251) has confirmed the association of the *Val66Met* with Extraversion, but not with anxiety-related traits (Terracciano et al., 2010a).

Hypothalamic-pituitary-adrenal system genes

The hypothalamic-pituitary-adrenal (HPA) axis plays the key role in pathogenesis of impulsivity, aggression, suicidal behaviour and affective disorders via activation within several seconds of corticotrophin-releasing hormone (CRH) as a response to stress (Gragnoli, 2014). Neuropeptides oxytocin (OXT) and arginine-vasopressin (AVP) are involved in regulation of social behaviour (e.g., attachment, aggression, anxiety). A higher oxytocin level in plasma was associated with higher confidence, diminished hormonal response to stress and lower anxiety in patients with depression (for review, see Meyer-Lindenberg et al., 2011). Genetic association studies have demonstrated that *rs53576* polymorphism in the oxytocin receptor gene (*OXTR*) was related to maternal sensitivity (Bakermans-Kranenburg et al., 2008), empathy (Rodrigues et al., 2009) and Reward dependence (Tost et al., 2010), as well as other behavioural characteristics (Prichard et al., 2007; Kawamura et al., 2010).

Research has also found the link between AVP and its receptors with variation in personality. Genetic association studies have provided evidence for the involvement of arginine-vasopressin receptor 1A (*AVPR1A*) gene in variation of social behaviour in chimpanzee (Hopkins et al., 2012) as well as in human personality traits including Novelty seeking, Harm avoidance and Extraversion (Ukkola et al., 2009; Meyer-Lindenberg et al., 2001 Kazantseva et al., 2014). Another AVP receptor gene *AVPR1B* (Cagliani et al., 2009) may also be involved in regulation of social behaviour (Wersinger et al., 2002; Carter et al., 2008). Genetic association studies have demonstrated the association of the *AVPR1B* gene with affective and panic psychopathologies characterized by high Harm avoidance (Cagliani et al., 2009). One study found an association of *AVPR1B G-A*-haplotype (*rs28632197-rs33911258*) and lower scores on Self-transcendence – a personality trait linked to spirituality, self-actualization and liability to psychopathologies (Kazantseva et al., 2014).

Sex hormone genes

Personality traits are characterized by sex differences, therefore polymorphic loci of genes that are involved in the functioning of sex hormones are of particular relevance. For example, higher self-perceived abilities have been reported in boys (Boudreault-Bouchard et al., 2013). Self-perceived abilities in boys have been linked with free testosterone level in the presence of the allele with a longer *CAG* length in the androgen receptor gene (*AR*) (Vermeersch et al., 2010). This gene has also been associated with sensation seeking (Aluja et al., 2011). The role of another sex hormone – oestrogen – is linked to emotional control and affective behaviour. In particular, a decrease in oestrogen level in women has been shown to result in irritability, emotional instability and depressive thoughts (Freeman et al., 2006). Genetic association studies have suggested the role of the oestrogen receptor alpha gene (*ESR1*) in personality traits such as Self-transcendence (Kutlumbetova et al., 2014) and Harm avoidance (Gade-Andavolu et al., 2009).

Other genes

Molecular genetic studies reported associations of the whole range of other candidate genes with individual differences in personality including: GABA-receptor genes (e.g., *GABRA6, GABRA2*) with Harm avoidance and Impulsivity (Arias et al., 2012; Villafuerte et al., 2012); a glutamatergic system gene (*GRIK3*) with Harm avoidance and exploratory excitability (Minelli et al., 2009); a signal pathway gene (*DARPP*-32) with Harm avoidance and Novelty seeking (Li et al., 2011); a cholinergic system gene (*CHRNA4*) with emotional instability and anxiety (Markett et al., 2011); a heat shock protein gene (*FKBP5*) with Harm avoidance and Cooperation (Shibuya et al., 2010); the nitric oxide synthase 1 gene (*NOS1*) with Reward dependence (Giegling et al., 2011); the glutathione peroxidase gene (*GPX-1*) with Openness to experience (Matsuzawa et al., 2005); the cannabinoid receptor 1 gene (*CNR1*) with Neuroticism and Agreeableness (Juhasz et al., 2009); the GTP cyclohydrolase 1 gene (*GCH1*) with Novelty seeking (Sadahiro et al., 2011); and the chitinase 3-like 1 gene (cartilage glycoprotein-39; *CHI3L1*) with Self-transcendence (Yamamori et al., 2012) .

Recently, genome-wide association studies (GWASs) have reported several new genes associated with personality traits including: the *MAMDC1* (or *MDGA2*), *PDE4D, SNAP25, LCE3C, POLR3A, LMAN1L, ULK3*, and *SCAMP2* genes with Neuroticism; the *MDGA2* gene with Harm avoidance; the *CDH13, CDH23* and *KIAA0802* genes with Extraversion; the *CNTNAP2* and *RASA1* genes with Openness to experience; the *CLOCK* gene with Agreeableness; the *DYRK1A* and *KATNAL2* genes with Conscientiousness; and the *ABLIM1* gene with Novelty seeking, Harm avoidance and Reward dependence (van den Oord et al., 2008; Shifman et al., 2008; Terracciano et al., 2010b; Luciano et al., 2012; De Moor et al., 2012; Wang et al., 2012). However, few of these associations have been replicated in independent samples (Terracciano et al., 2011).

Gene–environment interaction in individual differences in personality

Personality traits can be influenced by co-action of environmental and genetic factors (Reif & Lesch, 2003). Twin studies show that environmental factors explain ~50% of the observed variation in personality (Van Gestel et al., 2003). Several factors have been identified as potential environmental factors for personality including birth order, season of birth (Chotai et al., 2001), parenting style (Reti et al., 2002), parental psychopathology (Ravaja et al., 2001), place of residence and socio-economic status (Jokela, 2007b).

Inconsistencies in the results of genetic association studies can be partly explained by the potential presence of gene–environment interaction (GxE). The effects of the serotonin receptor 2A gene (*HTR2A*) on Harm avoidance have been shown to be modified by maternal overprotection (Nakamura et al., 2010); family socio-economic status (SES; Jokela et al., 2007) and place of residence (urban/rural residence; Jokela, Lehtimäki, & Keltikangas-Järvinen, 2007). For example, one study reported that people from families of high SES had lower scores on Harm avoidance than those from families of low SES, but only among those with the *T*-allele of the *102T>C* polymorphism of the *HTR2A* gene (Jokela et al., 2007). The interaction effect between the *CRHR1* gene and childhood maltreatment on individual differences in neuroticism has also been reported (DeYoung et al., 2011).

A novel approach to the study of GxE is analysis of genome-wide gene–environment interactions. This approach is based on the inclusion of environmental measures in a genome-wide association study making it possible to detect genetic effects that are otherwise masked by gene–environment interactions. This approach offers promises to identify new mechanisms underlying variation in human personality (Thomas, 2010; Winham & Biernacka, 2013).

Gene–gene interaction (epistasis) in individual differences in personality

In addition to additive genetic effects on personality, gene–gene interactions (epistasis) may also contribute to variation in personality. Epistasis represents suppression or enhancement of the effects of one gene by other genes (Benjamin et al., 2008; Comings et al., 2000a).

Several studies have demonstrated the interaction effect between the serotonin transporter gene (*5-HTT*) with the brain derived neurotrophic factor gene (*BDNF*) on anxiety-related traits (Arias et al., 2012; Terracciano et al., 2010a). For example, carriers of *L/L*-genotype of the *5-HTT* gene were characterized by lower neuroticism scores in the presence of the *BDNF Val*-allele, but by higher Neuroticism scores in the presence of *BDNF Met*-allele (Terracciano et al., 2010a). Another study has demonstrated lower conscientiousness scores in those with the *BDNF Met*-allele only in the presence of the *5-HTT S/S*-genotype (Hiio et al., 2011).

Other studies have pointed to the interaction between the serotonergic system (i.e., *5-HTT*) and noradrenergic system genes (i.e., *NET*) (Hall et al., 2009; Suzuki et al., 2008). For example, females with the *5-HTT S/S*-genotype demonstrated higher scores on Harm avoidance and lower scores on Novelty seeking only in the presence of the *NET* (*–3081A/T* polymorphism) low-active *T/T*-genotype (Suzuki et al., 2008). Interactions between serotonergic and dopaminergic system genes have also been demonstrated in several studies. For example, higher scores on Persistence were observed only in carriers of *COMT Val/Val*-genotype in the presence of the *5-HTT S*-allele. In the absence of the *5-HTT S*-allele and in the presence of the *COMT Val/Val*-genotype, Novelty seeking scores were higher in the presence of the *DRD4 7R*-allele (Benjamin et al., 2008).

A number of studies have reported the interaction effect of the vasopressin gene with the oxytocin receptor gene and other hormonal and neurotransmitter system genes. For example, one study showed an interaction effect of the *5-HTT* and *OXTR* genes on parenting: parents with low-activity alleles of both *5-HTT* and *OXTR* genes demonstrated lower sensitivity towards their children (Bakermans-Kranenburg et al., 2008). Another study also reported an interaction effect between the *5-HTT* and *OXTR* genes on negative emotionality (Montag et al., 2011).

Several studies have indicated the presence of epistasis for the genes from the same functional network. For example, the *REST* gene which encodes RE1 transcription factor and is involved in regulation of BDNF expression level was shown to moderate the association between the *BDNF Val*-allele and cognitive functions (Miyajima et al., 2008). An association of the *C*-allele of the *DRD4* (*–521C>T* polymorphism) with higher scores on extraversion was observed only in the presence of the *COMT Met/Met*-genotype (Golimbet et al., 2007).

We are a long way away from understanding the molecular mechanisms of gene–gene interactions that are detected at a statistical level. Future functional studies will be necessary to characterize biological mechanisms and pathways that bridge the gap between genes and behaviour.

Conclusions

The evidence reviewed in this chapter strongly suggests that individual differences in personality traits are influenced by multiple genes of very small effects. There is also evidence for both common and specific genetic effects on different personality traits (Terracciano et al., 2010b). To date, only a small proportion of heritability of personality traits has been explained with identified genetic variants (see Chapter 1 for more details on the missing heritability problem). Much of the remaining trait variation is likely to be due to the combined effect of genes, environmental factors and their interactions.

It is worth noting that most of the reviewed evidence comes from studies of adults or adolescents. Molecular genetic studies of temperament in infants and young children are scarce because of methodological difficulties of measuring temperament in this early period of life. Nevertheless, the existing studies suggest that individual differences in temperament in infants and young children can also be influenced by multiple genes with small effects. A number of studies have demonstrated associations of children's temperament with candidate genes that are involved in personality traits in adults (Ivorra et al., 2011; Auerbach et al., 1999; De Luca et al., 2001; De Luca et al., 2003; Lakatos et al., 2003). Despite the methodological problems, the search for molecular genetic bases of personality has already provided important insights into mechanisms underlying personality development across the life course.

Acknowledgements

This work is partly supported by the Russian foundation for humanities grant (no. 13-06-00583a) and the Russian foundation for basic research grant (no. 11-04-97032-r_povolzhye_a).

References

Aluja, A., García, L. F., Blanch, A., & Fibla, J. (2011). Association of androgen receptor gene, CAG and GGN repeat length polymorphism and impulsive-disinhibited personality traits in inmates: the role of short–long haplotype. *Psychiatric Genetics, 21*(5), 229–239.

Amin, N., Schuur, M., Gusareva, E. S., Isaacs, A., Aulchenko, Y. S., Kirichenko, A. V., ..., & van Duijn, C. M. (2012). A genome-wide linkage study of individuals with high scores on NEO personality traits. *Molecular Psychiatry, 17*(10), 1031–1041.

Anghelescu, I., Klawe, C., Fehr, C., Singer, P., Schleicher, A., Himmerich, H., ..., & Szegedi, A. (2005). The TPH intron 7 A218C polymorphism and TCI dimension scores in alcohol-dependent patients: hints to nonspecific psychopathology. *Addictive Behaviors, 30*(6), 1135–1143.

Arias, B., Aguilera, M., Moya, J., Sáiz, P. A., Villa, H., Ibáñez, M. I., …, & Fananas, L. (2012). The role of genetic variability in the SLC6A4, BDNF and GABRA6 genes in anxiety-related traits. *Acta Psychiatrica Scandinavica, 125*(3), 194–202.

Asghari, V., Sanyal, S., Buchwaldt, S., Paterson, A., Jovanovic, V., & Van Tol, H. H. (1995). Modulation of intracellular cyclic AMP levels by different human dopamine D4 receptor variants. *Journal of Neurochemistry, 65*(3), 1157–1165.

Auerbach, J., Geller, V., Lezer, S., Shinwell, E., Belmaker, R. H., Levine, J., & Ebstein, R. (1999). Dopamine D4 receptor (D4DR) and serotonin transporter promoter (5-HTTLPR) polymorphisms in the determination of temperament in 2-month-old infants. *Molecular Psychiatry, 4*(4), 369–373.

Aulchenko, Y. S. (2010). Razrabotka i primeneniye metodov polnogenomnogo analiza geneticheskikh assotsiatsiy slozhnykh priznakov [Development and application of genome-wide association study of complex traits] (Doctoral thesis, Novosibirsk). Retrieved from http://www.dissercat.com/content/razrabotka-i-primenenie-metodov-polnogenomnogo-analiza-geneticheskikh-assotsiatsii-slozhnykh

Bachner-Melman, R., Dina, C., Zohar, A. H., Constantini, N., Lerer, E., Hoch, S., …, & Ebstein, R. P. (2005). AVPR1a and SLC6A4 gene polymorphisms are associated with creative dance performance. *PLoS Genetics, 1*(3), e42.

Bakermans-Kranenburg, M. J., & van IJzendoorn, M. H. (2008). Oxytocin receptor (OXTR) and serotonin transporter (5-HTT) genes associated with observed parenting. *Social Cognitive and Affective Neuroscience*, nsn004.

Becker, J. B. (1999). Gender differences in dopaminergic function in striatum and nucleus accumbens. *Pharmacology Biochemistry and Behavior, 64*(4), 803–812.

Becker, K., Laucht, M., El-Faddagh, M., & Schmidt, M. H. (2005). The dopamine D4 receptor gene exon III polymorphism is associated with novelty seeking in 15-year-old males from a high-risk community sample. *Journal of Neural Transmission, 112*(6), 847–858.

Benjamin, J., Ebstein, R. P., & Belmaker, R. H. (Eds) (2008). *Molecular genetics and the human personality*. American Psychiatric Pub.

Benjamin, J., Osher, Y., Kotler, M., Gritsenko, I., Nemanov, L., Belmaker, R. H., & Ebstein, R. P. (2000). Association between tridimensional personality questionnaire (TPQ) traits and three functional polymorphisms: dopamine receptor D4 (DRD4), serotonin transporter promoter region (5-HTTLPR) and catechol O-methyltransferase (COMT). *Molecular Psychiatry, 5*(1), 96–100.

Benjamini, Y., & Hochberg, Y. (1995). Controlling the false discovery rate: a practical and powerful approach to multiple testing. *Journal of the Royal Statistical Society. Series B (Methodological)*, 289–300.

Bosker, F. J., Hartman, C. A., Nolte, I. M., Prins, B. P., Terpstra, P., Posthuma, D., …, & Nolen, W. A. (2011). Poor replication of candidate genes for major depressive disorder using genome-wide association data. *Molecular Psychiatry, 16*(5), 516–532.

Boudreault-Bouchard, A. M., Dion, J., Hains, J., Vandermeerschen, J., Laberge, L., & Perron, M. (2013). Impact of parental emotional support and coercive control on adolescents' self-esteem and psychological distress: Results of a four-year longitudinal study. *Journal of Adolescence, 36*(4), 695–704.

Cagliani, R., Fumagalli, M., Pozzoli, U., Riva, S., Cereda, M., Comi, G. P., …, & Sironi, M. (2009). A complex selection signature at the human AVPR1B gene. *BMC Evolutionary Biology, 9*(1), 123.

Carter, C. S., Grippo, A. J., Pournajafi-Nazarloo, H., Ruscio, M. G., & Porges, S. W. (2008). Oxytocin, vasopressin and sociality. *Progress in Brain Research, 170*, 331–336.

Chapman, B. P., Duberstein, P. R., Sörensen, S., & Lyness, J. M. (2007). Gender differences in Five Factor Model personality traits in an elderly cohort. *Personality and Individual Differences, 43*(6), 1594–1603.

Chen, J., Lipska, B. K., Halim, N., Ma, Q. D., Matsumoto, M., Melhem, S.,..., & Weinberger, D. R. (2004). Functional analysis of genetic variation in catechol-O-methyltransferase (COMT): effects on mRNA, protein, and enzyme activity in postmortem human brain. *American Journal of Human Genetics, 75*(5), 807–821.

Chen, C., Chen, C., Moyzis, R., Dong, Q., He, Q., Zhu, B.,..., & Lessard, J. (2011). Sex modulates the associations between the COMT gene and personality traits. *Neuropsychopharmacology, 36*(8), 1593–1598.

Chiao, J. Y., & Blizinsky, K. D. (2010). Culture–gene coevolution of individualism–collectivism and the serotonin transporter gene. *Proceedings of the Royal Society of London B: Biological Sciences,* 277(1681), 529–537.

Chotai, J., Forsgren, T., Nilsson, L. G., & Adolfsson, R. (2001). Season of birth variations in the temperament and character inventory of personality in a general population. *Neuropsychobiology, 44*(1), 19–26.

Cloninger, C. R., Svrakic, D. M., & Przybeck, T. R. (1993). A psychobiological model of temperament and character. *Archives of General Psychiatry, 50*(12), 975–990.

Cloninger, C. R., Van Eerdewegh, P., Goate, A., Edenberg, H. J., Blangero, J., Hesselbrock, V.,..., & Begleiter, H. (1998). Anxiety proneness linked to epistatic loci in genome scan of human personality traits. *American Journal of Medical Genetics, 81*(4), 313–317.

Comings, D. E., Gade-Andavolu, R., Gonzalez, N., Wu, S., Muhleman, D., Blake, H.,..., & MacMurray, J. P. (2000a). A multivariate analysis of 59 candidate genes in personality traits: the temperament and character inventory. *Clinical Genetics, 58*(5), 375–385.

Comings, D. E., Johnson, J. P., Gonzalez, N. S., Huss, M., Saucier, G., McGue, M., & MacMurray, J. (2000b). Association between the adrenergic [alpha] 2A receptor gene (ADRA2A) and measures of irritability, hostility, impulsivity and memory in normal subjects. *Psychiatric Genetics, 10*(1), 39–42.

Cosgrove, K. P., Mazure, C. M., & Staley, J. K. (2007). Evolving knowledge of sex differences in brain structure, function, and chemistry. *Biological Psychiatry, 62*(8), 847–855.

Costa Jr, P., Terracciano, A., & McCrae, R. R. (2001). Gender differences in personality traits across cultures: robust and surprising findings. *Journal of Personality and Social Psychology, 81*(2), 322.

Costa, P. T., & McCrae, R. R. (1992). *Revised NEO personality inventory (NEO PI-R) and NEO five-factor inventory (NEO FFI): Professional manual.* Psychological Assessment Resources.

De Luca, A., Rizzardi, M., Torrente, I., Alessandroni, R., Salvioli, G. P., Filograsso, N.,..., & Novelli, G. (2001). Dopamine D4 receptor (DRD4) polymorphism and adaptability trait during infancy: a longitudinal study in 1-to 5-month-old neonates. *Neurogenetics, 3*(2), 79–82.

De Luca, A., Rizzardi, M., Buccino, A., Alessandroni, R., Salvioli, G. P., Filograsso, N.,..., & Dallapiccola, B. (2003). Association of dopamine D4 receptor (DRD4) exon III repeat polymorphism with temperament in 3-year-old infants. *Neurogenetics, 4*(4), 207–212.

De Moor, M. H., Costa, P. T., Terracciano, A., Krueger, R. F., De Geus, E. J., Toshiko, T.,..., & Metspalu, A. (2012). Meta-analysis of genome-wide association studies for personality. *Molecular Psychiatry, 17*(3), 337–349.

De, R., Bush, W. S., & Moore, J. H. (2014). Bioinformatics Challenges in Genome-Wide Association Studies (GWAS). In *Clinical Bioinformatics* (pp. 63–81). Springer New York.

DeYoung, C. G., Cicchetti, D., & Rogosch, F. A. (2011). Moderation of the association between childhood maltreatment and neuroticism by the corticotropin-releasing hormone receptor 1 gene. *Journal of Child Psychology and Psychiatry, 52*(8), 898–906.

Dina, C., Nemanov, L., Gritsenko, I., Rosolio, N., Osher, Y., Heresco-Levy, U., ..., & Ebstein, R. P. (2005). Fine mapping of a region on chromosome 8p gives evidence for a QTL contributing to individual differences in an anxiety-related personality trait: TPQ harm avoidance. *American Journal of Medical Genetics Part B: Neuropsychiatric Genetics, 132*(1), 104–108.

Ding, Y. C., Chi, H. C., Grady, D. L., Morishima, A., Kidd, J. R., Kidd, K. K., ..., & Moyzis, R. K. (2002). Evidence of positive selection acting at the human dopamine receptor D4 gene locus. *Proceedings of the National Academy of Sciences, 99*(1), 309–314.

Dougherty, L. R., Klein, D. N., Congdon, E., Canli, T., & Hayden, E. P. (2010). Interaction between 5-HTTLPR and BDNF Val66Met polymorphisms on HPA axis reactivity in preschoolers. *Biological Psychology, 83*(2), 93–100.

Duncan, L. E., & Keller, M. C. (2011). A critical review of the first 10 years of candidate gene-by-environment interaction research in psychiatry. *American Journal of Psychiatry, 168*(10), 1041–1049.

Ebstein, R. P., & Belmaker, R. H. (1997). Saga of an adventure gene: novelty seeking, substance abuse and the dopamine D4 receptor (D4DR) exon III repeat polymorphism. *Molecular Psychiatry, 2*(5), 381–384.

Egan, M. F., Kojima, M., Callicott, J. H., Goldberg, T. E., Kolachana, B. S., Bertolino, A., ... & Weinberger, D. R. (2003). The BDNF val66met polymorphism affects activity-dependent secretion of BDNF and human memory and hippocampal function. *Cell, 112*(2), 257–269.

Else-Quest, N. M., Hyde, J. S., Goldsmith, H. H., & Van Hulle, C. A. (2006). Gender differences in temperament: a meta-analysis. *Psychological Bulletin, 132*(1), 33.

Feingold, A. (1994). Gender differences in personality: a meta-analysis. *Psychological Bulletin, 116*(3), 429.

Fernández-Teruel, A., Escorihuela, R. M., Gray, J. A., Aguilar, R., Gil, L., Giménez-Llort, L., ..., & Flint, J. (2002). A quantitative trait locus influencing anxiety in the laboratory rat. *Genome Research, 12*(4), 618–626.

Freeman, E. W., Sammel, M. D., Lin, H., & Nelson, D. B. (2006). Associations of hormones and menopausal status with depressed mood in women with no history of depression. *Archives of General Psychiatry, 63*(4), 375–382.

Fullerton, J., Cubin, M., Tiwari, H., Wang, C., Bomhra, A., Davidson, S., ..., & Flint, J. (2003). Linkage analysis of extremely discordant and concordant sibling pairs identifies quantitative-trait loci that influence variation in the human personality trait neuroticism. *American Journal of Human Genetics, 72*(4), 879–890.

Gade-Andavolu, R., Macmurray, J., Comings, D. E., Calati, R., Chiesa, A., & Serretti, A. (2009). Association between the estrogen receptor TA polymorphism and Harm avoidance. *Neuroscience Letters, 467*(2), 155–158.

Gallagher, J. J., Zhang, X., Hall, F. S., Uhl, G. R., Bearer, E. L., & Jacobs, R. E. (2013). Altered reward circuitry in the norepinephrine transporter knockout mouse. *PloS One, 8*(3), e57597.

Garvey, M. J., Noyes, R., Cook, B., & Blum, N. (1996). Preliminary confirmation of the proposed link between reward-dependence traits and norepinephrine. *Psychiatry Research, 65*(1), 61–64.

Gerra, G., Avanzini, P., Zaimovic, A., Sartori, R., Bocchi, C., Timpano, M.,..., & Brambilla, F. (1999). Neurotransmitters, neuroendocrine correlates of sensation-seeking temperament in normal humans. *Neuropsychobiology, 39*(4), 207–213.

Giegling, I., Calati, R., Porcelli, S., Hartmann, A. M., Möller, H. J., De Ronchi, D.,..., & Serretti, A. (2011). NCAM1, TACR1 and NOS genes and temperament: a study on suicide attempters and controls. *Neuropsychobiology, 64*(1), 32–37.

Gillespie, N. A., Zhu, G., Evans, D. M., Medland, S. E., Wright, M. J., & Martin, N. G. (2008). A Genome-Wide Scan for Eysenckian Personality Dimensions in Adolescent Twin Sibships: Psychoticism, Extraversion, Neuroticism, and Lie. *Journal of Personality, 76*(6), 1415–1446.

Golimbet, V. E., Alfimova, M. V., Gritsenko, I. K., & Ebstein, R. P. (2007). Relationship between dopamine system genes and extraversion and novelty seeking. *Neuroscience and Behavioral Physiology, 37*(6), 601–606.

Gowrishankar, R., Hahn, M. K., & Blakely, R. D. (2014). Good riddance to dopamine: Roles for the dopamine transporter in synaptic function and dopamine-associated brain disorders. *Neurochemistry international, 73*, 42–48.

Gragnoli, C. (2014). Hypothesis of the neuroendocrine cortisol pathway gene role in the comorbidity of depression, type 2 diabetes, and metabolic syndrome. *Application of Clinical Genetics, 7*, 43.

Hall, F. S., Li, X. F., Randall-Thompson, J., Sora, I., Murphy, D. L., Lesch, K. P.,..., & Uhl, G. R. (2009). Cocaine-conditioned locomotion in dopamine transporter, norepinephrine transporter and 5-HT transporter knockout mice. *Neuroscience, 162*(4), 870–880.

Ham, B. J., Choi, M. J., Lee, H. J., Kang, R. H., & Lee, M. S. (2005). Reward dependence is related to norepinephrine transporter T-182C gene polymorphism in a Korean population. *Psychiatric Genetics, 15*(2), 145–147.

Hamon, M., & Blier, P. (2013). Monoamine neurocircuitry in depression and strategies for new treatments. *Progress in Neuro-Psychopharmacology and Biological Psychiatry, 45*, 54–63.

Hashimoto, R., Noguchi, H., Hori, H., Ohi, K., Yasuda, Y., Takeda, M., & Kunugi, H. (2007). A possible association between the Val158Met polymorphism of the catechol-O-methyl transferase gene and the personality trait of harm avoidance in Japanese healthy subjects. *Neuroscience Letters, 428*(1), 17–20.

Heck, A., Lieb, R., Ellgas, A., Pfister, H., Lucae, S., Roeske, D.,..., & Ising, M. (2009). Investigation of 17 candidate genes for personality traits confirms effects of the HTR2A gene on novelty seeking. *Genes, Brain and Behavior, 8*(4), 464–472.

Hiio, K., Merenäkk, L., Nordquist, N., Parik, J., Oreland, L., Veidebaum, T., & Harro, J. (2011). Effects of serotonin transporter promoter and BDNF Val66Met genotype on personality traits in a population representative sample of adolescents. *Psychiatric Genetics, 21*(5), 261–264.

Hines, M., Golombok, S., Rust, J., Johnston, K. J., Golding, J., & The Avon Longitudinal Study of Parents and Children Study Team. (2002). Testosterone during pregnancy and gender role behavior of preschool children: a longitudinal, population study. *Child Development*, 1678–1687.

Hiroi, R., McDevitt, R. A., & Neumaier, J. F. (2006). Estrogen selectively increases tryptophan hydroxylase-2 mRNA expression in distinct subregions of rat midbrain raphe nucleus: association between gene expression and anxiety behavior in the open field. *Biological Psychiatry, 60*(3), 288–295.

Hopkins, W. D., Donaldson, Z. R., & Young, L. J. (2012). A polymorphic indel containing the RS3 microsatellite in the 5′ flanking region of the vasopressin V1a receptor gene is associated with chimpanzee (Pan troglodytes) personality. *Genes, Brain and Behavior, 11*(5), 552–558.

Hwang, J. P., Tsai, S. J., Hong, C. J., Yang, C. H., Lirng, J. F., & Yang, Y. M. (2006). The Val66Met polymorphism of the brain-derived neurotrophic-factor gene is associated with geriatric depression. *Neurobiology of Aging, 27*(12), 1834–1837.

Ivorra, J. L., D'Souza, U. M., Jover, M., Arranz, M. J., Williams, B. P., Henry, S. E...., & Molto, M. D. (2011). Association between neonatal temperament, SLC6A4, DRD4 and a functional polymorphism located in TFAP2B. *Genes, Brain and Behavior, 10*(5), 570–578.

Jokela, M., Keltikangas-Järvinen, L., Kivimäki, M., Puttonen, S., Elovainio, M., Rontu, R., & Lehtimäki, T. (2007). Serotonin receptor 2A gene and the influence of childhood maternal nurturance on adulthood depressive symptoms. *Archives of General Psychiatry, 64*(3), 356–360.

Jokela, M., Lehtimäki, T., & Keltikangas-Järvinen, L. (2007). The influence of urban/rural residency on depressive symptoms is moderated by the serotonin receptor 2A gene. *American Journal of Medical Genetics Part B: Neuropsychiatric Genetics, 144*(7), 918–922.

Jönsson, E. G., Nöthen, M. M., Gustavsson, J. P., Neidt, H., Bunzel, R., Propping, P., & Sedvall, G. C. (1998). Polymorphisms in the dopamine, serotonin, and norepinephrine transporter genes and their relationships to monoamine metabolite concentrations in CSF of healthy volunteers. *Psychiatry Research, 79*(1), 1–9.

Joo, Y. H., Oh, H. B., Kim, B., Jung, S. H., Chung, J. K., Hong, J. P., & Kim, C. Y. (2007). No association between 5-HTTLPR and harm avoidance in Korean college students. *Journal of Korean Medical Science, 22*(1), 138–141.

Jorm, A. F., Prior, M., Sanson, A., Smart, D., Zhang, Y., & Easteal, S. (2000). Association of a functional polymorphism of the serotonin transporter gene with anxiety-related temperament and behavior problems in children: a longitudinal study from infancy to the mid-teens. *Molecular Psychiatry, 5*(5), 542–547.

Juhasz, G., Chase, D., Pegg, E., Downey, D., Toth, Z. G., Stones, K., ..., & Deakin, J. W. (2009). CNR1 gene is associated with high neuroticism and low agreeableness and interacts with recent negative life events to predict current depressive symptoms. *Neuropsychopharmacology, 34*(8), 2019–2027.

Kaiser, R., Tremblay, P. B., Klufmöller, F., Roots, I., & Brockmöller, J. (2001). Relationship between adverse effects of antipsychotic treatment and dopamine D (2) receptor polymorphisms in patients with schizophrenia. *Molecular Psychiatry, 7*(7), 695–705.

Kato, N., Takeuchi, F., Tabara, Y., Kelly, T. N., Go, M. J., Sim, X., ..., & He, J. (2011). Meta-analysis of genome-wide association studies identifies common variants associated with blood pressure variation in east Asians. *Nature Genetics, 43*(6), 531–538.

Kawamura, Y., Liu, X., Akiyama, T., Shimada, T., Otowa, T., Sakai, Y., ..., & Akiskal, H. S. (2010). The association between oxytocin receptor gene (OXTR) polymorphisms and affective temperaments, as measured by TEMPS-A. *Journal of Affective Disorders, 127*(1), 31–37.

Khusnutdinova, E. K., Bermisheva, M. A., Kutuev, I. A., Yunusbayev, B. B., & Villems, R. (2008). Genetic landscape of the central Asia and Volga–Ural region. In *Biosphere Origin and Evolution* (pp. 373–381). Springer US.

Kazantseva, A. V., Gaysina, D. A., Faskhutdinova, G. G., Noskova, T., Malykh, S. B., & Khusnutdinova, E. K. (2008). Polymorphisms of the serotonin transporter gene (5-HTTLPR, A/G SNP in 5-HTTLPR, and STin2 VNTR) and their relation to personality traits in healthy individuals from Russia. *Psychiatric Genetics, 18*(4), 167–176.

Kazantseva, A., Gaysina, D., Malykh, S., & Khusnutdinova, E. (2011). The role of dopamine transporter (SLC6A3) and dopamine D2 receptor/ankyrin repeat and kinase domain containing 1 (DRD2/ANKK1) gene polymorphisms in personality traits. *Progress in Neuro-Psychopharmacology and Biological Psychiatry, 35*(4), 1033–1040.

Kazantseva, A., Kutlumbetova, Y., Malykh, S., Lobaskova, M., & Khusnutdinova, E. (2014). Assotsiatsiya polimorfnykh markerov genov arginin-vazopressinovykh retseptorov (AVPR1A, AVPR1B) s chertami lichnosti [The role of arginine-vasopressin receptor gene (AVPR1A, AVPR1B) polymorphisms in personality traits] *Journal of Genetics,* V. 50 (3), p.341–352.

Konno, A., Inoue-Murayama, M., & Hasegawa, T. (2011). Androgen receptor gene polymorphisms are associated with aggression in Japanese Akita Inu. *Biology Letters, 7*(5), 658–660.

Kubinyi, E., Vas, J., Hejjas, K., Ronai, Z., Brúder, I., Turcsán, B., ..., & Miklósi, Á. (2012). Polymorphism in the tyrosine hydroxylase (TH) gene is associated with activity-impulsivity in German Shepherd Dogs. *PloS One, 7*(1), e30271.

Kuo, P. H., Neale, M. C., Riley, B. P., Patterson, D. G., Walsh, D., Prescott, C. A., & Kendler, K. S. (2007). A genome-wide linkage analysis for the personality trait neuroticism in the Irish affected sib-pair study of alcohol dependence. *American Journal of Medical Genetics Part B: Neuropsychiatric Genetics, 144*(4), 463–468.

Kutlumbetova, Y., Kazantseva, A., Kanzafarova, R., Malykh, S., Lobaskova, M., & Khusnutdinova, E. (2014). Vovlechennost' polimorfnyh markerov gena receptora ehstrogena al'fa (ESR1) v variacii chert lichnosti[The role of estrogen receptor alpha gene (ESR1) polymorphisms in personality traits]. *Vestnik Bashkirskogo Gosudarstvennogo Universiteta,*V. 19(1), 49–53.

Lakatos, K., Nemoda, Z., Birkas, E., Ronai, Z., Kovacs, E., Ney, K., ..., & Gervai, J. (2003). Association of D4 dopamine receptor gene and serotonin transporter promoter polymorphisms with infants' response to novelty. *Molecular Psychiatry, 8*(1), 90–97.

Lang, U. E., Bajbouj, M., Wernicke, C., Rommelspacher, H., Danker-Hopfe, H., & Gallinat, J. U. R. (2004). No association of a functional polymorphism in the serotonin transporter gene promoter and anxiety-related personality traits. *Neuropsychobiology, 49*(4), 182–184.

Lang, U. E., Hellweg, R., Kalus, P., Bajbouj, M., Lenzen, K. P., Sander, T., ..., & Gallinat, J. (2005). Association of a functional BDNF polymorphism and anxiety-related personality traits. *Psychopharmacology, 180*(1), 95–99.

Lee, H. J., Lee, H. S., Kim, Y. K., Kim, L., Lee, M. S., Jung, I. K., ..., & Kim, S. (2003). D2 and D4 dopamine receptor gene polymorphisms and personality traits in a young Korean population. *American Journal of Medical Genetics Part B: Neuropsychiatric Genetics, 121*(1), 44–49.

Lee, S. H., Ham, B. J., Cho, Y. H., Lee, S. M., & Shim, S. H. (2007). Association study of dopamine receptor D2TaqI A polymorphism and reward-related personality traits in healthy Korean young females. *Neuropsychobiology, 56*(2–3), 146–151.

Lee, B. C., Yang, J. W., Lee, S. H., Kim, S. H., Joe, S. H., Jung, I. K., ..., & Ham, B. J. (2008). An interaction between the norepinephrine transporter and monoamine oxidase A polymorphisms, and novelty-seeking personality traits in Korean females. *Progress in Neuro-Psychopharmacology and Biological Psychiatry, 32*(1), 238–242.

Lesch, K. P., Bengel, D., Heils, A., Sabol, S. Z., Greenberg, B. D., Petri, S., ..., & Murphy, D. L. (1996). Association of anxiety-related traits with a polymorphism in the serotonin transporter gene regulatory region. *Science, 274*(5292), 1527–1531.

Li, J., Ma, H., Zhou, H., Huang, Y., Wu, L., Li, J., & Zhu, G. (2011). Association between DARPP-32 gene polymorphism and personality traits in healthy Chinese-Han subjects. *Journal of Molecular Neuroscience, 44*(1), 48–52.

Lin, S. C., Wu, P. L., Ko, H. C., Wu, J. Y. W., Huang, S. Y., Lin, W. W., & Lu, R. B. (2007). Specific personality traits and dopamine, serotonin genes in anxiety–depressive alcoholism among Han Chinese in Taiwan. *Progress in Neuro-Psychopharmacology and Biological Psychiatry, 31*(7), 1526–1534.

Luciano, M., Huffman, J. E., Arias-Vásquez, A., Vinkhuyzen, A. A., Middeldorp, C. M., Giegling, I., ..., & Deary, I. J. (2012). Genome-wide association uncovers shared genetic effects among personality traits and mood states. *American Journal of Medical Genetics part B: Neuropsychiatric Genetics, 159*(6), 684–695.

Mäestu, J., Allik, J., Merenäkk, L., Eensoo, D., Parik, J., Veidebaum, T., & Harro, J. (2008). Associations between an alpha 2a adrenergic receptor gene polymorphism and adolescent personality. *American Journal of Medical Genetics Part B: Neuropsychiatric Genetics, 147*(4), 418–423.

Mann, J. J., Brent, D. A., & Arango, V. (2001). The neurobiology and genetics of suicide and attempted suicide: a focus on the serotonergic system. *Neuropsychopharmacology, 24*(5), 467–477.

Markett, S., Montag, C., & Reuter, M. (2011). The nicotinic acetylcholine receptor gene CHRNA4 is associated with negative emotionality. *Emotion, 11*(2), 450.

Matsuzawa, D., Hashimoto, K., Shimizu, E., Fujisaki, M., & Iyo, M. (2005). Functional polymorphism of the glutathione peroxidase 1 gene is associated with personality traits in healthy subjects. *Neuropsychobiology, 52*(2), 68–70.

McCrae, R. R. (2001). Trait psychology and culture: Exploring intercultural comparisons. *Journal of Personality, 69*(6), 819–846.

McCrae, R. R., & Terracciano, A. (2005). Personality profiles of cultures: aggregate personality traits. *Journal of Personality and Social Psychology, 89*(3), 407.

McCrae, R. R., & Terracciano, A. (2008). The Five-Factor Model and its correlates in individuals and cultures. *Multilevel analysis of individuals and cultures*, 249–283.

Meyer-Lindenberg, A., Domes, G., Kirsch, P., & Heinrichs, M. (2011). Oxytocin and vasopressin in the human brain: social neuropeptides for translational medicine. *Nature Reviews Neuroscience, 12*(9), 524–538.

Miettunen, J., Kantojärvi, L., Veijola, J., Järvelin, M. R., & Joukamaa, M. (2006). International comparison of Cloninger's temperament dimensions. *Personality and Individual Differences, 41*(8), 1515–1526.

Minelli, A., Scassellati, C., Bonvicini, C., Perez, J., & Gennarelli, M. (2009). An association of GRIK3 Ser310Ala functional polymorphism with personality traits. *Neuropsychobiology, 59*(1), 28–33.

Miyajima, F., Quinn, J. P., Horan, M., Pickles, A., Ollier, W. E., Pendleton, N., & Payton, A. (2008). Additive effect of BDNF and REST polymorphisms is associated with improved general cognitive ability. *Genes, Brain and Behavior,7*(7), 714–719.

Montag, C., Fiebach, C. J., Kirsch, P., & Reuter, M. (2011). Interaction of 5-HTTLPR and a variation on the oxytocin receptor gene influences negative emotionality. *Biological Psychiatry, 69*(6), 601–603.

Morón, J. A., Brockington, A., Wise, R. A., Rocha, B. A., & Hope, B. T. (2002). Dopamine uptake through the norepinephrine transporter in brain regions with low levels of the dopamine transporter: evidence from knock-out mouse lines. *Journal of Neuroscience, 22*(2), 389–395.

Munafò, M. R., Clark, T. G., Moore, L. R., Payne, E., Walton, R., & Flint, J. (2003). Genetic polymorphisms and personality in healthy adults: a systematic review and meta-analysis. *Molecular Psychiatry, 8*(5), 471–484.

Munafò, M. R., Durrant, C., Lewis, G., & Flint, J. (2009a). Gene × environment interactions at the serotonin transporter locus. *Biological Psychiatry, 65*(3), 211–219.

Munafò, M. R., Freimer, N. B., Ng, W., Ophoff, R., Veijola, J., Miettunen, J., ..., & Flint, J. (2009b). 5-HTTLPR genotype and anxiety-related personality traits: A meta-analysis and new data. *American Journal of Medical Genetics Part B: Neuropsychiatric Genetics, 150*(2), 271–281.

Nakamura, Y., Ito, Y., Aleksic, B., Kushima, I., Yasui-Furukori, N., Inada, T., ..., & Ozaki, N. (2010). Influence of HTR2A polymorphisms and parental rearing on personality traits in healthy Japanese subjects. *Journal of Human Genetics, 55*(12), 838–841.

Nakamura, M., Ueno, S., Sano, A., & Tanabe, H. (2000). The human serotonin transporter gene linked polymorphism (5-HTTLPR) shows ten novel allelic variants. *Molecular Psychiatry, 5*(1), 32–38.

Nakazato, M., Hashimoto, K., Shimizu, E., Kumakiri, C., Koizumi, H., Okamura, N., ..., & Iyo, M. (2003). Decreased levels of serum brain-derived neurotrophic factor in female patients with eating disorders. *Biological Psychiatry, 54*(4), 485–490.

Nash, M. W., Huezo-Diaz, P., Williamson, R. J., Sterne, A., Purcell, S., Hoda, F., ..., & Sham, P. C. (2004). Genome-wide linkage analysis of a composite index of neuroticism and mood-related scales in extreme selected sibships. *Human Molecular Genetics, 13*(19), 2173–2182.

Nederhof, E., Creemers HE., Huizink AC., & Ormel, J., (2011). L-DRD4 genotype not associated with sensation seeking, gambling performance and startle reactivity in adolescents: the TRAILS study. *Neuropsychologia*, 49(5), 1359–1362.

Neville, M. J., Johnstone, E. C., & Walton, R. T. (2004). Identification and characterization of ANKK1: a novel kinase gene closely linked to DRD2 on chromosome band 11q23. 1. *Human Mutation, 23*(6), 540–545.

Nielsen, D. A., Jenkins, G. L., Stefanisko, K. M., Jefferson, K. K., & Goldman, D. (1997). Sequence, splice site and population frequency distribution analyses of the polymorphic human tryptophan hydroxylase intron 7. *Molecular Brain Research, 45*(1), 145–148.

Noble, E. P. (2003). D2 dopamine receptor gene in psychiatric and neurologic disorders and its phenotypes. *American Journal of Medical Genetics Part B: Neuropsychiatric Genetics, 116*(1), 103–125.

Noble, E. P., Ozkaragoz, T. Z., Ritchie, T. L., Zhang, X., Belin, T. R., & Sparkes, R. S. (1998). D 2 and D 4 dopamine receptor polymorphisms and personality. *American Journal of Medical Genetics, 81*(3), 257–267.

Nomura, M., & Nomura, Y. (2006). Psychological, Neuroimaging, and Biochemical Studies on Functional Association between Impulsive Behavior and the 5-HT2A Receptor Gene Polymorphism in Humans. *Annals of the New York Academy of Sciences, 1086*(1), 134–143.

Noskova, T., Pivac, N., Nedic, G., Kazantseva, A., Gaysina, D., Faskhutdinova, G., ..., & Seler, D. M. (2008). Ethnic differences in the serotonin transporter polymorphism (5-HTTLPR) in several European populations. *Progress in Neuro-Psychopharmacology and Biological Psychiatry, 32*(7), 1735–1739.

Pełka-Wysiecka, J., Ziętek, J., Grzywacz, A., Kucharska-Mazur, J., Bienkowski, P., & Samochowiec, J. (2012). Association of genetic polymorphisms with personality profile in individuals without psychiatric disorders. *Progress in Neuro-Psychopharmacology and Biological Psychiatry, 39*(1), 40–46.

Poortinga, Y. H., & Van Hemert, D. A. (2001). Personality and culture: Demarcating between the common and the unique. *Journal of Personality, 69*(6), 1033–1060.

Presutti, C., Rosati, J., Vincenti, S., & Nasi, S. (2006). Non coding RNA and brain. *BMC Neuroscience, 7*(Suppl 1), S5.

Prichard, Z. M., Mackinnon, A. J., Jorm, A. F., & Easteal, S. (2007). AVPR1A and OXTR polymorphisms are associated with sexual and reproductive behavioral phenotypes in humans. *Human Mutation, 28*(11), 1150.

Rasmusson, A. M., Shi, L., & Duman, R. (2002). Downregulation of BDNF mRNA in the hippocampal dentate gyrus after re-exposure to cues previously associated with footshock. *Neuropsychopharmacology, 27*(2), 133–142.

Ravaja, N., & Keltikangas-Järvinen, K. (2001). Cloninger's temperament and character dimensions in young adulthood and their relation to characteristics of parental alcohol use and smoking. *Journal of Studies on Alcohol, 62*(1), 98–104.

Reif, A., & Lesch, K. P. (2003). Toward a molecular architecture of personality. *Behavioural Brain Research, 139*(1), 1–20.

Reti, I. M., Samuels, J. F., Eaton, W. W., Bienvenu Iii, O. J., Costa Jr, P. T., & Nestadt, G. (2002). Influences of parenting on normal personality traits. *Psychiatry Research, 111*(1), 55–64.

Reuter, M., Roth, S., Holve, K., & Hennig, J. (2006). Identification of first candidate genes for creativity: a pilot study. *Brain Research, 1069*(1), 190–197.

Richter, J. O. R., Eisemann, M., & Richter, G. (2000). Temperament, character and perceived parental rearing in healthy adults: Two related concepts? *Psychopathology, 33*(1), 36–42.

Ritchie, T., & Noble, E. P. (2003). Association of seven polymorphisms of the D2 dopamine receptor gene with brain receptor-binding characteristics. *Neurochemical Research, 28*(1), 73–82.

Rodrigues, S. M., Saslow, L. R., Garcia, N., John, O. P., & Keltner, D. (2009). Oxytocin receptor genetic variation relates to empathy and stress reactivity in humans. *Proceedings of the National Academy of Sciences, 106*(50), 21437–21441.

Rouff, J. H., Sussman, R. W., & Strube, M. J. (2005). Personality traits in captive lion-tailed macaques (Macaca silenus). *American Journal of Primatology, 67*(2), 177–198.

Rybakowski, F., Slopien, A., Dmitrzak-Weglarz, M., Czerski, P., Rajewski, A., & Hauser, J. (2006). The 5-HT2A–1438 A/G and 5-HTTLPR polymorphisms and personality dimensions in adolescent anorexia nervosa: association study. *Neuropsychobiology, 53*(1), 33–39.

Sabol, S. Z., Nelson, M. L., Fisher, C., Gunzerath, L., Brody, C. L., Hu, S., ..., & Hamer, D. H. (1999). A genetic association for cigarette smoking behavior. *Health Psychology, 18*(1), 7.

Sadahiro, R., Suzuki, A., Matsumoto, Y., Shibuya, N., Enokido, M., Kamata, M., ..., & Otani, K. (2011). Functional polymorphism of the GTP cyclohydrolase 1 gene affects the personality trait of novelty seeking in healthy subjects. *Neuroscience Letters, 503*(3), 220–223.

Sadahiro, R., Suzuki, A., Shibuya, N., Kamata, M., Matsumoto, Y., Goto, K., & Otani, K. (2010). Association study between a functional polymorphism of tyrosine hydroxylase gene promoter and personality traits in healthy subjects. *Behavioural Brain Research, 208*(1), 209–212.

Samochowiec, J., Rybakowski, F., Czerski, P., Zakrzewska, M., Stepień, G., Pełka-Wysiecka, J., ..., & Hauser, J. (2001). Polymorphisms in the dopamine, serotonin, and norepinephrine transporter genes and their relationship to temperamental dimensions measured by the Temperament and Character Inventory in healthy volunteers. *Neuropsychobiology, 43*(4), 248–253.

Samochowiec, J., Kucharska-Mazur, J., Rybakowski, F., Ostapowicz, A., Horodnicki, J., Rozpara, M., ..., & Schmidt, L. G. (2002). Norepinephrine transporter polymorphism and personality trait of reward dependence in male alcoholics. *Pharmacopsychiatry,* 35(5), 195–196.

Saudino, K. J. (2005). Behavioral genetics and child temperament. *Journal of Developmental and Behavioral Pediatrics: JDBP, 26*(3), 214.

Schinka, J. A., Busch, R. M., & Robichaux-Keene, N. (2004). A meta-analysis of the association between the serotonin transporter gene polymorphism (5-HTTLPR) and trait anxiety. *Molecular Psychiatry, 9*(2), 197–202.

Schmidt, K. T., & Weinshenker, D. (2014). Adrenaline rush: the role of adrenergic receptors in stimulant-induced behaviors. *Molecular Pharmacology, 85*(4), 640–650.

Sen, S., Burmeister, M., & Ghosh, D. (2004). Meta-analysis of the association between a serotonin transporter promoter polymorphism (5-HTTLPR) and anxiety-related personality traits. *American Journal of Medical Genetics Part B: Neuropsychiatric Genetics, 127*(1), 85–89.

Serretti, A., Calati, R., Giegling, I., Hartmann, A. M., Möller, H. J., Colombo, C., & Rujescu, D. (2007). 5-HT 2A SNPs and the Temperament and Character Inventory. *Progress in Neuro-Psychopharmacology and Biological Psychiatry, 31*(6), 1275–1281.

Shibuya, N., Suzuki, A., Sadahiro, R., Kamata, M., Matsumoto, Y., Goto, K., ..., & Otani, K. (2010). Association study between a functional polymorphism of FK506-binding protein 51 (FKBP5) gene and personality traits in healthy subjects. *Neuroscience Letters, 485*(3), 194–197.

Shifman, S., Bhomra, A., Smiley, S., Wray, N. R., James, M. R., Martin, N. G., ..., & Flint, J. (2008). A whole genome association study of neuroticism using DNA pooling. *Molecular Psychiatry, 13*(3), 302–312.

Stein, M. B., Fallin, M. D., Schork, N. J., & Gelernter, J. (2005). COMT polymorphisms and anxiety-related personality traits. *Neuropsychopharmacology, 30*(11), 2092–2102.

Strobel, A., Lesch, K. P., Jatzke, S., Paetzold, F., & Brocke, B. (2003). Further evidence for a modulation of Novelty Seeking by DRD4 exon III, 5-HTTLPR, and COMT val/met variants. *Molecular Psychiatry, 8*(4), 371–372.

Suzuki, A., Matsumoto, Y., Oshino, S., Kamata, M., Goto, K., & Otani, K. (2008). Combination of the serotonin transporter and norepinephrine transporter gene promoter polymorphisms might influence harm avoidance and novelty seeking in healthy females. *Neuroscience Letters, 439*(1), 52–55.

Terracciano, A., Balaci, L., Thayer, J., Scally, M., Kokinos, S., Ferrucci, L., ..., & Costa, P. T. (2009). Variants of the serotonin transporter gene and NEO-PI-R Neuroticism: No association in the BLSA and SardiNIA samples. *American Journal of Medical Genetics Part B: Neuropsychiatric Genetics, 150*(8), 1070–1077.

Terracciano, A., Martin, B., Ansari, D., Tanaka, T., Ferrucci, L., Maudsley, S., ..., & Costa, P. T. (2010a). Plasma BDNF concentration, Val66Met genetic variant and depression-related personality traits. *Genes, Brain and Behavior, 9*(5), 512–518.

Terracciano, A., Sanna, S., Uda, M., Deiana, B., Usala, G., Busonero, F., ..., & Costa, P. T. (2010b). Genome-wide association scan for five major dimensions of personality. *Molecular Psychiatry, 15*(6), 647–656.

Terracciano, A., Esko, T., Sutin, A. R., De Moor, M. H. M., Meirelles, O., Zhu, G., ... & Uda, M. (2011). Meta-analysis of genome-wide association studies identifies common variants in CTNNA2 associated with excitement-seeking. *Translational Psychiatry, 1*(10), e49.

Thomas, D. (2010). Methods for investigating gene–environment interactions in candidate pathway and genome-wide association studies. *Annual Review of Public Health, 31*, 21.

Thompson, T. L., & Certain, M. E. (2005). Estrogen mediated inhibition of dopamine transport in the striatum: Regulation by Gα i/o. *European Journal of Pharmacology, 511*(2), 121–126.

Tost, H., Kolachana, B., Hakimi, S., Lemaitre, H., Verchinski, B. A., Mattay, V. S., ..., & Meyer-Lindenberg, A. (2010). A common allele in the oxytocin receptor gene (OXTR) impacts prosocial temperament and human hypothalamic-limbic structure and function. *Proceedings of the National Academy of Sciences, 107*(31), 13936–13941.

Tsai, S. J., Hong, C. J., Yu, Y. Y., & Chen, T. J. (2004). Association study of catechol-O-methyltransferase gene and dopamine D4 receptor gene polymorphisms and personality traits in healthy young chinese females. *Neuropsychobiology, 50*(2), 153–156.

Turri, M. G., Talbot, C. J., Radcliffe, R. A., Wehner, J. M., & Flint, J. (1999). High-resolution mapping of quantitative trait loci for emotionality in selected strains of mice. *Mammalian Genome, 10*(11), 1098–1101.

Ukkola, L. T., Onkamo, P., Raijas, P., Karma, K., & Järvelä, I. (2009). Musical aptitude is associated with AVPR1A-haplotypes. *PLoS One, 4*(5), e5534.

Vaidya, J. G., Grippo, A. J., Johnson, A. K., & Watson, D. (2004). A comparative developmental study of impulsivity in rats and humans: the role of reward sensitivity. *Annals of the New York Academy of Sciences, 1021*(1), 395–398.

van den Oord, E. J., Kuo, P. H., Hartmann, A. M., Webb, B. T., Möller, H. J., Hettema, J. M., ..., & Rujescu, D. (2008). Genomewide association analysis followed by a replication study implicates a novel candidate gene for neuroticism. *Archives of General Psychiatry, 65*(9), 1062–1071.

Van Gestel, S., & Van Broeckhoven, C. (2003). Genetics of personality: are we making progress? *Molecular Psychiatry, 8*(10), 840–852.

Van Heeringen, C., Audenaert, K., Van Laere, K., Dumont, F., Slegers, G., Mertens, J., & Dierckx, R. A. (2003). Prefrontal 5-HT 2a receptor binding index, hopelessness and personality characteristics in attempted suicide. *Journal of Affective Disorders, 74*(2), 149–158.

van Oers, K., & Mueller, J. C. (2010). Evolutionary genomics of animal personality. *Philosophical Transactions of the Royal Society B: Biological Sciences, 365*(1560), 3991–4000.

Vermeersch, H., T'sjoen, G., Kaufman, J. M., Vincke, J., & Van Houtte, M. (2010). Testosterone, androgen receptor gene CAG repeat length, mood and behaviour in adolescent males. *European Journal of Endocrinology, 163*(2), 319–328.

Villafuerte, S., Heitzeg, M. M., Foley, S., Yau, W. W., Majczenko, K., Zubieta, J. K., ..., & Burmeister, M. (2012). Impulsiveness and insula activation during reward anticipation are associated with genetic variants in GABRA2 in a family sample enriched for alcoholism. *Molecular Psychiatry, 17*(5), 511–519.

Vormfelde, S. V., Hoell, I., Tzvetkov, M., Jamrozinski, K., Sehrt, D., Brockmöller, J., & Leibing, E. (2006). Anxiety-and novelty seeking-related personality traits and serotonin transporter gene polymorphisms. *Journal of Psychiatric Research, 40*(6), 568–576.

Vriend, C., Pattij, T., van der Werf, Y. D., Voorn, P., Booij, J., Rutten, S., ..., & van den Heuvel, O. A. (2014). Depression and impulse control disorders in Parkinson's disease: two sides of the same coin? *Neuroscience & Biobehavioral Reviews, 38*, 60–71.

Walther, D. J., & Bader, M. (2003). A unique central tryptophan hydroxylase isoform. *Biochemical Pharmacology, 66*(9), 1673–1680.

Wang, E., Ding, Y. C., Flodman, P., Kidd, J. R., Kidd, K. K., Grady, D. L., ... & Moyzis, R. K. (2004). The genetic architecture of selection at the human dopamine receptor D4 (DRD4) gene locus. *American Journal of Human Genetics, 74*(5), 931–944.

Wang, K. S., Liu, X., Aragam, N., Mullersman, J. E., Jian, X., Pan, Y., & Liu, Y. (2012). Polymorphisms in ABLIM1 are associated with personality traits and alcohol dependence. *Journal of Molecular Neuroscience, 46*(2), 265–271.

Wersinger, S. R., Ginns, E. I., O'Carroll, A. M., Lolait, S. J., & Young III, W. S. (2002). Vasopressin V1b receptor knockout reduces aggressive behavior in male mice. *Molecular Psychiatry, 7*(9), 975–984.

Willis-Owen, S. A., Turri, M. G., Munafò, M. R., Surtees, P. G., Wainwright, N. W., Brixey, R. D., & Flint, J. (2005). The serotonin transporter length polymorphism, neuroticism, and depression: a comprehensive assessment of association. *Biological Psychiatry, 58*(6), 451–456.

Winham, S. J., & Biernacka, J. M. (2013). Gene–environment interactions in genome-wide association studies: current approaches and new directions. *Journal of Child Psychology and Psychiatry, 54*(10), 1120–1134.

Wray, N. R., Middeldorp, C. M., Birley, A. J., Gordon, S. D., Sullivan, P. F., Visscher, P. M., ... & Boomsma, D. I. (2008). Genome-wide linkage analysis of multiple measures of neuroticism of 2 large cohorts from Australia and the Netherlands. *Archives of General Psychiatry, 65*(6), 649–658.

Yamamori, H., Hashimoto, R., Ohi, K., Yasuda, Y., Fukumoto, M., Kasahara, E., ..., & Takeda, M. (2012). A promoter variant in the chitinase 3-like 1 gene is associated with serum YKL-40 level and personality trait. *Neuroscience Letters, 513*(2), 204–208.

Yang, J., Lee, M. S., Lee, S. H., Lee, B. C., Kim, S. H., Joe, S. H., ... & Ham, B. J. (2010). Association between tryptophan hydroxylase 2 polymorphism and anger-related personality traits among young Korean women. *Neuropsychobiology, 62*(3), 158–163.

Zhang, G., Goldblatt, J., & Lesouëf, P. (2009). The era of genome-wide association studies: opportunities and challenges for asthma genetics. *Journal of Human Genetics, 54*(11), 624–628.

Zohar, A. H., Dina, C., Rosolio, N., Osher, Y., Gritsenko, I., Bachner-Melman, R., ..., & Ebstein, R. P. (2003). Tridimensional personality questionnaire trait of harm avoidance (anxiety proneness) is linked to a locus on chromosome 8p21. *American Journal of Medical Genetics Part B: Neuropsychiatric Genetics, 117*(1), 66–69.

9

Behavioural Genetic Studies of Child and Adolescent Psychopathology

Elena Gindina and Darya Gaysina

The role of genetic and environmental factors in the development of psychopathology has been investigated for almost a century. However, most of the earlier research focused on the aetiology of psychopathology in adults. With the exception of severe psychiatric disorders, such as autism and Tourette syndrome, systematic behavioural genetic investigations of child and adolescent psychopathology (i.e., emotional and behavioural problems) are relatively recent (Rutter et al., 1990).

Results obtained from studies in adults cannot be directly generalized to children and adolescents (Caspi et al., 2008). Therefore, the need to investigate the aetiology of emotional and behavioural problems in children and adolescents cannot be overestimated. Firstly, epidemiological studies show that emotional and behavioural disorders in children have a relatively high prevalence (about 12%); varying from approximately 3% for attention deficit hyperactivity disorder (ADHD) to 9% for anxiety (Costello, Egger, & Angold, 2005). There is also evidence that prevalence of child and adolescent psychopathology has been increasing (Costello, Foley, & Angold, 2006). Secondly, emotional and behavioural problems are frequently comorbid: they can co-occur or one can precede another (Campbell, 1995; Hofstra, Van der Ende, & Verhulst, 2002; Laitinen-Krispijn, Van der Ende, Wierdsma, & Verhulst, 1999; Verhulst & van der Ende, 1993; 1995). Thirdly, emotional and behavioural problems can lead to negative outcomes, such as low academic achievement, parent–child conflict, peer rejection, antisocial behaviour and chronic somatic disorders (Angold et al., 1999; Caspi, Moffitt, Newman, & Silva, 1996; Gaysina et al., 2011; Hofstra, Van der Ende, & Verhulst, 2002; Patterson, DeBaryshe, & Ramsey, 1989).

One of the main methodological problems in investigating the nature of psychopathology is diagnosing and operationalizing the phenotypes. This problem stems from unclear boundaries between separate taxonomic categories, as well as from the absence of independent criteria that are used to validate different taxonomic categories (Taylor & Rutter, 2002). Nevertheless, recent decades have witnessed significant progress in developing the standardized diagnostic procedures in child psychopathology (Achenbach & McConaughy,1987; Angold, 2002; Simonoff et al., 1997; Verhulst & Van der Ende, 2006). Nowadays, two approaches – clinical and empirical – to defining child and adolescent psychopathology are used.

In the **clinical diagnostic approach**, psychopathologies are considered to be discrete categories defined on the basis of criteria proposed by experts. The criteria are represented in some classifications, such as the Diagnostic and Statistical Manual of Mental Disorders (DSM; American Psychiatric Association, 2000) and the International Classification of Diseases (ICD; World Health Organization, 1992; 1993). As a rule, standardized interviews, which aim to identify the presence or absence of a particular set of symptoms, are used to diagnose psychopathologies (Angold, 2002; Simonoff et al., 1997).

In the **empirical approach**, psychopathology symptoms are assessed on a continuous scale. This quantitative approach is based on the idea of a continuum from the norm to psychopathology, with the quantity of symptoms measuring the deviation from the norm. In population-based research, parental, teacher and self-reports are used to assess the degree to which certain behaviour is exhibited in a child. It has been shown that the most frequent psychological problems in children and adolescents can be grouped into eight clusters (syndromes): withdrawn/depressed, anxious/depressed, somatic complaints, social problems, thought problems, inattention and hyperactivity, delinquent behaviour and aggressive behaviour. These clusters can be combined into larger dimensions – internalizing and externalizing problems. Internalizing problems are characterized by withdrawal, somatic complaints, anxiety and depression. Externalizing problems are characterized by reactions oriented outwards, such as delinquent behaviour and aggression (Achenbach & McConaughy, 1987; Achenbach & Rescorla, 2001; Verhulst & Van der Ende, 2006).

In order to gain a comprehensive knowledge of child psychological development, behavioural genetic studies use both categorical and quantitative approaches (Ferdinand et al., 2004). Most common psychopathology symptoms, such as aggression, delinquency, thought problems, hyperactivity, anxiety, withdrawal and depression, represent continuous quantitative variations in behaviour in the population. Therefore, the quantitative approach has an advantage because it enables one to assess symptoms of psychopathology on a continuous scale instead of using a particular diagnostic threshold. Besides, the empirical approach provides more differentiated data on different symptoms within a particular psychiatric disorder (Hudziak, 2001).

Quantitative behavioural genetics allow us to study the effects of genetic and environmental factors on individual differences in developmental psychopathologies using twin, sibling or adoption methods. Molecular genetic methods allow the investigation of specific genetic factors influencing the development of psychopathology.

In **family studies**, the rate of disorder in first- (and, at times, second-) degree relatives of probands is compared to the rate in the control group (healthy individuals or those with a different psychiatric diagnosis). Some family studies analyse the psychopathology risk in relatives of children with psychiatric disorders, while others assess the psychopathology risk in children of parents suffering from psychiatric disorders. All child disorders bear evidence of familiality that can be associated both with common genetic factors and environmental factors that contribute to similarities among family members (i.e., common, shared environmental factors). Family risk is usually expressed in the form of relative risk and assessed as a ratio between the probability of developing the disorder for the relative of a proband and this probability for the relative of a non-affected person. For child and adolescent depression, the relative risk is from 2 to 4 (Rice & Thapar, 2009); for ADHD risk is from 4 to 9 (Chen et al., 2008; Faraone, Biederman, & Friedman, 2000); and for autism risk is over 50 (Rutter, 2000). Since familiality may stem from both genetic and common environmental factors, the data from twin or adoption samples are required in order to estimate the relative contribution of genetic and environmental factors.

Twin studies (see Box 1) of developmental psychopathology have shown that psychopathology in childhood and adolescence is influenced by both genetic and environmental factors. The highest heritability (up to 90%) has been found for autism and ADHD (Derks et al., 2006; Rutter, 2000). Moderate heritability (30–60%) has been found for depressive disorders, oppositional defiant disorder, conduct disorder without hyperactivity, anxiety disorders, eating disorders, substance use disorders and obsessive disorders (Klein & Pine, 2002; Klump et al., 2001; Price et al., 1985; Rhee et al., 2003; Rutter, Giller, & Hagell, 1998; Sullivan, Neale, & Kendler, 2000; van Grootheest et al., 2005). It is important to remember that heritability statistics are specific to the studied population, and therefore can vary as a function of sample, methods and other aspects of studies. Non-shared environment has been shown as important in practically all twin studies of child and adolescent psychopathology, whereas the effects of shared environment factors seem to be weaker. However, the majority of twin studies have low statistical power to identify common environmental effects, because these effects are assessed as a residual component of twin similarity (that cannot be explained by common genetic factors). Adoption studies have more power to estimate shared environmental effects.

Adoption studies are based on comparisons of adoptees with biological and adoptive (non-biological) relatives. Similarity between adopted away children and their biological relatives is taken as evidence of genetic influences, whilst their similarity with adoptive relatives provides evidence of postnatal shared environmental influences. Details of adoption study methods can be found elsewhere (e.g., Plomin et al., 2013). Adoption studies have also been applied to testing environmental effects of specific risk factors, such as rearing style in a family (see Chapter 2 for more details). Adoption studies have some limitations: adoptive families are on average more educated and have a higher socioeconomic status as compared with general population; placement to adoptive families may not be random (selective adoption placement); the role of prenatal environment is usually not taken into account (Stoolmiller, 1999). Nevertheless, strong convergence between results of twin and adoption studies validates the results obtained using both approaches.

Molecular genetic studies of psychopathology have been carried out worldwide for clinically diagnosed disorders (e.g., unipolar depression) as well as quantitative traits normally distributed in the general population (e.g., depressiveness). At present, the genetic association method is the most commonly used approach to study molecular genetic bases of complex behavioural traits. This method is particularly suitable for identifying multiple genetic variants that have small effects and that are frequently observed in the general population (Cordell & Clayton, 2005). Traditionally, in genetic association studies, the frequency of a specific genetic variant is compared between a group of people with a specific disorder (i.e., cases) and a group of people without this disorder (i.e., controls). If a genetic variant is more frequently present in the group with the disorder than in the control group, the given variant is considered to be related (associated) to the risk of this disorder. Two main genetic association methods are the candidate gene approach and genome-wide association study (GWAS) approach. Candidate gene association approach enables researchers to test hypotheses about the potential role of a specific gene in neurobiological processes involved in the development of a specific psychopathology. This approach has led to the identification of several genetic variants involved in the development of psychopathology (Jia, Kao, Kuo, & Zhao, 2011). Nevertheless, a large number of inconsistent results and a low replication rate of such studies have given rise to widespread criticism of this method as well as initiated the search for new approaches to genetic studies of complex disorders.

With the development of the HapMap international project and with the improved genotyping technologies, GWASs have become widely used. Usually hundreds of thousands of bi-allelic single nucleotide polymorphisms (SNPs), located across the entire genome, are examined in GWAS. Due to multiple testing (that leads to an increase of type I error), the level of statistical significance for GWAS is set at $p \leq 5.0 \times 10^{-8}$ (Shyn & Hamilton, 2010).

To date, GWASs of many developmental psychopathologies, including autism, ADHD and child anxiety have been conducted. Overall, only very few associations of genetic variants with these disorders have been replicated in independent studies. Similar to candidate gene association studies, possible reasons for the lack of replication include: low statistical power associated with small sample sizes; phenotypic heterogeneity; and sample differences. In addition, technical characteristics of genotyping methods may also contribute to non-replication. Most GWASs have focused on common (rather than rare) genetic variants, such as common SNPs. However, rare deletions and insertions in the genome have also been studied in some GWASs. For example, the effects of rare copy number variants (CNVs) have been shown in a number of studies (Elia et al., 2012; Hawi et al., 2015; Jarick et al., 2014; Levy et al., 2011). More recently, high-throughput sequencing methods have been employed to study rare point mutations across the whole genome (Iossifov et al., 2014; Wang et al., 2013; Yuen et al., 2015). It is increasingly evident that the genetic architecture of developmental psychopathologies involves both common and rare variants and their impact on hundreds of genes.

Below, we present the review of behavioural genetic studies of developmental psychopathology and discuss such important issues as the nature of comorbidity of psychopathologies, sex and age differences in heritability, and the role of gene–environment interplay.

Review of behavioural genetic studies of developmental psychopathology

Specific developmental psychopathologies

Autism spectrum disorder (ASD). The first twin study of autism, conducted in 1977, showed a high heritability of this disorder (Rutter, 2000). Further twin studies supported this initial result, with MZ concordance ranging from 36% to 91%, and DZ concordance from 0% to 24%, leading to heritability estimates of about 90% (Bailey et al., 1995; Ritvo et al., 1985; Steffenburg et al., 1989). A very low concordance for DZ twin pairs as compared to MZ twin pairs indicates a possible effect of gene–gene interactive processes (i.e., epistasis). Studies also demonstrated that autism should be considered as representing a wide spectrum of phenotypes (i.e., autistic spectrum disorder, ASD). About 90% of MZ twin pairs and 10% of DZ twin pairs are concordant for some cognitive and social difficulties that are qualitatively similar to symptomatology associated with autism, but expressed to a lesser degree. There is evidence for common aetiology across the whole autistic spectrum, from the normal variation to the extreme end (Robinson et al., 2010). However, the results of twin studies showed that the triad of impairments in ASD (impairment of social interaction; impairments in communication; restricted interests and repetitive behaviours) is genetically heterogeneous (Ronald et al., 2006).

More than 100 associations of common genetic variants with ASD have been found, however, only a few associations have been consistently replicated (for review, see Xu et al., 2012). Several rare CNVs have also been implicated in ASD, with approximately 10% of all cases of autism associated with such rare mutations (Levy et al., 2011). Recent studies using whole genome sequencing have identified rare inherited and de novo point mutations for ASD (Iossifov et al., 2014; Yuen et al., 2015).

Attention deficit hyperactivity disorder (ADHD). ADHD includes three subtypes: attention deficit, hyperactivity, and a combination of attention deficit with hyperactivity. Both twin and adoption studies have consistently found high heritability of ADHD (up to 100%), and a negligible effect of common environment, irrespective of methodology, sex or age of participants, and other characteristics of the studies. Interestingly, data obtained from parent, but not teacher, reports, showed that MZ twin correlations are more than twice as high as DZ twin correlations. This could be due to a number of reasons: a possible presence of the contrast effect, where parents rate DZ twins more differently than they really are; the sibling interaction effect, where siblings behave more differently to contrast each other – an effect to which parents are particularly sensitive; or the increased reported twin similarity, where teachers tend to rate twins in the same class more similar than they really are (Alberts-Corush, Firestone, & Goodman, 1986; Derks, Hudziak, & Boomsma, 2009; Kovas et al., 2007; Sprich et al., 2000).

Similarly to autism, the same aetiology underlies the whole spectrum of ADHD symptoms (Levy et al., 1997; Price et al., 2001; Willcutt, Pennington, & DeFries, 2000). However, there is some evidence that heritability is slightly lower in the group of those with extreme hyperactivity but without attention deficit problems (Gjone, Stevenson, & Sundet, 1996).

A meta-analysis of a large number of candidate gene studies showed that the dopamine transporter gene (*SLC6A3*) and the dopamine D4 receptor gene (*DRD4*) were reliably associated with ADHD. However, the effects of the discovered genetic variants account for only a small percent of ADHD phenotypic variation (Li et al., 2006). Another meta-analysis that included results of several GWASs of ADHD has failed to find statistically significant associations of any genetic variants with ADHD (Neale et al., 2010). A more recent GWAS, that used a quantitative approach to attention deficit and hyperactivity, has also failed to find any reliable significant associations (Ebejer et al., 2013). Recent molecular genetic investigations that focused on rare deletions and duplications have implicated the role of the *PARK2* gene (involved in the aetiology of Parkinson's disease; Jarick et al., 2014); as well as genes related to the metabotropic glutamate receptor (Elia et al., 2012). It is possible that the search for reliable genetic effects is complicated by the presence of gene–gene and/or gene–environment interactions. For example, some authors reported that genetic influences on ADHD manifest only under the impact of specific environmental factors (Brookes et al., 2006; Kahn, Khoury, Nichols, & Lanphear, 2003; Seeger, Schloss, Schmidt, Ruter-Jungfleisch, & Henn, 2004). A recently created database integrating the results of all existing molecular genetic studies of ADHD represents a valuable resource for further research of this psychopathology (Elia et al., 2012).

Antisocial behaviour. Twin studies of antisocial behaviour, which is characterized by aggression, delinquency and oppositional behaviour, have shown that genetic factors account for approximately 45% of the variance in this trait in childhood and adolescence (Burt, 2002). Antisocial behaviour is a heterogeneous phenotype; some authors differentiate an *aggressive disorder* that manifests itself in early periods (childhood) and a *delinquent disorder* that manifests itself in later periods (adolescence) (Moffitt, 1993). Ratings of aggression correlate with a diagnosis of oppositional disorder, whereas scores on the delinquent behaviour scale correlate with a diagnosis of conduct disorder (Hudziak, Copeland, Stanger, & Wadsworth, 2004). It is possible that the two types of antisocial behaviour disorders have different aetiologies. Genetic factors account for about 75% of phenotypic variance in aggression, and these genetic influences are stable across ages (Bartels et al., 2003; Burt, 2009; Eley, Lichtenstein, & Stevenson, 1999; Schmitz, Fulker, & Mrazek, 1995; van Beijsterveldt, Bartels, Hudziak, & Boomsma, 2003; van den Oord, Boomsma, & Verhulst, 1994). Delinquent behaviour has been found to be less heritable (30–40%), with less stability in genetic influences across ages (Bartels et al., 2003; Burt & Neiderhiser, 2009; Eley, Lichtenstein, & Moffitt, 2003; Eley, Lichtenstein & Stevenson, 1999). However, different types of conduct disorder may themselves have different aetiologies. For example, conduct disorder which is comorbid with ADHD has been found to be more heritable than other forms of conduct disorder (Thapar, Harrington, & McGuffin, 2001).

Adoption studies have also demonstrated the influence of genetic factors on the development of antisocial behaviour. Studies of externalizing problems in adopted children aged 10–15 years, and their biological and non-biological siblings, showed that externalizing behaviour was mainly explained by genetic factors (65%). Similarly to the results of twin studies, heritability of aggressive behaviour (70%) was higher than heritability of delinquent behaviour (39%; van den Oord et al., 1994). However, in the Colorado Adoption Project, which also compared biological and non-biological siblings, heritability of aggression (24–49%) and delinquent behaviour (17–36%), assessed by parental and teachers' reports, was lower (Deater-Deckard & Plomin, 1999).

A recently published systematic meta-analysis of 185 studies of 31 candidate genes showed the absence of reliable associations between analysed genetic variants and aggression. Sub-group analysis, which took into account the severity of disorder and sample characteristics, did not find any significant associations (Vassos, Collier, & Fazel, 2014). No reliable associations were revealed in a single GWAS of aggression in adults (Tielbeek et al., 2012).

Depression and anxiety. Twin studies of internalizing problems, such as depression, anxiety and withdrawal, are sufficiently consistent and show a moderate heritability (from 30 to 40%; Bartels et al., 2003; Deater-Deckard, Reiss, Hetherington, & Plomin, 1997; Gjone & Stevenson, 1997; Schmitz et al., 1995) that does not change across the period from 7 to 12 years of age (Boomsma et al., 2005). In contrast, adoption studies did not reveal any considerable contribution of genetic factors to the variance of internalizing symptoms (Eley et al., 1998; van den Oord et al., 1994).

Studies of child and adolescent depression have reported a wide range of heritability estimates (11–72%; Rice & Thapar, 2009). This inconsistency in heritability estimates can be due to sampling differences as well as the use of different methods to diagnose depressive symptomatology (for example, reports of different respondents). Cross-rater agreement on depression ratings is low (Verhulst, Dekker, & van der Ende, 1997), which probably reflects the fact that different respondents assess different aspects of depressiveness (Boomsma, van Beijsterveldt, & Hudziak, 2005; Eaves et al., 1997; Rice, Harold, & Thapar, 2002; Thapar & McGuffin, 1995). Another possible explanation for differences in heritability estimates is differences in aetiology of depression at different ages: genetic factors seem to be more important in adolescence than in childhood (Eley & Stevenson, 1999; Rice, Harold, & Thapar, 2002; Thapar & McGuffin, 1995). An increase in genetic influences with age may be partially attributed to active and reactive gene–environment correlations associated with an increase in the number of dependent stressful life events during adolescence (Rice et al., 2003). Such a combined action of genetic and environmental factors is included in the component of heritability in most traditional behavioural genetic analyses like twin studies (Rutter, 2006).

There is evidence of discontinuity in the aetiology of depressive symptoms with various degrees of severity. In some studies of adolescent depression, the effect of shared environment was stronger in the groups with severe depression (extreme groups) than in the groups with a normal range of depressive symptoms (Eley, 1997; Rende, Slomkowski, Stocker, Fulker, & Plomin, 1992; Rice, Harold, & Thapar, 2002).

A high aetiological heterogeneity has been found across anxiety disorders in children (Eaves et al., 1997; Eley et al., 2003a; Legrand, McGue, & Iacono, 1999; Rice, van den Bree, & Thapar, 2004; Thapar & McGuffin, 1995). For example, individual differences in separation anxiety were mainly due to the effect of shared environment (Eaves et al., 1997; Eley et al., 2003a; Silberg, Rutter, Neale, & Eaves, 2001), whereas individual differences in generalized anxiety disorder were mainly explained by genetic factors (Silberg et al., 2001).

To date, no genetic variants that contribute to heritability of depression, depressiveness or anxiety have been reliably identified in molecular genetic studies (Hek et al., 2013; Shi et al., 2011; Trzaskowski et al., 2013).

Comorbidity of developmental psychopathologies

Different emotional and behavioural problems are often comorbid (Angold, Costello, & Erkanli, 1999; Caron & Rutter, 1991; Rutter et al., 1997). Mechanisms of this co-occurrence remain poorly understood. Twin studies that use multivariate structural modelling can clarify to what extent genetic and environmental factors contribute to comorbidity of two or more disorders.

Twin research has shown that comorbidity of ADHD and ASD, observed in 30% of cases, with ADHD often preceding ASD, stems largely from overlapping genetic factors (Lichtenstein et al., 2010; Ronald, Larsson, Anckarsater, & Lichtenstein, 2011; Ronald et al., 2008; Taylor & Rutter, 2002; Willcutt et al., 2012). ADHD is also often comorbid with other behavioural disorders (30–50% of cases), with this comorbidity largely explained by common genetic vulnerability (Biederman, Newcorn, & Sprich, 1991; Burt, Krueger, McGue, & Iacono, 2001; Dick et al., 2005).

Common genetic factors have also been implicated in comorbidity between conduct disorders and substance use and abuse, observed in approximately 50% of cases (Reebye, Moretti, & Lessard, 1995), which is also explained by common genetic factors (Button et al., 2006; Miles, van den Bree, & Pickens, 2002; Silberg, Rutter, D'Onofrio, & Eaves, 2003). Comorbidity of delinquent and aggressive behaviour symptoms is also largely explained by common genetic factors. For instance, in a study of 1481 pairs of 12-year old twins from the Netherlands Twin Register, 80% of covariance between aggressive and delinquent behaviour was explained by genetic factors (Bartels et al., 2003).

Common genetic factors have also been implicated in comorbidity between depression and anxiety (from 20% to 70% of cases) in childhood and adulthood, with anxiety disorders often preceding depressive disorders (Angold, Costello, & Erkanli, 1999; Eley & Stevenson, 1999; Goodwin, Fergusson, & Horwood, 2004; Rice et al., 2004; Silberg et al., 2001; Thapar & McGuffin, 1997).

More research is needed in order to clarify the extent of comorbidity and its aetiology between internalizing and externalizing problems as studies to date have produced inconsistent results. For example, the comorbidity rate for conduct disorder and depression varies from 21% to 83% (Angold & Costello, 1993). One study demonstrated that 50% of covariance in antisocial behaviour and depression in adolescence was explained by genetic factors (O'Connor et al., 1998b). Another study of covariance between extreme scores on internalizing and externalizing scales (51–58%) in 5–15 year-old twins showed that this covariance was mainly explained by shared environment, especially in the younger age group (5–9 years old; Gjone & Stevenson, 1997).

Sex differences in the aetiology of developmental psychopathologies

Epidemiological studies have provided evidence for sex differences in the incidence of emotional and behavioural problems (Rutter, Caspi, & Moffitt,

2003). Boys more frequently exhibit externalizing problems, ADHD, aggression and conduct disorders, whereas girls have higher incidence of anxiety and depression (Cohen et al., 1993; Vierikko et al., 2003; Zahn-Waxler et al., 1996). Behavioural genetic research can: 1) assess the relative contribution of genetic and environmental factors to phenotypic variance for both sexes (quantitative differences); and 2) find out whether the same genetic and environment factors affect trait variance in boys and girls (qualitative differences).

The results of behavioural genetic studies of sex differences in the aetiology of emotional and behavioural problems are rather inconsistent (Cohen et al., 1993; Eley & Stevenson, 1999). For example, one study showed a higher heritability of hyperactivity in girls (Kuo et al., 2004; van den Oord et al., 1996), whilst the other demonstrated a higher heritability in boys based on fathers' interviews, and no differences in heritability between boys and girls based on parental and teachers' questionnaires (Eaves et al., 1997). Studies of externalizing problems also showed discrepancies in results: no significant sex differences were identified in some studies (van den Oord, Verhulst, & Boomsma, 1996), whereas other studies showed larger genetic influences in boys (Silberg et al., 1994; van den Oord et al., 1994). Studies of aggression also revealed inconsistent results, with several studies finding higher heritability in boys (Hudziak et al., 2003; Miles & Carey, 1997), whilst other studies indicated higher heritability in girls (Bartels et al., 2003; Kuo et al., 2004; Vierikko et al., 2003), or no significant sex differences (Eley, Lichtenstein, & Stevenson, 1999). A similar inconsistency was observed for depression: results based on parental assessments did not show any sex differences (Nadder et al., 2002; Rice, Harold, & Thapar, 2002); whereas results based on teachers' ratings showed higher heritability for girls (Happonen et al., 2002); and results based on self-report indicated higher heritability for boys (Eley & Stevenson, 1999; Rice et al., 2002).

Several studies reported qualitative sex differences in the aetiology of ADHD (Derks et al., 2007; Saudino, Ronald, & Plomin, 2005; Vierikko, Pulkkinen, Kaprio, & Rose, 2004), oppositional defiant disorder (Derks et al., 2007) and aggression (Vierikko et al., 2003), suggesting that, to some extent, different genetic factors are involved in the aetiology of this condition in girls and boys.

Inconsistent results can be explained by differences in sample characteristics (for instance, differences in subjects' age), respondents and diagnostic tools (Eley & Stevenson, 1999; Silberg et al., 1994). Moreover, many studies are likely to be underpowered to detect small sex differences (Eaves et al., 1997).

Stability and change in developmental psychopathology

Longitudinal genetically informative studies can examine the role of genetic and environmental factors in stability and variability of psychopathology. Longitudinal methods can help to address whether heritability changes across development and the reasons for any such changes. For example, increasing heritability at a certain age may result from expression of new genes or from an amplification of existing genetic influences. Longitudinal studies can also provide insights into the aetiology underlying phenotypic stability, which can result from the contribution of the same genes or from the same environmental factors acting at different ages.

Most longitudinal studies of externalizing problems have shown that genetic factors are the main source of symptomatic stability, whereas nonshared environmental factors and new genetic influences contribute to changes in symptomatology (O'Connor et al., 1998a; Schmitz et al., 1995). For example, for ADHD, genetic factors explain the largest part of age stability, whereas both genetic and environmental factors contribute to change (Larsson, Larsson, & Lichtenstein, 2004; Rietveld et al., 2004). Another externalizing problem – aggressive behaviour – is a stable and heritable phenotype. This stability from childhood to adolescence is mainly due to genetic factors. Genetic effects for stability of aggression seem stronger for boys, while common environmental effects are more important for stability of aggressive behaviour in girls (van Beijsterveldt et al., 2003). Compared with aggressive behaviour, delinquency is less stable and heritable, with its relative stability largely explained by environmental factors (Eley, Lichtenstein, & Moffitt, 2003).

Some evidence suggests that environmental factors, rather than genetic factors, influence age-related stability of depression and anxiety. Although in some studies, age-related stability of depression and anxiety were attributed to genetic factors (Lau & Eley, 2006; O'Connor et al., 1998b); in the Dutch Twins Registry, stability of internalizing symptoms was explained by both genetic (43%) and common environmental (47%) factors (Bartels et al., 2004). In a number of twin studies, stability of internalizing symptoms was mainly due to common environmental factors (Schmitz, Fulker, & Mrazek, 1995; Scourfield et al., 2003). The results of longitudinal studies of adopted children attributed stability of internalizing problems to nonshared environmental factors (van der Valk, Verhulst, Neale, & Boomsma, 1998).

Gene–environment interplay in developmental psychopathology

Genetic and environmental factors co-act in their influence on the development of individual differences in behavioural traits. This phenomenon is discussed in detail in Chapter 2, and here we present some examples of gene–environment interplay in development of psychopathology in children and adolescents.

Gene–environment correlation. Genes may increase (or decrease) the probability of exposure to environmental risk factors, through the processes of passive, evocative and active gene–environment correlations (Plomin,

DeFries, & McClearn, 2008; Rutter et al., 1997). For example, in a study of children of twins, a passive gene–environment correlation partially explained the impact of parental depression on behaviour problems in adolescents (Silberg, Maes, & Eaves, 2010). A study of families, where one of the parents was suffering from a mental disorder, found that parental psychopathology (partially genetically driven) was associated with arguments and conflicts in the family, and, in turn, with child psychopathology (Quinton & Rutter, 1988).

Several studies of adopted children indicated the presence of an evocative gene–environment correlation, in which genetically determined characteristics of children provoked harsh parental discipline (Ge et al., 1996; O'Connor et al., 1998a). Behavioural genetic studies also indicated the presence of an active gene–environment correlation (Jaffee & Price, 2007; Plomin, DeFries, & Loehlin, 1977). For example, a study of MZ twins showed that within a pair, the twin with a higher level of antisocial behaviour in early adolescence was more likely to have deviant friends in late adolescence (Burt, 2009). Similarly, another twin study demonstrated that antisocial behaviour in boys was a predictor of selection of deviant peers (the so-called "social selection"). This study also found the environmental impact of deviant peers on antisocial behaviour (so-called "social conditioning"). It was shown that social conditioning was more important during childhood, while social selection became important during adolescence (Kendler, Jacobson, Myers, & Eaves, 2008).

Research has shown that gene–environment correlation plays a role in the development of antisocial behaviour and depression in children and adolescents. For example, genetic factors accounted for 20–30% of the variability in corporal punishment (Jaffee et al., 2004; Wade & Kendler, 2000) and harsh discipline in the family (Button, Lau, Maughan, & Eley, 2008; Pike, McGuire, Hetherington, Reiss, & Plomin, 1996). In addition, the covariance of harsh discipline and children's antisocial behaviour was mainly attributed to shared genetic factors (Button et al., 2008; Jaffee et al., 2004; Neiderhiser, Reiss, Hetherington, & Plomin, 1999). These results are consistent with the "child effect" hypothesis, according to which child behaviour provokes harsh parenting. Adoption studies also showed that children with high genetic risk of antisocial behaviour received harsh discipline from adoptive parents more often than children with low genetic risk (Ge et al., 1996; O'Connor et al., 1998a; Riggins-Caspers, Cadoret, Knutson, & Langbehn, 2003).

Studies have demonstrated that active and evocative gene–environment correlation with stressful life events plays an important role in the development of depression in adolescence (Silberg, Rutter, Neale, & Eaves, 2001; Thapar, Harold, & McGuffin, 1998). A reactive gene–environment correlation was found for negative family environment (anger/hostility, coercion, conflict) in the development of depressive symptoms in children (Pike et al., 1996). A three-year longitudinal study indicated the presence of gene–environment correlation with parental conflict in the development of depression in children (Neiderhiser, Reiss, Hetherington, & Plomin, 1999).

Gene–environment interaction. This type of gene–environment co-action describes a situation where genetic factors influence behaviour through the control of sensitivity to environmental factors (Moffitt, Caspi, & Rutter, 2005; Rutter, 2006; Rutter, Moffitt, & Caspi, 2006). This phenomenon should be taken into account because genetic effects may remain undetected in the absence of a specific environmental factor.

A study of adopted children showed that harsh discipline was associated with antisocial behaviour in adulthood only when children had a genetic predisposition to antisocial behaviour, whereas this parenting style had a negligible effect on children without a genetic predisposition to antisocial behaviour (Riggins-Caspers, Cadoret, Knutson, & Langbehn, 2003). In the Colorado Adoption Project, the interaction between genetic risk and parental divorce had a significant impact on behaviour and emotional problems in adopted children at 12 years of age. In this study, the index of genetic risk was the level of biological parents' negative emotionality. The results showed that in families without divorce, the level of problems in adopted children was independent from genetic risk, whereas in the group of adopted children whose adoptive parents went through a divorce, significant correlations of genetic predisposition and behaviour problems were identified (Connor et al., 2003). Similar evidence was found for conduct problems using the twin design: the effect of physical maltreatment on risk for conduct problems was strongest among those at high genetic risk as compared to those at low genetic risk (Jaffee et al., 2005).

Gene–environment interaction was also found in twin studies that investigated changes in heritability as a function of environmental risk exposure. The results of these studies are inconsistent and indicate both higher and lower heritability estimates in the presence of environmental risk. For example, studies showed lower heritability of internalizing problems in adolescents in the presence of environmental risk factors, such as parental divorce (Hicks et al., 2009); and the frequency of conflicts and arguments between parents and children (Feinberg et al., 2007; Rice, Harold, Shelton, & Thapar, 2006). Such lower heritability estimates were found for aggression and antisocial behaviour in families with lower levels of emotional warmth and higher levels of family dysfunction (Button, Thapar, & McGuffin, 2005; Rowe, Almeida, & Jacobson, 1999;). Another study found lower heritability of behaviour problems in boys from families with low socio-economic status (Tuvblad, Grann, & Lichtenstein, 2006). However, it was reported that heritability of depressive symptoms was higher in adolescents who experienced more stressful life events (Hicks et al., 2009; Lau & Eley, 2008). Heritability of antisocial behaviour in children was higher in families with lower level of parental warmth (Feinberg et al., 2007).

New developments in molecular genetics make it possible to test the effects of environmental risk factors in interaction with specific genetic variants. For example, the research has suggested that the effect of maltreatment in childhood on antisocial behaviour might be stronger in people with a less active form of the monoamine oxidase A (*MAOA*) gene (Caspi et al., 2002). Research has also demonstrated that the serotonin transporter gene (*5-HTT*) interacts with stressful life events in the development of depression: a less active form of the *5-HTT* gene (*S*-allele) increases the risk of depression in response to stressful life events more than twofold (Caspi et al., 2003). These findings have been replicated in a number of studies (Karg, Burmeister, Shedden, & Sen, 2011; Kim-Cohen et al., 2006). Studies have also shown the interaction between dopamine system genes (e.g., *DRD4, SLC6A3*) and adverse environmental conditions in the development of antisocial behaviour (Bakermans-Kranenburg & van Ijzendoorn, 2011). For example, three-year old children were more likely to exhibit externalizing problems, if they experienced lower maternal sensitivity at 10 months and were carriers of the *7R*-allele of *DRD4* gene (Bakermans-Kranenburg & van Ijzendoorn, 2006).

Disentangling environmental and genetic effects on developmental psychopathology. The observed associations between environmental factors and child psychological outcomes can results from a direct (causal) effect of environment, or alternatively, they can stem from gene–environment correlations. Behavioural genetic methods (adoption studies, twin studies and the studies of families with children born as a result of in vitro fertilization, IVF) allow us to investigate the effects of specific environmental factors while controlling for the effects of genes (Jaffee, Strait, & Odgers, 2012; Rice et al., 2010). Moreover, the combination of a classical adoption study (adoption at birth) and IVF (adoption at conception) allows for investigation of the effect of prenatal environmental factors (e.g., mother's smoking during pregnancy) on the psychological development of the child (e.g., antisocial behaviour) while controlling for the effects of common genetic factors and factors of postnatal environment (Gaysina et al., 2013).

Several **adoption studies** have been used to assess the extent to which passive gene–environment correlation is involved in the development of antisocial behaviour. Association between parental divorce (a known risk factor of antisocial behaviour in children; Emery, 1999) and antisocial behaviour in children was present in both biological and non-biological (adoptive) families suggesting a non-genetic (environmental) effect of parental divorce (Burt, Barnes, McGue, & Iacono, 2008; O'Connor, Caspi, DeFries, & Plomin, 2000). Maternal depression has also been identified as an environmental risk factor for the development of depression and antisocial behaviour in children. The risk of depression and antisocial behaviour in adolescents whose mothers suffered from depression was higher in both biological and non-biological families, indicating an environmental mechanism of this link

(Tully, Iacono, & McGue, 2008). Another study of externalizing problems in children found that depressive symptoms in adoptive mothers were associated with behaviour problems in children, even when depressive symptoms in biological mothers were taken into account (Pemberton et al., 2010; Kerr et al., 2013). Paternal depression also had an environmental effect on externalizing problems in children (Pemberton et al., 2010).

Another useful approach to studying environmental influences is **the adoption at conception design**, in which families with children born through assisted reproduction (in vitro fertilization, IVF) are studied, with biological relatives compared with non-biological relatives (mother–child and father–child). IVF allows for different degrees of relatedness between parents and children: both parents are genetically related to the child; only one parent is genetically related to the child; or both parents are genetically unrelated to the child (Rice et al., 2010). Using this method to study the effect of maternal depression on internalizing problems in children, it was shown that the correlation between the mother's and the child's internalizing problems was not higher for biological relatives as compared to non-biological relatives, indicating that intergenerational effect of maternal depression was environmental rather than genetic (Lewis, Rice, Harold, Collishaw, & Thapar, 2011). Similarly, parental and child antisocial behaviour scores were significantly correlated in both biologically related and biologically unrelated parent–child dyads (Harold et al., 2011).

The twin method can be used to disentangle environmental from genetic effects in several ways. **Multivariate analysis of twin data** allows us to identify genetic and environmental effects when the environmental variables are included in the model as a phenotype. Using this approach, the effect of maternal negativity towards the child on adolescent depression was shown to be both genetically and environmentally mediated (Pike et al., 1996). Other studies showed an environmental effect of stressful life events (e.g., abuse, neglect) on child psychopathology (Rice, Harold, & Thapar, 2003). Children who have experienced maltreatment demonstrated higher levels of antisocial behaviour than children who have not experienced maltreatment (Wilson, Stover, & Berkowitz, 2009), with very little evidence for contribution of genetic factors in variability of maltreatment (6–7%) (Jaffee, Caspi, Moffitt, & Taylor, 2004; Schulz-Heik et al., 2009). The absence of genetic effects on maltreatment rules out the effect of gene–environment correlation in this link and is consistent with a causal (environmental) effect of maltreatment on child antisocial behaviour. In another study, the effect of neighbourhood deprivation on behaviour of 2 -year old twins was tested: children from more deprived neighbourhoods were at higher risk for emotional and behaviour problems regardless of their genetic predisposition. Common environment explained 20% of the variance in children's mental health, and neighbourhood deprivation explained 5% of the shared environmental effects (Caspi, Taylor, Moffitt, & Plomin, 2000).

The children of twins design, which includes adult twins and their children, makes it possible to study how differences between twins (when they are themselves parents) are related to differences in the behaviour of cousins. This method is based on the fact that the mother who is an MZ twin is genetically as close to her own children as to her nephews. In DZ twin pairs, a mother is genetically closer to her children than to her nephews. This model helps to estimate to what extent differences between cousins experiencing different environmental influences (e.g., different parenting styles) are attributed to the genotype or environment of their parents (who are twins).

Data from the classical twin design, when twins and their parents are studied, can be complemented by data on twins when they are parents and their children. Such expanded models can be used to identify even small passive and evocative gene–environment correlations. The study of Australian Children of Twins found that children whose parents were divorced typically had more conduct problems, and this effect was independent from whether their parents, aunts or uncles were MZ or DZ twins. Therefore, the effect of divorce was not associated with genetic factors or common environmental factors of parental families, i.e., parental divorce had an environmentally mediated effect on conduct problems in children (D'Onofrio et al., 2005). Further research is needed to understand whether this effect is related to divorce itself or to the related factors (e.g., family characteristics, such as excessive conflict, or a low socioeconomic status). Another study of the same sample found an environmentally mediated effect of parental antisocial behaviour on the development of conduct problems in boys. Association between behaviour problems of children and parents was significant, and the effects did not differ between the families of MZ and DZ twins (D'Onofrio et al., 2007). In the same sample, adolescent motherhood was an environmental risk factor for mental disorders in children: children born to adolescent mothers manifested more behaviour problems than their siblings born when their mothers reached adult age (Harden et al., 2007). In addition, a study of children of twins provided evidence for significant environmentally mediated effects of parental depression on behaviour problems in adolescents (Silberg, Maes, & Eaves, 2010).

The longitudinal twins and parents design estimates the magnitude of genetic and environmental pathways, through which non-twin parental behaviour is associated with twin offspring behaviour. For example, a longitudinal twins and parents study provided evidence for an environmentally mediated effect of maltreatment (i.e., neglect, violence, inconsistent discipline) on adolescents' antisocial behaviour over and above any genetically mediated effects. Maltreatment directly accounted for approximately 30% of shared environmental effects (that explained ~40% of the total phenotypic variance) on adolescents' antisocial behaviour (Eaves, Prom, & Silberg, 2010).

Finally, environmentally mediated effects of putative risk factors can be studied using the **discordant MZ twins design.** For example, maternal negativity was found to have an environmentally mediated effect on the development of externalizing problems in children (Caspi et al., 2004). Within MZ pairs, the twin receiving more maternal negativity and less warmth had more antisocial behaviour problems as assessed by parental and teachers' reports. Maternal attitude when the child was 5 years old predicted child antisocial behaviour at the age of 7. In another study, it was shown that harsh discipline had an environmentally mediated effect on the development of children's conduct problems. MZ twins who experienced more harsh discipline also showed more conduct problems. Although a child effect was present (indicating an evocative gene–environment correlation), harsh discipline aggravated children's antisocial behaviour (Asbury, 2003).

Conclusions

Findings of behavioural genetic studies indicate that genetic factors have an important influence on the development of psychopathology in children and adolescents. Genetic factors can also be responsible for the ontogenetic (life course) stability of emotional and behavioural problems. It has been shown that common genetic factors are largely responsible for the observed comorbidity between symptoms of: ADHD and autism; conduct disorders and ADHD; anxiety and depression; and aggression and delinquency. The role of genetic and environmental factors in the development of emotional and behaviour problems may depend on age and sex. To date, molecular genetic studies of emotional and behaviour problems in childhood and adolescence have focused on common genetic variants. A search for rare mutations in the genome has just begun and represents a promising direction. These studies have already led to the identification of several rare deletions and duplications associated with ASD and ADHD. Rapid development of high-throughput DNA technologies, such as genome-wide sequencing, will lead to new large-scale studies of molecular genetic underpinning of developmental psychopathologies. The paths from DNA to manifestation of individual characteristics are complex and involve action, interaction and correlation of many genes and environmental factors. The results of the research reviewed in this chapter show that genetically informative designs are valuable for disentangling environmental and genetic effects to further our understanding of the complex processes involved in developmental psychopathology.

References

Achenbach, T. M., & McConaughy, S. H. (1987). *Empirically based assessment of child and adolescent psychopathology: Practical applications*. Sage Publications, Inc.

Achenbach, T. M., & Rescorla, L. A. (2001). ASEBA School Age Forms and Profiles. Burlington, Vt.: ASEBA.

Alberts-Corush, J., Firestone, P., & Goodman, J. T. (1986). Attention and impulsivity characteristics of the biological and adoptive parents of hyperactive and normal control children. *American Journal of Orthopsychiatry, 56*(3), 413–423.

American Psychiatric Association (2000). Diagnostic and statistical manual of mental disorders: DSM-IV-TR: American Psychiatric Publishing, Inc.

Angold, A. (2002). Diagnostic interviews with parents and children. In M. J. Rutter & E. Taylor (Ed.), *Child and adolescent psychiatry* (pp. 32–51). Oxford, UK: Blackwell Publisher.

Angold, A., & Costello, E. J. (1993). Depressive Comorbidity in Children and Adolescents – Empirical, Theoretical, and Methodological Issues. *American Journal of Psychiatry, 150*(12), 1779–1791.

Angold, A., Costello, E. J., & Erkanli, A. (1999). Comorbidity. *Journal of Child Psychology and Psychiatry and Allied Disciplines, 40*(1), 57–87.

Angold, A., Costello, E. J., Farmer, E. M. Z., Burns, B. J., & Erkanli, A. (1999). Impaired but undiagnosed. *Journal of the American Academy of Child and Adolescent Psychiatry, 38*(2), 129–137.

Asbury, K., Dunn, J. F., Pike, A., & Plomin, R. (2003). Nonshared environmental influences on individual differences in early behavioral development: A monozygotic twin differences study. *Child Development, 74*(3), 933–943.

Bailey, A., Lecouteur, A., Gottesman, I., Bolton, P., Simonoff, E., Yuzda, E., & Rutter, M. (1995). Autism as a Strongly Genetic Disorder – Evidence from a British Twin Study. *Psychological Medicine, 25*(1), 63–77.

Bakermans-Kranenburg, M. J., & van Ijzendoorn, M. H. (2006). Gene–environment interaction of the dopamine D4 receptor (DRD4) and observed maternal insensitivity predicting externalizing behaviour in preschoolers. *Developmental Psychobiology, 48*(5), 406–409.

Bakermans-Kranenburg, M. J., & van Ijzendoorn, M. H. (2011). Differential susceptibility to rearing environment depending on dopamine-related genes: New evidence and a meta-analysis. *Development and Psychopathology, 23*(1), 39–52.

Bartels, M., Hudziak, J. J., Boomsma, D. I., Rietveld, M. J. H., Van Beijsterveldt, T. C. E. M., & Van den Oord, E. J. C. G. (2003). A study of parent ratings of internalizing and externalizing problem behaviour in 12-year-old twins. *Journal of the American Academy of Child and Adolescent Psychiatry, 42*(11), 1351–1359.

Bartels, M., van den Oord, E. J., Hudziak, J. J., Rietveld, M. J., van Beijsterveldt, C. E., & Boomsma, D. I. (2004). Genetic and environmental mechanisms underlying stability and change in problem behaviours at ages 3, 7, 10, and 12. *Developmental Psychology, 40*(5), 852–867.

Biederman, J., Newcorn, J., & Sprich, S. (1991). Comorbidity of Attention-Deficit Hyperactivity Disorder with Conduct, Depressive, Anxiety, and Other Disorders. *American Journal of Psychiatry, 148*(5), 564–577.

Boomsma, D. I., van Beijsterveldt, C. E. M., & Hudziak, J. J. (2005). Genetic and environmental influences on Anxious/Depression during childhood: a study from the Netherlands Twin Register. *Genes, Brain and Behaviour, 4*(8), 466–481.

Brookes, K. J., Mill, J., Guindalini, C., Curran, S., Xu, X. H., Knight, J., ... Asherson, P. (2006). A common haplotype of the dopamine transporter gene associated with attention-deficit/hyperactivity disorder and interacting with maternal use of alcohol during pregnancy. *Archives of General Psychiatry, 63*(1), 74–81.

Burt, S. A. (2009). Rethinking Environmental Contributions to Child and Adolescent Psychopathology: A Meta-Analysis of Shared Environmental Influences. *Psychological Bulletin, 135*(4), 608–637.

Burt, S. A., Barnes, A. R., McGue, M., & Iacono, W. G. (2008). Parental Divorce and Adolescent Delinquency: Ruling Out the Impact of Common Genes. *Developmental Psychology, 44*(6), 1668–1677.

Burt, S. A., Krueger, R. F., McGue, M., & Iacono, W. G. (2001). Sources of covariation among attention-deficit/hyperactivity disorder, oppositional defiant disorder, and conduct disorder: The importance of shared environment. *Journal of Abnormal Psychology, 110*(4), 516–525.

Burt, S. A., & Neiderhiser, J. M. (2009). Aggressive Versus Nonaggressive Antisocial Behaviour: Distinctive Etiological Moderation by Age. *Developmental Psychology, 45*(4), 1164–1176.

Button, T., Thapar, A., & McGuffin, P. (2005). Relationship between antisocial behaviour, attention-deficit hyperactivity disorder and maternal prenatal smoking. *British Journal of Psychiatry, 187*(2), 155–160.

Button, T. M., Lau, J. Y., Maughan, B., & Eley, T. C. (2008). Parental punitive discipline, negative life events and gene–environment interplay in the development of externalizing behaviour. *Psychological Medicine, 38*(1), 29–39.

Button, T. M. M., Hewitt, J. K., Rhee, S. H., Young, S. E., Corley, R. P., & Stallings, M. C. (2006). Examination of the causes of covariation between conduct disorder symptoms and vulnerability to drug dependence. *Twin Research and Human Genetics, 9*(1), 38–45.

Campbell, S. B. (1995). Behaviour problems in preschool children: a review of recent research. *Journal of Child Psychology and Psychiatry, 36*(1), 113–149.

Caron, C., & Rutter, M. (1991). Comorbidity in child psychopathology: Concepts, issues and research strategies. *Journal of Child Psychology and Psychiatry, 32*(7), 1063–1080.

Caspi, A., Langley, K., Milne, B., Moffitt, T. E., O'Donovan, M., Owen, M. J., ... Thapar, A. (2008). A replicated molecular genetic basis for subtyping antisocial behaviour in children with attention-deficit/hyperactivity disorder. *Archives of General Psychiatry, 65*(2), 203–210.

Caspi, A., McClay, J., Moffitt, T. E., Mill, J., Martin, J., Craig, I. W., ... Poulton, R. (2002). Role of genotype in the cycle of violence in maltreated children. *Science, 297*(5582), 851–854.

Caspi, A., Moffitt, T. E., Morgan, J., Rutter, M., Taylor, A., Arseneault, L., ... Polo-Tomas, M. (2004). Maternal expressed emotion predicts children's antisocial behaviour problems: Using monozygotic-twin differences to identify environmental effects on behavioural development. *Developmental Psychology, 40*(2), 149–161.

Caspi, A., Moffitt, T. E., Newman, D. L., & Silva, P. A. (1996). Behavioural observations at age 3 years predict adult psychiatric disorders. Longitudinal evidence from a birth cohort. *Archives of General Psychiatry, 53*(11), 1033–1039.

Caspi, A., Sugden, K., Moffitt, T. E., Taylor, A., Craig, I. W., Harrington, H., ... Poulton, R. (2003). Influence of life stress on depression: Moderation by a polymorphism in the 5-HTT gene. *Science, 301*(5631), 386–389.

Caspi, A., Taylor, A., Moffitt, T. E., & Plomin, R. (2000). Neighborhood deprivation affects children's mental health: Environmental risks identified in a genetic design. *Psychological Science, 11*(4), 338–342.

Chen, W., Zhou, K. X., Sham, P., Franke, B., Kuntsi, J., Campbell, D., ... Asherson, P. (2008). DSM-IV Combined Type ADHD Shows Familial Association With Sibling Trait Scores: A Sampling Strategy For QTL Linkage. *American Journal of Medical Genetics Part B-Neuropsychiatric Genetics, 147B*(8), 1450–1460.

Cohen, P., Cohen, J., Kasen, S., Velez, C. N., Hartmark, C., Johnson, J., ... Streuning, E. L. (1993). An Epidemiologic-Study of Disorders in Late Childhood and Adolescence .1. Age-Specific and Gender-Specific Prevalence. *Journal of Child Psychology and Psychiatry and Allied Disciplines, 34*(6), 851–867.

O'Connor, T. G., Caspi, A., DeFries, J. C., & Plomin, R. (2003). Genotype–environment interaction in children's adjustment to parental separation. *Journal of Child Psychology and Psychiatry, 44*(6), 849–856.

Cordell, H. J., & Clayton, D. G. (2005). Genetic association studies. *Lancet, 366*(9491), 1121–1131.

Costello, E. J., Egger, H., & Angold, A. (2005). 10-year research update review: the epidemiology of child and adolescent psychiatric disorders: I. Methods and public health burden. *Journal of the American Academy of Child & Adolescent Psychiatry, 44*(10), 972–986.

Costello, E. J., Foley, D. L., & Angold, A. (2006). 10-year research update review: the epidemiology of child and adolescent psychiatric disorders: II. Developmental epidemiology. *Journal of the American Academy of Child & Adolescent Psychiatry, 45*(1), 8–25.

D'Onofrio, B. M., Slutske, W. S., Turkheimer, E., Emery, R. E., Harden, K. P., Heath, A. C., ... Martin, N. G. (2007). Intergenerational transmission of childhood conduct problems – A children of twins study. *Archives of General Psychiatry, 64*(7), 820–829.

D'Onofrio, B. M., Turkheimer, E., Emery, R. E., Slutske, W. S., Heath, A. C., Madden, P. A., & Martin, N. G. (2005). A genetically informed study of marital instability and its association with offspring psychopathology. *Journal of Abnormal Psychology, 114*(4), 570–586.

Deater-Deckard, K., & Plomin, R. (1999). An adoption study of the etiology of teacher and parent reports of externalizing behaviour problems in middle childhood. *Child Development, 70*(1), 144–154.

Deater-Deckard, K., Reiss, D., Hetherington, E. M., & Plomin, R. (1997). Dimensions and disorders of adolescent adjustment: a quantitative genetic analysis of unselected samples and selected extremes. *Journal of Child Psychology and Psychiatry, 38*(5), 515–525.

Deater-Deckard, K., Reiss, D., Hetherington, E. M., & Plomin, R. (1997). Dimensions and disorders of adolescent adjustment: A quantitative genetic analysis of unselected samples and selected extremes. *Journal of Child Psychology and Psychiatry, 38*(5), 515–525.

Derks, E. M., Dolan, C. V., Hudziak, J. J., Neale, M. C., & Boomsma, D. I. (2007). Assessment and etiology of attention deficit hyperactivity disorder and oppositional defiant disorder in boys and girls. *Behavior genetics, 37*(4), 559–566.

Derks, E. M., Hudziak, J. J., & Boomsma, D. I. (2009). Genetics of ADHD, hyperactivity, and attention problems. In *Handbook of behavior genetics* (pp. 361–378). Springer New York.

Derks, E. M., Hudziak, J. J., Van Beijsterveldt, C. E. M., Dolan, C. V., & Boomsma, D. I. (2006). Genetic analyses of maternal and teacher ratings on attention problems in 7-year-old Dutch twins. *Behavior Genetics, 36*(6), 833–844.

Dick, D. M., Viken, R. J., Kaprio, J., Pulkkinen, L., & Rose, R. J. (2005). Understanding the covariation among childhood externalizing symptoms: Genetic and environmental influences on conduct disorder, attention deficit hyperactivity disorder, and oppositional defiant disorder symptoms. *Journal of Abnormal Child Psychology, 33*(2), 219–229.

Eaves, L., Silberg, J., & Erkanli, A. (2003). Resolving multiple epigenetic pathways to adolescent depression. *Journal of Child Psychology and Psychiatry and Allied Disciplines, 44*(7), 1006–1014.

Eaves, L. J., Prom, E. C., & Silberg, J. L. (2010). The Mediating Effect of Parental Neglect on Adolescent and Young Adult Anti-Sociality: A Longitudinal Study of Twins and Their Parents. *Behavior Genetics, 40*(4), 425–437.

Eaves, L. J., Silberg, J. L., Meyer, J. M., Maes, H. H., Simonoff, E., Pickles, A., ... & Hewitt, J. K. (1997). Genetics and developmental psychopathology: 2. The main effects of genes and environment on behavioral problems in the Virginia Twin Study of Adolescent Behavioral Development. *Journal of Child Psychology and Psychiatry, 38*(8), 965–980.

Ebejer, J. L., Duffy, D. L., van der Werf, J., Wright, M. J., Montgomery, G., Gillespie, N. A., ..., & Medland, S. E. (2013). Genome-wide association study of inattention and hyperactivity–impulsivity measured as quantitative traits. *Twin Research and Human Genetics, 16*(02), 560–574.

Eley, T. C. (1997). Depressive symptoms in children and adolescents: Etiological links between normality and abnormality: A research note. *Journal of Child Psychology and Psychiatry and Allied Disciplines, 38*(7), 861–865.

Eley, T. C., Bolton, D., O'Connor, T. G., Perrin, S., Smith, P., & Plomin, R. (2003a). A twin study of anxiety-related behaviours in pre-school children. *Journal of Child Psychology and Psychiatry and Allied Disciplines, 44*(7), 945–960.

Eley, T. C., Deater-Deckard, K., Fombonne, E., Fulker, D. W., & Plomin, R. (1998). An adoption study of depressive symptoms in middle childhood. *Journal of Child Psychology and Psychiatry and Allied Disciplines, 39*(3), 337–345.

Eley, T. C., Lichtenstein, P., & Moffitt, T. E. (2003b). A longitudinal behavioural genetic analysis of the etiology of aggressive and nonaggressive antisocial behaviour. *Development and Psychopathology, 15*(2), 383–402.

Eley, T. C., Lichtenstein, P., & Stevenson, J. (1999). Sex differences in the etiology of aggressive and nonaggressive antisocial behaviour: Results from two twin studies. *Child Development, 70*(1), 155–168.

Eley, T. C., & Stevenson, J. (1999). Exploring the covariation between anxiety and depression symptoms: A genetic analysis of the effects of age and sex. *Journal of Child Psychology and Psychiatry and Allied Disciplines, 40*(8), 1273–1282.

Eley, T. C., Sugden, K., Corsico, A., Gregory, A. M., Sham, P., McGuffin, P., ..., & Craig, I. W. (2004). Gene–environment interaction analysis of serotonin system markers with adolescent depression. *Molecular Psychiatry, 9*(10), 908–915.

Elia, J., Glessner, J. T., Wang, K., Takahashi, N., Shtir, C. J., Hadley, D., ..., & Hakonarson, H. (2012). Genome-wide copy number variation study associates metabotropic glutamate receptor gene networks with attention deficit hyperactivity disorder. *Nature Genetics, 44*(1), 78–84.

Emery, R. E. (1999). *Marriage, divorce, and children's adjustment* (Vol. 14): SAGE Publications, Incorporated.

Faraone, S. V., Biederman, J., & Friedman, D. (2000). Validity of DSM-IV subtypes of attention-deficit/hyperactivity disorder: a family study perspective. *Journal of the American Academy of Child & Adolescent Psychiatry, 39*(3), 300–307.

Feinberg, M. E., Button, T. M., Neiderhiser, J. M., Reiss, D., & Hetherington, E. M. (2007). Parenting and adolescent antisocial behaviour and depression: evidence of genotype x parenting environment interaction. *Archives of General Psychiatry, 64*(4), 457–465.

Ferdinand, R. F., Visser, J. H., Hoogerheide, K. N., van der Ende, J., Kasius, M. C., Koot, H. M., & Verhulst, F. C. (2004). Improving estimation of the prognosis of childhood psychopathology; combination of DSM-III-R/DISC diagnoses and CBCL scores. *Journal of Child Psychology and Psychiatry, 45*(3), 599–608.

Gaysina, D., Fergusson, D. M., Leve, L. D., Horwood, J., Reiss, D., Shaw, D. S., ... & Harold, G. T. (2013). Maternal smoking during pregnancy and offspring conduct problems: evidence from 3 independent genetically sensitive research designs. *JAMA Psychiatry, 70*(9), 956–963.

Gaysina, D., Pierce, M., Richards, M., Hotopf, M., Kuh, D., & Hardy, R. (2011). Association between adolescent emotional problems and metabolic syndrome: The modifying effect of C-reactive protein gene (CRP) polymorphisms. *Brain, Behaviour, and Immunity, 25*(4), 750–758.

Ge, X., Conger, R. D., Cadoret, R. J., Neiderhiser, J. M., Yates, W., Troughton, E., & Stewart, M. A. (1996). The developmental interface between nature and nurture: a mutual influence model of child antisocial behaviour and parent behaviours. *Developmental Psychology, 32*(4), 574–589.

Gjone, H., & Stevenson, J. (1997). The association between internalizing and externalizing behaviour in childhood and early adolescence: genetic of environmental common influences? *Journal of Abnormal Child Psychology, 25*(4), 277–286.

Gjone, H., Stevenson, J., & Sundet, J. M. (1996). Genetic influence on parent-reported attention-related problems in a Norwegian general population twin sample. *Journal of the American Academy of Child & Adolescent Psychiatry, 35*(5), 588–598.

Gjone, H., Stevenson, J., Sundet, J. M., & Eilertsen, D. E. (1996). Changes in heritability across increasing levels of behaviour problems in young twins. *Behavior Genetics, 26*(4), 419–426.

Goodwin, R. D., Fergusson, D. M., & Horwood, L. J. (2004). Early anxious/withdrawn behaviours predict later internalising disorders. *Journal of Child Psychology and Psychiatry, 45*(4), 874–883.

Happonen, M., Pulkkinen, L., Kaprio, J., Van der Meere, J., Viken, R. J., & Rose, R. J. (2002). The heritability of depressive symptoms: multiple informants and multiple measures. *Journal of Child Psychology and Psychiatry, 43*(4), 471–479.

Harden, K. P., Lynch, S. K., Turkheimer, E., Emery, R. E., D'Onofrio, B. M., Slutske, W. S., ... Martin, N. G. (2007). A behaviour genetic investigation of adolescent motherhood and offspring mental health problems. *Journal of Abnormal Psychology, 116*(4), 667–683.

Harold, G. T., Rice, F., Hay, D. F., Boivin, J., van den Bree, M., & Thapar, A. (2011). Familial transmission of depression and antisocial behaviour symptoms: disentangling the contribution of inherited and environmental factors and testing the mediating role of parenting. *Psychological Medicine, 41*(6), 1175–1185.

Hawi, Z., Cummins, T. D. R., Tong, J., Johnson, B., Lau, R., Samarrai, W., & Bellgrove, M. A. (2015). The molecular genetic architecture of attention deficit hyperactivity disorder. *Molecular Psychiatry, 20*, 289–297.

Hek, K., Demirkan, A., Lahti, J., Terracciano, A., Teumer, A., Cornelis, M. C., ... Murabito, J. (2013). A genome-wide association study of depressive symptoms. *Biological Psychiatry, 73*(7), 667–678.

Hicks, B. M., South, S. C., Dirago, A. C., Iacono, W. G., & McGue, M. (2009). Environmental adversity and increasing genetic risk for externalizing disorders. *Archives of General Psychiatry, 66*(6), 640–648.

Hofstra, M. B., & Verhulst, F. C. (2000). Continuity and change of psychopathology from childhood into adulthood: a 14-year follow-up study. *Journal of the American Academy of Child & Adolescent Psychiatry, 39*(7), 850–858.

Hofstra, M. B., Van der Ende, J., & Verhulst, F. C. (2002). Pathways of self-reported problem behaviours from adolescence into adulthood. *American Journal of Psychiatry, 159*(3), 401–407.

Hudziak, J. J. (2001). The role of phenotypes (diagnoses) in genetic studies of attention-deficit/hyperactivity disorder and related child psychopathology. *Child and Adolescent Psychiatric Clinics of North America, 10*(2), 279–297.

Hudziak, J. J., Copeland, W., Stanger, C., & Wadsworth, M. (2004). Screening for DSM-IV externalizing disorders with the Child Behaviour Checklist: a receiver-operating characteristic analysis. *Journal of Child Psychology and Psychiatry, 45*(7), 1299–1307.

Hudziak, J. J., van Beijsterveldt, C. E., Bartels, M., Rietveld, M. J., Rettew, D. C., Derks, E. M., & Boomsma, D. I. (2003). Individual differences in aggression: genetic analyses by age, gender, and informant in 3-, 7-, and 10-year-old Dutch twins. *Behavior Genetics, 33*(5), 575–589.

Iossifov, I., O'Roak, B. J., Sanders, S. J., Ronemus, M., Krumm, N., Levy, D., ... & Wigler, M. (2014). The contribution of de novo coding mutations to autism spectrum disorder. *Nature, 515*(7526), 216–221.

Jaffee, S. R., Caspi, A., Moffitt, T. E., Dodge, K. A., Rutter, M., Taylor, A., & Tully, L. A. (2005). Nature X nurture: genetic vulnerabilities interact with physical maltreatment to promote conduct problems. *Developmental Psychopathology, 17*(1), 67–84.

Jaffee, S. R., Caspi, A., Moffitt, T. E., Polo-Tomas, M., Price, T. S., & Taylor, A. (2004). The limits of child effects: evidence for genetically mediated child effects on corporal punishment but not on physical maltreatment. *Developmental Psychology, 40*(6), 1047–1058.

Jaffee, S. R., Caspi, A., Moffitt, T. E., & Taylor, A. (2004). Physical maltreatment victim to antisocial child: evidence of an environmentally mediated process. *Journal of Abnormal Psychology, 113*(1), 44–55.

Jaffee, S. R., & Price, T. S. (2007). Gene–environment correlations: a review of the evidence and implications for prevention of mental illness. *Molecular Psychiatry, 12*(5), 432–442.

Jaffee, S. R., Strait, L. B., & Odgers, C. L. (2012). From correlates to causes: can quasi-experimental studies and statistical innovations bring us closer to identifying the causes of antisocial behavior? *Psychological Bulletin, 138*(2), 272–295.

Jarick, I., Volckmar, A. L., Putter, C., Pechlivanis, S., Nguyen, T. T., Dauvermann, M. R., ... Hinney, A. (2014). Genome-wide analysis of rare copy number variations reveals PARK2 as a candidate gene for attention-deficit/hyperactivity disorder. *Molecular Psychiatry, 19*(1), 115–121.

Jia, P., Kao, C. F., Kuo, P. H., & Zhao, Z. (2011). A comprehensive network and pathway analysis of candidate genes in major depressive disorder. *BMC Systems Biology, 5*(Suppl 3), S12.

Kahn, R. S., Khoury, J., Nichols, W. C., & Lanphear, B. P. (2003). Role of dopamine transporter genotype and maternal prenatal smoking in childhood hyperactive-impulsive, inattentive, and oppositional behaviors. *Journal of Pediatrics, 143*(1), 104–110.

Karg, K., Burmeister, M., Shedden, K., & Sen, S. (2011). The serotonin transporter promoter variant (5-HTTLPR), stress, and depression meta-analysis revisited: evidence of genetic moderation. *Archives of General Psychiatry, 68*(5), 444–454.

Kendler, K. S., Jacobson, K., Myers, J. M., & Eaves, L. J. (2008). A genetically informative developmental study of the relationship between conduct disorder and peer deviance in males. *Psychological Medicine, 38*(7), 1001–1011.

Kerr, D. C., Leve, L. D., Harold, G. T., Natsuaki, M. N., Neiderhiser, J. M., Shaw, D. S., & Reiss, D. (2013). Influences of biological and adoptive mothers' depression and antisocial behavior on adoptees' early behavior trajectories. *Journal of Abnormal Child Psychology, 41*(5), 723–734.

Kim-Cohen, J., Caspi, A., Taylor, A., Williams, B., Newcombe, R., Craig, I. W., & Moffitt, T. E. (2006). MAOA, maltreatment, and gene–environment interaction predicting children's mental health: new evidence and a meta-analysis. *Molecular Psychiatry, 11*(10), 903–913.

Kim-Cohen, J., Moffitt, T. E., Taylor, A., Pawlby, S. J., & Caspi, A. (2005). Maternal depression and children's antisocial behaviour: nature and nurture effects. *Archives of General Psychiatry, 62*(2), 173–181.

Klein, R. G., Pine, D. S. (2002). Anxiety disorders. In M. J. Rutter & E. Taylor (Ed.), *Child and Adolescent Psychiatry* (pp. 485–509). Oxford, UK: Blackwell Publishers.

Klump, K. L., Miller, K. B., Keel, P. K., McGue, M., & Iacono, W. G. (2001). Genetic and environmental influences on anorexia nervosa syndromes in a population–based twin sample. *Psychological Medicine, 31*, 737–740.

Kovas, Y., Haworth, C. M., Dale, P. S., Plomin, R., Weinberg, R. A., Thomson, J. M., & Fischer, K. W. (2007). The genetic and environmental origins of learning abilities and disabilities in the early school years. *Monographs of the Society for Research in Child Development, 72*(3), 1–144.

Kuo, P. H., Lin, C. C., Yang, H. J., Soong, W. T., & Chen, W. J. (2004). A twin study of competence and behavioural/emotional problems among adolescents in Taiwan. *Behavior Genetics, 34*(1), 63–74.

Laitinen-Krispijn, S., Van der Ende, J., Wierdsma, A. I., & Verhulst, F. C. (1999). Predicting Adolescent Mental Health Service Use in a Prospective Record-Linkage Study. *Journal of the American Academy of Child & Adolescent Psychiatry, 38*(9), 1073–1080.

Larsson, J. O., Larsson, H., & Lichtenstein, P. (2004). Genetic and environmental contributions to stability and change of ADHD symptoms between 8 and 13 years of age: a longitudinal twin study. *Journal of the American Academy of Child and Adolescent Psychiatry, 43*(10), 1267–1275.

Lau, J. Y., & Eley, T. C. (2006). Changes in genetic and environmental influences on depressive symptoms across adolescence and young adulthood. *British Journal of Psychiatry, 189*, 422–427.

Lau, J. Y., & Eley, T. C. (2008). Disentangling gene–environment correlations and interactions on adolescent depressive symptoms. *Journal of Child Psychology and Psychiatry, 49*(2), 142–150.

Laucht, M., Skowronek, M. H., Becker, K., Schmidt, M. H., Esser, G., Schulze, T. G., & Rietschel, M. (2007). Interacting effects of the dopamine transporter gene and psychosocial adversity on attention-deficit/hyperactivity disorder symptoms among 15-year-olds from a high-risk community sample. *Archives of General Psychiatry, 64*(5), 585–590.

Legrand, L. N., McGue, M., & Iacono, W. G. (1999). A twin study of state and trait anxiety in childhood and adolescence. *Journal of Child Psychology and Psychiatry, 40*(6), 953–958.

Levy, D., Ronemus, M., Yamrom, B., Lee, Y.-h., Leotta, A., Kendall, J., ... Ye, K. (2011). Rare de novo and transmitted copy-number variation in autistic spectrum disorders. *Neuron, 70*(5), 886–897.

Levy, F., Hay, D. A., McStephen, M., Wood, C., & Waldman, I. (1997). Attention-deficit hyperactivity disorder: a category or a continuum? Genetic analysis of a large-scale twin study. *Journal of the American Academy of Child and Adolescent Psychiatry, 36*(6), 737–744.

Lewis, G., Rice, F., Harold, G. T., Collishaw, S., & Thapar, A. (2011). Investigating environmental links between parent depression and child depressive/anxiety symptoms using an assisted conception design. *Journal of the American Academy of Child and Adolescent Psychiatry, 50*(5), 451–459.

Lichtenstein, P., Carlstrom, E., Rastam, M., Gillberg, C., & Anckarsater, H. (2010). The genetics of autism spectrum disorders and related neuropsychiatric disorders in childhood. *American Journal of Psychiatry, 167*(11), 1357–1363.

Miles, D. R., & Carey, G. (1997). Genetic and environmental architecture on human aggression. *Journal of Personality and Social Psychology, 72*(1), 207–217.

Miles, D. R., van den Bree, M., & Pickens, R. W. (2002). Sex differences in shared genetic and environmental influences between conduct disorder symptoms and marijuana use in adolescents. *American journal of medical genetics, 114*(2), 159–168.

Moffitt, T. E. (1993). Adolescence-limited and life-course-persistent antisocial behavior: a developmental taxonomy. *Psychological Review, 100*(4), 674–701.

Moffitt, T. E., Caspi, A., & Rutter, M. (2005). Strategy for investigating interactions between measured genes and measured environments. *Archives of General Psychiatry, 62*(5), 473–481.

Murray, K. T., & Sines, J. O. (1996). Parsing the genetic and nongenetic variance in children's depressive behavior. *Journal of Affective Disorders, 38*(1), 23–34.

Nadder, T. S., Rutter, M., Silberg, J. L., Maes, H. H., & Eaves, L. J. (2002). Genetic effects on the variation and covariation of attention deficit-hyperactivity disorder (ADHD) and oppositional-defiant disorder/conduct disorder (Odd/CD) symptomatologies across informant and occasion of measurement. *Psychological Medicine, 32*(1), 39–53.

Neale, B. M., Medland, S. E., Ripke, S., Asherson, P., Franke, B., Lesch, K.-P., ... Holmans, P. (2010). Meta-analysis of genome-wide association studies of attention-deficit/hyperactivity disorder. *Journal of the American Academy of Child & Adolescent Psychiatry, 49*(9), 884–897.

Neale, M. C., & Cardon, L. R. (1992). *Methodology for genetic studies of twins and families*. Springer.

Neiderhiser, J. M., Reiss, D., Hetherington, E. M., & Plomin, R. (1999). Relationships between parenting and adolescent adjustment over time: genetic and environmental contributions. *Developmental Psychology, 35*(3), 680–692.

O'Connor, T. G., Caspi, A., DeFries, J. C., & Plomin, R. (2000). Are associations between parental divorce and children's adjustment genetically mediated? An adoption study. *Developmental Psychology, 36*(4), 429–437.

O'Connor, T. G., Deater-Deckard, K., Fulker, D., Rutter, M., & Plomin, R. (1998a). Genotype-environment correlations in late childhood and early adolescence: antisocial behavioural problems and coercive parenting. *Developmental Psychology, 34*(5), 970–981.

O'Connor, T. G., Neiderhiser, J. M., Reiss, D., Hetherington, E. M., & Plomin, R. (1998b). Genetic contributions to continuity, change, and co-occurrence of antisocial and depressive symptoms in adolescence. *Journal of Child Psychology and Psychiatry, 39*(3), 323–336.

Patterson, G. R., DeBaryshe, B. D., & Ramsey, E. (1989). A developmental perspective on antisocial behaviour. *American Psychologist, 44*(2), 329–335.

Pemberton, C. K., Neiderhiser, J. M., Leve, L. D., Natsuaki, M. N., Shaw, D. S., Reiss, D., & Ge, X. (2010). Influence of parental depressive symptoms on adopted toddler behaviours: an emerging developmental cascade of genetic and environmental effects. *Developmental Psychopathology, 22*(4), 803–818.

Pike, A., McGuire, S., Hetherington, E. M., Reiss, D., & Plomin, R. (1996). Family environment and adolescent depressive symptoms and antisocial behaviour: a multivariate genetic analysis. *Developmental Psychology, 32*(4), 590–603.

Plomin, R., DeFries, J. C., & Loehlin, J. C. (1977). Genotype-environment interaction and correlation in the analysis of human behaviour. *Psychological Bulletin, 84*(2), 309–322.

Plomin, R., DeFries, J. C., Knopik, V. S., & Neiderhiser, J. M. (2013). *Behavioral genetics* (6th edition). New York: Worth.

Price, R. A., Kidd, K. K., Cohen, D. J., Pauls, D. L., & Leckman, J. F. (1985). A twin study of Tourette symdrome. *Archives of General Psychiatry, 42*, 815–820.

Price, T. S., Simonoff, E., Waldman, I., Asherson, P., & Plomin, R. (2001). Hyperactivity in preschool children is highly heritable. *Journal of the American Academy of Child & Adolescent Psychiatry, 40*(12), 1362–1364.

Quinton, D., & Rutter, M. (1988). *Parenting breakdown: The making and breaking of inter-generational links*. Avebury Aldershot, UK.

Reebye, P., Moretti, M. M., & Lessard, J. C. (1995). Conduct disorder and substance use disorder: Comorbidity in a clinical sample of preadolescents and adolescents. *The Canadian Journal of Psychiatry/La Revue canadienne de psychiatrie, 40*(6), 313–319.

Rende, R. D., Slomkowski, C. L., Stocker, C., Fulker, D. W., & Plomin, R. (1992). Genetic and environmental influences on maternal and sibling interaction in middle childhood: A sibling adoption study. *Developmental Psychology, 28*(3), 484–490.

Rhee, S., Hewitt, J. K., Young, S. E., Corley, R. P., Crowley, T. J., Stallings, M. C. (2003). Genetic and Environmental Influences on Substance Initiation, Use, and Problem Use in Adolescents. *Archives of General Psychiatry, 60*(12), 1256–1264.

Rice, F., Harold, G. T., Boivin, J., van den Bree, M., Hay, D. F., & Thapar, A. (2010). The links between prenatal stress and offspring development and psychopathology: disentangling environmental and inherited influences. *Psychological Medicine, 40*(2), 335–345.

Rice, F., Harold, G. T., Shelton, K. H., & Thapar, A. (2006). Family conflict interacts with genetic liability in predicting childhood and adolescent depression. *Journal of the American Academy of Child and Adolescent Psychiatry, 45*(7), 841–848.

Rice, F., Harold, G. T., & Thapar, A. (2002). Assessing the effects of age, sex and shared environment on the genetic aetiology of depression in childhood and adolescence. *Journal of Child Psychology and Psychiatry, 43*(8), 1039–1051.

Rice, F., Harold, G. T., & Thapar, A. (2003). Negative life events as an account of age-related differences in the genetic aetiology of depression in childhood and adolescence. *Journal of Child Psychology and Psychiatry, 44*(7), 977–987.

Rice, F., & Thapar, A. (2009). Depression and anxiety in childhood and adolescence: Developmental pathways, genes and environment. In Y. Kim (Ed.), *Handbook of behavioural genetics* (pp. 379–396). Springer.

Rice, F., van den Bree, M. B., & Thapar, A. (2004). A population-based study of anxiety as a precursor for depression in childhood and adolescence. *BMC Psychiatry, 4*, 43.

Rietveld, M. J., Hudziak, J. J., Bartels, M., van Beijsterveldt, C. E., & Boomsma, D. I. (2004). Heritability of attention problems in children: longitudinal results from a study of twins, age 3 to 12. *Journal of Child Psychology and Psychiatry, 45*(3), 577–588.

Riggins-Caspers, K. M., Cadoret, R. J., Knutson, J. F., & Langbehn, D. (2003). Biology-environment interaction and evocative biology-environment correlation:

contributions of harsh discipline and parental psychopathology to problem adolescent behaviours. *Behavior Genetics, 33*(3), 205–220.

Ritvo, E. R., Freeman, B. J., Mason-Brothers, A., Mo, A., & Ritvo, A. M. (1985). Concordance for the syndrome of autism in 40 pairs of afflicted twins. *American Journal of Psychiatry, 142*(1), 74–77.

Robinson, M., McLean, N. J., Oddy, W. H., Mattes, E., Bulsara, M., Li, J. H., ... Newnham, J. P. (2010). Smoking cessation in pregnancy and the risk of child behavioural problems: a longitudinal prospective cohort study. *Journal of Epidemiology and Community Health, 64*(7), 622–629.

Ronald, A., Happe, F., Bolton, P., Butcher, L. M., Price, T. S., Wheelwright, S., ... Plomin, R. (2006). Genetic heterogeneity between the three components of the autism spectrum: a twin study. *Journal of the American Academy of Child and Adolescent Psychiatry, 45*(6), 691–699.

Ronald, A., Larsson, H., Anckarsater, H., & Lichtenstein, P. (2011). A twin study of autism symptoms in Sweden. *Molecular Psychiatry, 16*(10), 1039–1047.

Ronald, A., Simonoff, E., Kuntsi, J., Asherson, P., & Plomin, R. (2008). Evidence for overlapping genetic influences on autistic and ADHD behaviours in a community twin sample. *Journal of Child Psychology and Psychiatry, 49*(5), 535–542.

Rowe, D. C., Almeida, D. M., & Jacobson, K. C. (1999). School context and genetic influences on aggression in adolescence. *Psychological Science, 10*(3), 277–280.

Rutter, M. (2000). Genetic studies of autism: from the 1970s into the millennium. *Journal of Abnormal Child Psychology, 28*(1), 3–14.

Rutter, M. (2006). *Genes and behaviour: Nature-nurture interplay explained*. UK: Blackwell Publishing.

Rutter, M., Caspi, A., & Moffitt, T. E. (2003). Using sex differences in psychopathology to study causal mechanisms: unifying issues and research strategies. *Journal of Child Psychology and Psychiatry, 44*(8), 1092–1115.

Rutter, M., Dunn, J., Plomin, R., Simonoff, E., Pickles, A., Maughan, B., ... Eaves, L. (1997). Integrating nature and nurture: Implications of person–environment correlations and interactions for developmental psychopathology. *Developmental Psychopathology, 9*(02), 335–364.

Rutter, M., Giller, H., & Hagell, A. (1998). *Antisocial behaviour by young people: The main messages from a major new review of the research*. Social Information Systems.

Rutter, M., Macdonald, H., Le Couteur, A., Harrington, R., Bolton, P., & Bailey, A. (1990). Genetic factors in child psychiatric disorders – II. Empirical findings. *Journal of Child Psychology and Psychiatry, 31*(1), 39–83.

Rutter, M., Moffitt, T. E., & Caspi, A. (2006). Gene–environment interplay and psychopathology: multiple varieties but real effects. *Journal of Child Psychology and Psychiatry, 47*(3–4), 226–261.

Saudino, K. J., Ronald, A., & Plomin, R. (2005). The etiology of behaviour problems in 7-year-old twins: substantial genetic influence and negligible shared environmental influence for parent ratings and ratings by same and different teachers. *Journal of Abnormal Child Psychology, 33*(1), 113–130.

Schmitz, S., Fulker, D. W., & Mrazek, D. A. (1995). Problem behaviour in early and middle childhood: an initial behaviour genetic analysis. *Journal of Child Psychology and Psychiatry, 36*(8), 1443–1458.

Schulz-Heik, R. J., Rhee, S. H., Silvern, L., Lessem, J. M., Haberstick, B. C., Hopfer, C., & Hewitt, J. K. (2009). Investigation of genetically mediated child effects on maltreatment. *Behavior Genetics, 39*, 265–276.

Scourfield, J., Rice, F., Thapar, A., Harold, G. T., Martin, N., & McGuffin, P. (2003). Depressive symptoms in children and adolescents: changing aetiological influences with development. *Journal of Child Psychology and Psychiatry, 44*(7), 968–976.

Seeger, G., Schloss, P., Schmidt, M. H., Rüter-Jungfleisch, A., & Henn, F. A. (2004). Gene–environment interaction in hyperkinetic conduct disorder (HD+ CD) as indicated by season of birth variations in dopamine receptor (DRD4) gene polymorphism. *Neuroscience Letters, 366*(3), 282–286.

Shi, J., Potash, J. B., Knowles, J. A., Weissman, M. M., Coryell, W., Scheftner, W. A., ... Levinson, D. F. (2011). Genome-wide association study of recurrent early-onset major depressive disorder. *Molecular Psychiatry, 16*(2), 193–201.

Shyn, S. I., & Hamilton, S. P. (2010). The genetics of major depression: moving beyond the monoamine hypothesis. *Psychiatric Clinics of North America, 33*(1), 125–140.

Silberg, J., Rutter, M., D'Onofrio, B., & Eaves, L. (2003). Genetic and environmental risk factors in adolescent substance use. *Journal of Child Psychology and Psychiatry, 44*(5), 664–676.

Silberg, J., Rutter, M., Neale, M., & Eaves, L. (2001). Genetic moderation of environmental risk for depression and anxiety in adolescent girls. *British Journal of Psychiatry, 179,* 116–121.

Silberg, J. L., Erickson, M. T., Meyer, J. M., Eaves, L. J., Rutter, M. L., & Hewitt, J. K. (1994). The application of structural equation modeling to maternal ratings of twins' behavioral and emotional problems. *Journal of Consulting and Clinical Psychology, 62*(3), 510.

Silberg, J. L., Maes, H., & Eaves, L. J. (2010). Genetic and environmental influences on the transmission of parental depression to children's depression and conduct disturbance: an extended Children of Twins study. *Journal of Child Psychology and Psychiatry, 51*(6), 734–744.

Simonoff, E., Pickles, A., Meyer, J. M., Silberg, J. L., Maes, H. H., Loeber, R., ... Eaves, L. J. (1997). The Virginia Twin Study of Adolescent Behavioural Development. Influences of age, sex, and impairment on rates of disorder. *Archives of General Psychiatry, 54*(9), 801–808.

Sprich, S., Biederman, J., Crawford, M. H., Mundy, E., & Faraone, S. V. (2000). Adoptive and biological families of children and adolescents with ADHD. *Journal of the American Academy of Child and Adolescent Psychiatry, 39*(11), 1432–1437.

Steffenburg, S., Gillberg, C., Hellgren, L., Andersson, L., Gillberg, I. C., Jakobsson, G., & Bohman, M. (1989). A twin study of autism in Denmark, Finland, Iceland, Norway and Sweden. *Journal of Child Psychology and Psychiatry, 30*(3), 405–416.

Stoolmiller, M. (1999). Implications of the restricted range of family environments for estimates of heritability and nonshared environment in behaviour-genetic adoption studies. *Psychological Bulletin, 125*(4), 392–409.

Sullivan, P. F., Neale, M. C., & Kendler, K. S. (2000). Genetic epidemiology of major depression: review and meta-analysis. *American Journal of Psychiatry, 157*(10), 1552–1562.

Taylor, E., & Rutter, M. (2002). Classification: conceptual issues and substantive findings. *Child and Adolescent Psychiatry, 4,* 3–17.

Thapar, A., Harold, G., & McGuffin, P. (1998). Life events and depressive symptoms in childhood – shared genes or shared adversity? A research note. *Journal of Child Psychology and Psychiatry, 39*(8), 1153–1158.

Thapar, A., Harrington, R., & McGuffin, P. (2001). Examining the comorbidity of ADHD-related behaviours and conduct problems using a twin study design. *British Journal of Psychiatry, 179*, 224–229.

Thapar, A., & McGuffin, P. (1994). A twin study of depressive symptoms in childhood. *British Journal of Psychiatry, 165*(2), 259–265.

Thapar, A., & McGuffin, P. (1995). Are anxiety symptoms in childhood heritable? *Journal of Child Psychology and Psychiatry, 36*(3), 439–447.

Thapar, A., & McGuffin, P. (1997). Anxiety and depressive symptoms in childhood – a genetic study of comorbidity. *Journal of Child Psychology and Psychiatry, 38*(6), 651–656.

Tielbeek, J. J., Medland, S. E., Benyamin, B., Byrne, E. M., Heath, A. C., Madden, P. A., ... Verweij, K. J. (2012). Unraveling the genetic etiology of adult antisocial behaviour: a genome-wide association study. *PloS One, 7*(10), e45086.

Trzaskowski, M., Eley, T. C., Davis, O. S., Docherty, S. J., Hanscombe, K. B., Meaburn, E. L., ... Plomin, R. (2013). First genome-wide association study on anxiety-related behaviours in childhood. *PloS One, 8*(4), e58676.

Tully, E. C., Iacono, W. G., & McGue, M. (2008). An adoption study of parental depression as an environmental liability for adolescent depression and childhood disruptive disorders. *American Journal of Psychiatry, 165*(9), 1148–1154.

Tuvblad, C., Grann, M., & Lichtenstein, P. (2006). Heritability for adolescent antisocial behaviour differs with socioeconomic status: gene–environment interaction. *Journal of Child Psychology and Psychiatry, 47*(7), 734–743.

van Beijsterveldt, C. E., Bartels, M., Hudziak, J. J., & Boomsma, D. I. (2003). Causes of stability of aggression from early childhood to adolescence: a longitudinal genetic analysis in Dutch twins. *Behavior Genetics, 33*(5), 591–605.

van Grootheest, D. S., Cath, D. C., Beekman, A. T., & Boomsma, D. I. (2005). Twin studies on obsessive–compulsive disorder: a review. *Twin Research and Human Genetics, 8*(05), 450–458.

van den Oord, E. J., Boomsma, D. I., & Verhulst, F. C. (1994). A study of problem behaviours in 10- to 15-year-old biologically related and unrelated international adoptees. *Behavior Genetics, 24*(3), 193–205.

van den Oord, E. J., Verhulst, F. C., & Boomsma, D. I. (1996). A genetic study of maternal and paternal ratings of problem behaviours in 3-year-old twins. *Journal of Abnormal Psychology, 105*(3), 349–357.

van der Valk, J. C., Verhulst, F. C., Neale, M. C., & Boomsma, D. I. (1998). Longitudinal genetic analysis of problem behaviours in biologically related and unrelated adoptees. *Behavior Genetics, 28*(5), 365–380.

Vassos, E., Collier, D. A., & Fazel, S. (2014). Systematic meta-analyses and field synopsis of genetic association studies of violence and aggression. *Molecular Psychiatry, 19*(4), 471–477.

Verhulst, F. C., & van der Ende, J. (2006). *Assessment scales in child and adolescent psychiatry*. CRC Press.

Verhulst, F. C., Dekker, M. C., & Ende, J. V. D. (1997). Parent, teacher and self-reports as predictors of signs of disturbance in adolescents: whose information carries the most weight?. *Acta Psychiatrica Scandinavica, 96*(1), 75–81.

Verhulst, F. C., & van der Ende, J. (1993). "Comorbidity" in an epidemiological sample: a longitudinal perspective. *Journal of Child Psychology and Psychiatry, 34*(5), 767–783.

Verhulst, F. C., & Van Der Ende, J. (1995). The eight-year stability of problem behavior in an epidemiologic sample. *Pediatric Research, 38*(4), 612–617.

Vierikko, E., Pulkkinen, L., Kaprio, J., & Rose, R. J. (2004). Genetic and environmental influences on the relationship between aggression and hyperactivity-impulsivity as rated by teachers and parents. *Twin Research, 7*(3), 261–274.
Vierikko, E., Pulkkinen, L., Kaprio, J., Viken, R., & Rose, R. J. (2003). Sex differences in genetic and environmental effects on aggression. *Aggressive Behaviour, 29*(1), 55–68.
Wade, T. D., & Kendler, K. S. (2000). The genetic epidemiology of parental discipline. *Psychological Medicine, 30*(6), 1303–1313.
Wang, M., Fan, X., Wang, T., & Wu, J. (2013). High-throughput sequencing of autism spectrum disorders comes of age. *Genetics Research, 95*(04), 121–129.
Willcutt, E. G., Pennington, B. F., & DeFries, J. C. (2000). Etiology of inattention and hyperactivity/impulsivity in a community sample of twins with learning difficulties. *Journal of Abnormal Child Psychology, 28*(2), 149–159.
Willcutt, E. G., Nigg, J. T., Pennington, B. F., Solanto, M. V., Rohde, L. A., Tannock, R., et al. (2012). Validity of DSM-IV attention deficit/hyperactivity disorder symptom dimensions and subtypes. *Journal of Abnormal Psychology, 121*, 991–1010.
Wilson, H. W., Stover, C. S., & Berkowitz, S. J. (2009). Research review: The relationship between childhood violence exposure and juvenile antisocial behaviour: a meta-analytic review. *Journal of Child Psychology and Psychiatry, 50*(7), 769–779.
World Health Organization (1992). *The ICD-10 classification of mental and behavioural disorders: clinical descriptions and diagnostic guidelines*: World Health Organization.
World Health Organization (1993). *The ICD-10 classification of mental and behavioural disorders: diagnostic criteria for research*: World Health Organization.
Xu, L.-M., Li, J.-R., Huang, Y., Zhao, M., Tang, X., & Wei, L. (2012). AutismKB: an evidence-based knowledgebase of autism genetics. *Nucleic Acids Research, 40*(D1), D1016-D1022.
Yuen, R. K., Thiruvahindrapuram, B., Merico, D., Walker, S., Tammimies, K., Hoang, N., ... & Scherer, S. W. (2015). Whole-genome sequencing of quartet families with autism spectrum disorder. *Nature Medicine, 21*(2), 185–191.
Zahn-Waxler, C., Schmitz, S., Fulker, D., Robinson, J., & Emde, R. (1996). Behaviour problems in 5-year-old monozygotic and dizygotic twins: Genetic and environmental influences, patterns of regulation, and internalization of control. *Developmental Psychopathology, 8*, 103–122.

10
Genetic Research on Sleep, Sleep Disturbances and Associated Difficulties

Alice M. Gregory, Michael J. Parsons, Nicola L. Barclay, Philip Gehrman, and Rachael E. O'Leary

Sleep is a fundamental state. We spend approximately one third of our lives asleep and our dedication to this activity is likely to reflect its profound significance in many areas of our functioning and well-being. Whereas sleep is often taken for granted and berated by those who are keen to achieve more in their waking hours; its disruption can be profound. Indeed, any new parent can testify to the draining effects of sleep loss; and those suffering from insomnia can often spend many desperate hours lying awake. Those who have been touched by horrific accidents caused by someone "falling asleep on the job" know of the sometimes catastrophic effects of its disruption.

Understanding more about individual differences with regards to sleep is essential. Indeed, behavioural genetics can go some way towards answering questions such as: 1) "why do some people sleep for shorter periods than others?"; 2) "to what extent do genetic and environmental influences account for individual differences with regards to sleep disturbances such as insomnia?"; and 3) "why are certain sleep phenotypes associated with psychological problems such as anxiety and depression (e.g., Ford & Kamerow, 1989), as well as physical health problems, such as obesity (Cappuccio et al., 2008)"?

Behavioural genetic research focusing on "sleep" encompasses diverse areas of research (from research investigating homeostasis in fruit flies, Shaw et al., 2002; to self-reports of problems getting to sleep in children, Gregory, Rijsdijk, & Eley, 2006b) and multiple recent reviews have been written on this topic (e.g., Barclay & Gregory, 2013; Dauvilliers, Maret, & Tafti, 2005a; Gregory & Franken, 2009). Accordingly, this chapter does not aim to provide a comprehensive review of this entire field, but instead, focuses on a handful of interesting research findings to date and provides speculation as to where the field is going.

Traditional family studies of sleep

Traditional "family studies" investigate whether a trait, such as insomnia, runs in families. Demonstrating that it does provides a starting point from which to ask further questions about the extent to which genetic or environmental influences are of key importance in explaining this familiality. Such family studies have shown that a variety of sleep phenotypes run in families. These phenotypes include the sleep apnoea-hypopnoea syndrome (Mathur & Douglas, 1995), narcolepsy (see Mignot, 1998), restless legs syndrome (Winkelmann et al., 2000) and insomnia (Bastien & Morin, 2000; Beaulieu-Bonneau et al., 2007; Dauvilliers et al., 2005b).

With regards to studies focusing on insomnia, one report found that 73% of those with primary insomnia (as compared to 24% of those without insomnia) had a family history of insomnia (Dauvilliers et al., 2005b). In certain studies, familiality appears to be strongest in those with early-onset insomnia (Bastien & Morin, 2000). Interestingly, for other phenotypes such as depression, familiality also appears to be greatest when there is early onset (for a review, see Rice, 2012). Should this finding be further substantiated, this would suggest that focusing on those with early-onset insomnia in molecular genetic studies could prove particularly fruitful. Family studies have also suggested that mothers appear to be the first-degree relative most commonly affected by insomnia (Beaulieu-Bonneau et al., 2007; Dauvilliers et al., 2005b). This is perhaps unsurprising given that insomnia occurs more frequently in females than males (see Ohayon, 2002) – although this could also suggest that X-linked genes play a role in insomnia; there are parent of origin effects on imprinting; or that mothers provide particularly important role models for sleep behaviour.

Traditional family studies are sometimes dismissed as unimportant as they are not able to provide information about the extent to which genetic or environmental factors are important in influencing a trait. Nevertheless, these studies remain useful in helping to refine phenotypes for use in molecular genetic studies based on consideration of the conditions by which familiality are greatest; and providing information on the upper bound for heritability estimates (Rice, 2012). Societal changes, with increased rates of reconstituted families, provide a novel way of using families to aid understanding of the genetic and environmental processes influencing traits.

Twin studies of sleep

Much of what is known about genetic and environmental influences on sleep, sleep disturbances and associations with other phenotypes comes from twin studies. Twin studies compare the similarity of monozygotic (MZ) twins, who for the purposes of this design are considered genetic clones of one another, with dizygotic (DZ) twins, who share on average half of their segregating genes like any other full sibling pair. When using the twin design, it is standard to partition the variance of a trait (or the covariance between traits) into three components: additive genetics (where the effects of genes "add up" to influence a behaviour); common, or shared environmental influences (where environmental factors act to make individuals within a family more alike); and environment specific to each twin, or nonshared environment (which acts to make individuals within a family less alike). For further information and a discussion of the assumptions and limitations levelled at this design, see elsewhere (Plomin et al., 2013).

Over 100 twin studies have provided information on the aetiology of variation in both normal sleep-wake characteristics and sleep difficulties and disorders (for a more comprehensive discussion of this topic see Barclay & Gregory, 2013). Some of the key findings from these studies are outlined further in this chapter.

Twin studies using polysomnography

A number of twin studies have focused on the topography of the sleep electroencephalogram (EEG), which varies between individuals (De Gennaro et al., 2005). Some studies of this type have indicated high heritability for certain measures. For example, one small study reported the EEG composition of non-Rapid Eye Movement (REM) sleep to be one of the most heritable human traits (heritability estimate = 96% in a sample of 10 MZ and 10 DZ twin pairs; (De Gennaro et al., 2008). Other studies have highlighted the importance of environmental factors in these variables, with a study of 26 pairs of young adult male MZ and DZ twins showing that for the quantity of stage 1 and REM sleep, genetic influences appeared to be confounded by other factors such as cohabitation (Linkowski et al., 1989). Overall, twin studies on this topic have been insightful in emphasizing the heritability of EEG measurements – while flagging the need to control for possible environmental confounds. Despite this, it is also important to note that some studies within this field have focused on MZ twins without the inclusion of DZ twins (e.g., Chouvet et al., 1983) – meaning that it is impossible to estimate the extent to which familiality is due to shared genetic or environmental factors. Furthermore, studies of this type are typically based on very small sample sizes, given the expense of using EEG (as compared to questionnaires for example) to assess sleep.

Twin studies using questionnaire measures

Given the value of large sample sizes in twin studies, most sleep studies of this type have employed questionnaire measures. These studies have been informative about a variety of sleep phenotypes including sleep length (e.g., Barclay et al., 2010b; Brescianini et al., 2011; Gregory et al., 2006b; Partinen et al., 1983); diurnal preference – i.e. whether one has a preference for and functions best during mornings or evenings (Barclay et al., 2010a;

Hur et al., 1998; Hur, 2007; Koskenvuo et al., 2007; Vink et al., 2001); as well as a range of dyssomnias (e.g., Gregory, 2008) including insomnia (Drake et al., 2011; Hublin et al., 2011; McCarren et al., 1994; Watson et al., 2006), narcolepsy (Imlah, 1961; Kaprio et al., 1996; for a review, see Mignot, 1998), symptoms of obstructive sleep apnoea (Carmelli et al., 2001; Carmelli et al., 2004; Desai et al., 2004) and restless legs syndrome (Desai et al., 2004; Ondo et al., 2000). Twin studies have also provided information about the aetiology of parasomnias, including sleep walking (Bakwin, 1970; Hublin et al., 1997), sleep talking (Hublin et al., 1998c), bruxism (Hublin et al., 1998b) and enuresis (Bakwin, 1971; Hublin et al., 1998a).

Certain studies have suggested interesting trends, such as those related to age. For example, with regards to sleep length, results from a number of studies suggest that shared environmental influences may be particularly salient on sleep length in childhood (e.g., see, Brescianini et al., 2011; Gedda & Brenci, 1979; although for mixed results see Gregory et al., 2006b). Interestingly, there appear to be developmental changes with regards to genetic and environmental influences on sleep length, with genetic influences becoming more important in older as compared to younger participants. Indeed, in a study of 2238 MZ and 4545 DZ twins it was found that the heritability estimate for sleep duration was lower in participants aged 18–24 years compared to those aged over 25 years (Partinen et al., 1983). This could in part be linked to the fact that sleep architecture changes with age, and that possibly the mechanisms underlying individual differences in sleep length are also altered (see Dahl, 1996).

As well as flagging possible changes in the relative importance of genetic and environmental influences with age, studies also suggest that there are important differences related to who is providing information about sleep. Indeed, a number of studies have indicated that when examining sleep in a community sample of children, parents tend to report fewer sleep difficulties for their children as compared to the children themselves (e.g., see Gehrman et al., 2011; Gregory et al., 2006b; Owens et al., 2000) – possibly because parents are not aware of the extent of their children's difficulties or because children are not good reporters of their symptoms. Further exploring this rater discrepancy, we investigated genetic and environmental influences on child- and parent-reporting of children's sleep problems, finding that genetic estimates based on parent-reporting were greater than those based on child self-reporting (Gregory et al., 2006b). It may be that parental reports reflect more severe cases of sleep problems that might have stronger genetic influences, while milder cases go undetected.

Such results from quantitative genetic studies have potential implications for molecular genetic studies. For example, developmental changes in the heritability of sleep length would suggest that it could be more fruitful to examine older as compared to younger participants when attempting to specify genes for this phenotype; and the greater heritability of parent-report as compared to child self-report of childhood sleep difficulties suggests that parental reported data may be of particular value in molecular genetic studies.

Multivariate twin studies

Given the known co-occurrence of sleep disturbances with a variety of other difficulties, researchers have used twin studies to try to elucidate reasons for these associations. Multivariate twin studies typically attempt to answer two main questions. First, to what extent are genetic and environmental influences on one phenotype (e.g., sleep quality) the same as those on another (e.g., depression)? Second, to what extent is the association between phenotypes (e.g., sleep quality and depression) explained by genetic and environmental factors?

Emotional difficulties such as anxiety and depression symptoms are commonly associated with sleep disturbances in both children (Gregory & O'Connor, 2002; Johnson et al., 2000) and adults (American Psychiatric Association, 1994). Exploring reasons for these associations in 3-year old twins, van den Oord and colleagues reported that associations between sleep and emotional difficulties were mainly explained by shared environmental influences (Van den Oord et al., 2000). A further study examined reasons for the overlap between sleep disturbances and depression in 100 MZ and 199 DZ twins aged 8 years (Gregory et al., 2006a). The association between sleep difficulties and depression symptoms was largely accounted for by genes. These twins were followed up for two years, which enabled examination of the longitudinal associations between sleep and depression (Gregory et al., 2009b). Sleep problems at age 8 years predicted depression at 10 years (while the converse was not found). Furthermore, this association (although of a small and non-significant effect) was largely explained by shared genetic influences. In line with this research, a further study of the genetic covariation between insomnia symptoms, anxiety and depression in a sample of 689 MZ and 666 DZ 8–16-year old twins found that there were modest genetic influences on sleep disturbances which overlapped substantially with depression and anxiety (Gehrman et al., 2011). Moving into young adulthood, an additional study highlighted substantial overlap in the genes influencing sleep disturbance and anxiety (genetic correlation [rA] = .58); and sleep disturbance and depression (rA = .68) and that these phenotypic (observed) associations were primarily explained by genetic factors (58% and 74% respectively, Gregory et al., 2011).

There has been less research attention to the overlap between sleep and behavioural difficulties – such as conduct problems – but there is clear indication that these difficulties are associated (e.g., see Chervin et al., 2003; Goodnight et al., 2007; Gregory et al., 2008a; Semiz et al., 2008). Examining reasons for these links in 3-year olds, associations appeared to be mainly due to shared environmental factors (Van den Oord et al., 2000).

The associations between diurnal preference, sleep quality and externalizing behaviours (assessed by aggression and rule breaking subscales of the Adult Self Report) was examined in a sample of young adults (Barclay et al., 2011a). Evening types and those with poorer sleep reported greater externalizing problems as compared to morning types and those with better sleep quality. These associations were largely accounted for by genes (accounting for approximately 80% of the phenotypic correlations). Whereas there was some evidence for shared genes between the sleep and externalizing phenotypes, there was little evidence for common environmental influences.

Twin studies have also examined the overlap between sleep phenotypes including daytime sleepiness, insomnia and short sleep duration and other phenotypes such as snoring and BMI/obesity (Carmelli et al., 2001; Watson et al., 2006; Watson et al., 2010). One study reported that the bivariate associations between daytime sleepiness and snoring; sleepiness and obesity; and sleepiness and snoring in elderly male twin pairs were largely accounted for by genes (Carmelli et al., 2001); whereas another study suggested that the association between short sleep and BMI in adult participants (average age = 36.9 years) was mainly due to nonshared environmental influences (Watson et al., 2010).

Further quantitative genetic studies of sleep

Although estimates of genetic and environmental influences on sleep-related variables and associated phenotypes largely come from twin studies conducted on western participants, we note that heritability is a population statistic which may therefore vary between the populations under study and highlights the need to replicate findings in a wider variety of populations. Furthermore, we note the value of other methodologies, such as adoption studies as well as newly emerging designs such as the children of twins design and the IVF design – which have both been used successfully to inform about other phenotypes (e.g., anxiety and depression, Harold et al., 2011; Lewis et al., 2011; Silberg et al., 2010). It is hoped that these methodologies will be used to generate and corroborate findings about genetic and environmental influences on sleep-related phenotypes.

Specifying genes involved in sleep

Research in humans

Having established from quantitative genetic studies a substantial genetic influence on a variety of sleep phenotypes, research has turned to specifying which genes exert these influences. There has been a particular focus on circadian genes, given the role of the circadian clock on sleep behaviours and the considerable extent to which the molecular basis for circadian rhythms has been elucidated (see Dijk & von Schantz, 2005). For example, associations have been reported between *PER3* and extreme diurnal preference (Archer et al., 2003) as well as sleep homeostasis markers (Viola et al., 2007). Interestingly, research has also suggested that the link between *PER3* and diurnal preference may be influenced by age. Indeed, in one study, amongst participants aged 18–29 years, the *PER3* 4-repeat allele was more common in evening-types whereas the 5-repeat allele was more common in morning-types (Jones et al., 2007). Notably, this pattern of results was not significant in participants aged 30+ years. Polymorphisms in *CLOCK* have been associated with sleep duration (Allebrandt et al., 2010), diurnal preference (Katzenberg et al., 1998), sleep onset and duration in patients suffering from bipolar disorder (Benedetti et al., 2007), and with recurrence of insomnia in those suffering from depression (Serretti et al., 2003). The circadian gene *DEC2* has been associated with sleep length in mammals (He et al., 2009); and a recent genome-wide association study (GWAS) suggested linkage peaks which included the circadian genes *CSNK2A2* and *PROK2* and usual bedtime (Gottlieb et al., 2007). A further GWAS revealed an association between an *ABCC9* gene variant and sleep duration (Allebrandt et al., 2013). Despite these interesting associations, it is important to note that, as with other complex traits, there have also been some failures to replicate (e.g., see Barclay et al., 2011b).

In addition to focusing on the circadian genes, there has also been interest in genes involved in neurotransmitter systems that are considered important in the regulation of sleep. With regards to the serotonergic system, one study focused on a genetic polymorphism in the serotonin transporter gene (*5HTTLPR*), finding a greater frequency of short *5HTTLPR* alleles in a group of insomnia patients as compared to controls (Deuschle et al., 2010). In contrast, another study found that long-long (*"LL"*) homozygotes for *5HTTLPR* had poorer sleep when compared to carriers of at least one short (*"S"*) allele (Barclay et al., 2011b). Similarly conflicting results have been found for other phenotypes such as anxiety (Jorm et al., 2000; Lesch et al., 1996).

A gene in the GABAergic system has been associated with insomnia in one case study (Buhr et al., 2002). Furthermore, a functional genetic polymorphism in the adenosine systems has been associated with deep sleep (Retey et al., 2005). An association has also been reported between a genetic polymorphism involved in glutamatergic neurotransmission and sleep duration in women (Utge et al., 2011). Interestingly, many of the aforementioned genes have also been associated with psychiatric difficulties associated with sleep disturbances and may therefore help to explain their associations. For example, a glutamatergic polymorphism was also found to be associated with depression in females (Utge et al., 2010).

One sleep disorder for which there has been particular progress in terms of specifying genes is narcolepsy. For example, it has been reported that a very high proportion (85%+) of those suffering narcolepsy with cataplexy carry a specific human leukocyte antigen (HLA) allele (which compares to just 12–38% in the general population, for a review, see Mignot, 1998). More recently a GWAS revealed that a polymorphism in the TRA@ locus was associated with narcolepsy (Hallmayer et al., 2009). Research focusing on animals has supported and further elucidated the mechanisms underlying these associations seen in humans (see below for a further discussion).

Overall, progress in this domain has been slow and the genes specified to date explain only a small proportion of the variance in sleep disturbances such as insomnia. This is in line with what is known about specific genes involved in other complex phenotypes, and this "missing heritability" may be in part due to multiple factors such as a relative neglect of: structural variation such as the role of Copy Number Variations in influencing a trait; and rare genetic variants (for a discussion of missing heritability, see Manolio et al., 2009).

Research in animals

The use of a wide number of animal models has greatly aided in determining the mechanisms underlying sleep (for a review, see Cirelli, 2009). Traditionally, mammalian models (e.g., mouse, rat and dog) were used. These animals were used as their sleep architecture more closely mimics what is seen in humans. The mouse in particular has been widely used to investigate the mechanisms of sleep due to the ease of manipulating the mouse genome using knock-out or transgenic approaches. Recently, simpler model systems have also been used (e.g., drosophila, *C. Elegans* and zebra fish). Although the sleep architecture seen in these model systems is not as complex as that seen in humans, they still experience sleep-like states.

Animal models have implicated various neurotransmitter systems in the regulation of sleep. The neurotransmitter adenosine is generally considered to have sleep promoting effects (Bjorness & Greene, 2009). Indeed, mice with altered adenosine receptors have decreased slow wave sleep patterns (Bjorness et al., 2009). Similarly, much of what we know about the promoting neurotransmitters including histamine, dopamine, acetylcholine and norepinephrine; the sleep promoting neurotransmitter gamma-aminobutyric acid (GABA); and the REM sleep suppressing neurotransmitter serotonin, is due to studies conducted using animal models (Cirelli, 2009).

The use of animal models has also allowed us to better understand certain sleep disorders, such as narcolepsy. Two independent models implicated the hypocretin (orexin) system in narcolepsy. A knock-out mouse that was missing the hypocretin gene displayed narcolepsy-like symptoms (Chemelli et al., 1999). Almost simultaneously, another group identified a hypocretin receptor as the gene underlying a canine model of narcolepsy (Lin et al., 1999). It is now believed that the majority of human cases of narcolepsy are due to a loss of hypocretin-producing neurons in the hypothalamus (Chabas et al., 2003). The loss of these neurons is likely caused by a combination of environmental factors and a genetic susceptibility at least partially caused by the HLA and TCR genes (Mignot et al., 2001; Hallmayer et al., 2009).

Advanced sleep phase syndrome (ASPS) is a rare sleep disorder involving individuals who experience an advanced sleep phase (i.e. they may fall asleep earlier than others and wake very early in the morning). It is thought that the advanced sleep phase is caused by a shortened circadian cycle (Jones et al., 1999). A familial form of ASPS was found to be caused by a mutation in the *PER2* gene, a gene involved in the core circadian clock (Toh et al., 2001). Mutations in the gene in drosophila had previously been shown to shorten the circadian cycle, thus supporting this gene involvement in ASPS (Marrus et al., 1996). A mutation in another circadian gene, *Casein Kinase* δ gene, was found to underlie another familial form of ASPS. When this functional mutation was put into mice it caused a shortened circadian cycle (Xu et al., 2005). It has long been known that the circadian clock partially controls sleep timing, thus it is not surprising that mutations in any of the large number of core circadian genes, reviewed elsewhere (Zhang & Kay, 2010), underlie sleep timing disorders.

As mentioned in the previous section, both genome wide and candidate gene association studies for subjective sleep traits and sleep disorders have revealed new genes that are potentially involved in sleep. It is often difficult to determine the functional variant underlying an association, thus animal models can be used to validate these findings. The *ABCC9* gene, which codes for a potassium channel, has been found to be a GWAS loci for self-reported sleep duration (Allebrandt et al., 2013). The same study validated this association by knocking down the corresponding orthologue of this gene in drosophila and was found to alter sleep duration (Allebrandt et al., 2013). Similarly, the *DEC2* gene was found to affect sleep duration in a familial sleep pedigree, which was verified using both mouse and drosophila models (He et al., 2009). With development of genome-wide or next generation sequencing approaches to determining the genetic variants underlying various sleep-related phenotypes, animal models will likely play more of a role in validation of these findings.

Specifying environmental influences involved in sleep

Using behavioural genetic designs such as twin studies, it has become apparent that just as genetic factors are likely to influence almost every sleep phenotype, the environment clearly plays a role too. Indeed, it is clear from twin studies that genetic factors do not account for 100% of the variance in sleep phenotypes – clearly suggesting that environmental influences (and/ or measurement error) are important. If, for example, we take subjective sleep quality, heritability estimates have been consistently placed at roughly 40% (Barclay et al., 2010b; Heath et al., 1990; Partinen et al.,

1983) suggesting that around 60% of the variance is accounted for by environmental factors (or measurement error).

The importance of the environment in influencing sleep phenotypes is perhaps most clearly demonstrated by examining the sleep of MZ twins who are often considered "genetic clones" of each other (for further information about the twin design, see Plomin et al., 2013). Despite the fact that recent research has found that even MZ twins are not genetically identical – and differ slightly in certain areas such as in their Copy Number Variations (Bruder et al., 2008) – these differences are considered likely to have a small influence on traits. Based on the assumption that it is not genetic factors that account for differences between MZ twins, this suggests that it is the environment (and possibly to some extent measurement error) which account for these differences. Based on this logic, a study recently investigated within-pair differences in sleep quality (Barclay et al., 2012). Adjusting for genetic and shared environmental effects, MZ differences in sleep quality were significantly linked to within-pair differences in health for males; and relationship satisfaction in females. It is noteworthy, that most of the "environmental" influences investigated (e.g., negative life events, employment status, relationship status) did not appear to be purely "environmental" influences and may therefore intertwine with genetic factors to influence sleep quality. MZ differences in diurnal preference have also been examined and it was found that MZ twin pair differences in diurnal preference are associated with MZ differences in dependent negative life events, educational attainment, smoking and drug use – suggesting that all of these variables have a truly environmental influence on diurnal preference (Barclay et al., 2013).

Despite the value of the MZ differences approach to specifying environmental influences on sleep, most of the knowledge we have about assumed "environmental" influences on sleep to date comes from non-genetic studies – and such studies have highlighted a plethora of risk factors for poor sleep quality including relationship status, employment status, socioeconomic status as well as life-style factors such as engagement in exercise (Driver & Taylor, 2000; Ford & Kamerow, 1989; Grandner et al., 2010; Hale, 2005). Of note, some of these seemingly "environmental" influences may be more likely to be experienced or to have a negative impact on sleep in part for genetic reasons.

Studies of interactions between genes and the environment

Twin studies typically split the variance of a phenotype into genetic and environmental components without consideration of the interplay between these influences (other chapters of this book provide further information about interplay between genetic and environmental influences). However, it is now clear that genetic and environmental factors act in concert. Indeed, genotype may make certain people more likely to be *exposed* to certain environments than others (a process known as gene–environment correlation, for reviews see Jaffee & Price, 2007; Kendler & Baker, 2007). Twin studies have shown genetic influence on life events (Button et al., 2008) and the home environment (Saudino & Plomin, 1997), as well as being a victim of bullying (Ball et al., 2008) – all factors which have been associated with sleep disturbances (Gregory et al., 2012; Spilsbury et al., 2005; Vahtera et al., 2007; Williams et al., 1996). Molecular genetic studies have also suggested that exposure to certain environmental experiences, such as marital status, which has been previously linked to sleep quality and length (Ford & Kamerow, 1989; Grandner et al., 2010; Hale, 2005) may be influenced by genes (Dick et al., 2006).

Another process of gene–environment interplay is gene–environment interaction, which refers to the finding that certain people are more *sensitive* to environmental influences than others based on their genetic propensities. Gene–environment interaction can be estimated using the twin design (see Purcell, 2002). One such study examined the interplay between genes and negative life events on sleep quality (Barclay et al., 2011c). Those experiencing a greater number of negative life events in the past year had poorer sleep quality. There was also a substantial genetic correlation between sleep quality and dependent negative life events (i.e., those considered to be dependent on one's own behaviour such as a relationship split) suggesting gene–environment correlation. In contrast, genetic liability to sleep quality did not appear to be moderated by these life events (i.e., there was no evidence of gene–environment interaction in this study).

A handful of molecular genetic studies have also investigated gene–environment interaction in relation to sleep. One study of this type found that a *5HTTLPR* polymorphism is associated with poor sleep quality in those experiencing the environmental stressor of caring for a parent or spouse with dementia (Brummett et al., 2007). A further study found that young adults who have two copies of the low expressing *5HTTLPR* polymorphism experienced depression symptoms in response to short sleep (Carskadon et al., 2012). This study highlights the feasibility of conceptualizing short sleep as an environmental stressor in studies of gene–environment interaction.

Further understanding of gene–environment interplay is likely to come from focusing on epigenetics – which provides a mechanism by which the environment can change the properties of DNA and hence stably influence gene expression and individual differences in a trait (see Meaney, 2010). With this in mind, our group is currently examining epigenetic differences between MZ twins in relation to diurnal preference and sleep quality and we predict an increase in interest in this area of research over the next decade.

Specifying the pathways by which genes and the environment influence sleep

It is often assumed that when associations are found between specific genes and a behavioural trait (such as sleep quality) that the former must *cause* the latter. Of course this is not necessarily the case, as the specific genetic polymorphism highlighted may have no functional significance. It is thus important to further characterize the nature of the causality between polymorphisms and behaviour (for further discussion, see Meaney, 2010). Furthermore, even functionally significant genes may not have direct effects (e.g., influencing neuronal functioning/ plasticity) but may exert their effects indirectly (via complex pathways involving hormonal, neural and psychological processes). Hormonally, melatonin and cortisol, amongst other hormones, affect sleep (Born et al., 1989; Cajochen et al., 2003), so genes involved in the regulation of these hormones could be significant. At the neural level, it has been shown that certain genes such as the *5HTTLPR* polymorphism are associated with amygdala response to fearful stimuli (Hariri et al., 2002; see also Munafo et al., 2008). This could be relevant to sleep, given links between stress reactivity and sleep (Fenzl et al., 2011). Psychologically, it is known that multiple processes such as catastrophizing about the effects of sleeplessness, holding dysfunctional beliefs about sleep and presleep cognitive and somatic arousal are all associated with poor sleep (Gregory et al., 2008b; Gregory et al., 2009a; Gregory et al., 2010; Harvey, 2002; Harvey & Greenall, 2003; Morin et al., 1993; Nicassio et al., 1985) and inevitably genetic and environmental risk will influence these processes.

Conclusion

The plethora of work to date highlights the importance of both genetic and environmental factors on a host of sleep phenotypes. This is not surprising considering most quantitative traits appear to be influenced by both genetic and environmental factors (see Plomin et al., 2013). What is more interesting are the trends that have appeared within the literature (e.g., the increased heritability of sleep length with age) and the implications in terms of specifying genes. Understanding pathways and the ways in which genes and the environment interact has rightly become a popular/important source of enquiry and capitalizing on novel techniques and routes of investigation (such as epigenetics) provides an important way forward. In addition to fuelling a basic human need to understand why we are the way we are, behavioural genetics is likely to evolve to have useful implications in terms of predicting, preventing and treating difficulties. Indeed, research into treatment response has found that selective serotonin reuptake inhibitors (SSRIs) can have adverse effects, including insomnia-like symptoms and that differential response to treatment appears to be due in part to *5HTTLPR* genotype – with the *"S"* allele associated with an increased chance of developing new/worse insomnia symptoms (Perlis et al., 2003). Similarly, it was found that *"LL"* homozygotes appear to show improved mood following sleep deprivation in bipolar depression patients as compared to *"S"* carriers (Benedetti et al., 1999). This type of research needs to be taken further (e.g., examining the role of genotype on response to psychological treatment for insomnia) in order to unleash the potential of this field of enquiry. We are excited to see what the future holds.

References

Allebrandt, K. V., Amin, N., Muller-Myhsok, B., Esko, T., Teder-Laving, M., Azevedo, R. V. D. M. et al. (2013). A K(ATP) channel gene effect on sleep duration: from genome-wide association studies to function in Drosophila. *Molecular Psychiatry, 18,* 122–132.

Allebrandt, K. V., Teder-Laving, M., Akyol, M., Pichler, I., Muller-Myhsok, B., Pramstaller, P. et al. (2010). CLOCK gene variants associate with sleep duration in two independent populations. *Biological Psychiatry, 67,* 1040–1047.

American Psychiatric Association (1994). *Diagnostic and statistical manual of mental disorders: DSM IV.* Washington, DC: American Psychiatric Association.

Archer, S. N., Robilliard, D. L., Skene, D. J., Smits, M., Williams, A., Arendt, J. et al. (2003). A length polymorphism in the circadian clock gene Per3 is linked to delayed sleep phase syndrome and extreme diurnal preference. *Sleep, 26,* 413–415.

Bakwin, H. (1970). Sleep-walking in twins. *The Lancet, 29,* 446–447.

Bakwin, H. (1971). Enuresis in twins. *American Journal of Diseases in Children, 21,* 222–225.

Ball, H. A., Arseneault, L., Taylor, A., Maughan, B., Caspi, A., & Moffitt, T. E. (2008). Genetic and environmental influences on victims, bullies and bully-victims in childhood. *Journal of Child Psychology and Psychiatry, 49,* 104–112.

Barclay, N. L., Eley, T. E., Maughan, B., Rowe, R., & Gregory, A. M. (2011a). Associations between diurnal preference, sleep quality and externalizing behaviours: a behavioural genetic analysis. *Psychological Medicine, 41,* 1029–1040.

Barclay, N. L., Eley, T. C., Buysse, D. J., Maughan, B., & Gregory, A. M. (2012). Nonshared environmental influences on sleep quality: a study of monozygotic twin differences. *Behavioral Genetics, 42,* 234–244.

Barclay, N. L., Eley, T. C., Mill, J., Wong, C. C. Y., Zavos, H. M. S., Archer, S. N. et al. (2011b). Sleep quality and diurnal preference in a sample of young adults: associations with 5HTTLPR, PER3, and CLOCK 3111. *American Journal of Medical Genetics – Part B – Neuropsychiatric Genetics, 156,* 681–690.

Barclay, N. L., Eley, T. C., Parsons, M. J., Willis, T. A., & Gregory, A. M. (2013). Monozygotic twin differences in nonshared environmental factors associated with chronotype. *Journal of Biological Rhythms, 2013,* 51–61.

Barclay, N. L., Eley, T. C., Rijsdijk, F. V., & Gregory, A. M. (2011c). Dependent negative life events and sleep quality: an examination of gene-environment interplay. *Sleep Medicine, 12,* 403–409.

Barclay, N. L., & Gregory, A. M. (2013). Quantitative genetic research on sleep: A review of normal sleep, sleep disturbances and associated emotional, behavioural, and health-related difficulties. *Sleep Medicine Reviews, 17,* 29–40.

Barclay, N. L., Eley, T. C., Buysse, D. J., Archer, S. N., & Gregory, A. M. (2010a). Diurnal preference and sleep quality: same genes? A study of young adult twins. *Chronobiology International, 27,* 278–296.

Barclay, N. L., Eley, T. C., Buysse, D. J., Rijsdijk, F. V., & Gregory, A. M. (2010b). Genetic and environmental influences on different components of the Pittsburgh Sleep Quality Index and their overlap. *Sleep, 33,* 659–668.

Bastien, C. H., & Morin, C. M. (2000). Familial incidence of insomnia. *Journal of Sleep Research, 9,* 49–54.

Beaulieu-Bonneau, S., LeBlanc, M., Merette, C., Dauvilliers, Y., & Morin, C. M. (2007). Family history of insomnia in a population-based sample. *Sleep, 30,* 1739–1745.

Benedetti, F., Dallaspezia, S., Fulgosi, M. C., Lorenzi, C., Serretti, A., Barbini, B. et al. (2007). Actimetric evidence that CLOCK 3111 T/C SNP influences sleep and activity patterns in patients affected by bipolar depression. *American Journal of Medical Genetics Part B-Neuropsychiatric Genetics, 144B,* 631–635.

Benedetti, F., Serretti, A., Colombo, C., Campori, E., Barbini, B., Di Bella, D. et al. (1999). Influence of a functional polymorphism within the promoter of the serotonin transporter gene on the effects of total sleep deprivation in bipolar depression. *American Journal of Psychiatry, 156,* 1450–1452.

Bjorness, T. E., & Greene, R. W. (2009). Adenosine and sleep. *Current Neuropharmacology, 7,* 238–245.

Bjorness, T. E., Kelly, C. L., Gao, T., Poffenberger, V., & Greene, R. W. (2009). Control and function of the homeostatic sleep response by adenosine A(1) receptors. *Journal of Neuroscience, 29,* 1267–1276.

Born, J., Spathschwalbe, E., Schwakenhofer, H., Kern, W., & Fehm, H. L. (1989). Influences of corticotropin-releasing hormone, adrenocorticotropin, and cortisol on sleep in normal man. *Journal of Clinical Endocrinology & Metabolism, 68,* 904–911.

Brescianini, S., Volzone, A., Fagnani, C., Patriarca, V., Grimaldi, V., Lanni, R. et al. (2011). Genetic and environmental factors shape infant sleep patterns: A study of 18-month-old twins. *Pediatrics, 127,* E1296-E1302.

Bruder, C. E. G., Piotrowski, A., Gijsbers, A. A. C. J., Andersson, R., Erickson, S., Diaz de Stahl, T. et al. (2008). Phenotypically concordant and discordant monozygotic twins display different DNA copy-number-variation profiles. *The American Journal of Human Genetics, 82,* 763–771.

Brummett, B. H., Krystal, A. D., Ashley-Koch, A., Kuhn, C. M., Zuchner, S., Siegler, I. C. et al. (2007). Sleep quality varies as a function of 5-HTTLPR genotype and stress. *Psychosomatic Medicine, 69,* 621–624.

Buhr, A., Bianchi, M. T., Baur, R., Courtet, P., Pignay, V., Boulenger, J. P. et al. (2002). Functional characterization of the new human GABA(A) receptor mutation beta 3(R192H). *Human Genetics, 111,* 154–160.

Button, T. M. M., Lau, J. Y. F., Maughan, B., & Eley, T. C. (2008). Parental punitive discipline, negative life events and gene-environment interplay in the development of externalizing behavior. *Psychological Medicine, 38,* 29–39.

Cajochen, C., Krauchi, K., & Wirz-Justice, A. (2003). Role of melatonin in the regulation of human circadian rhythms and sleep. *Journal of Neuroendocrinology, 15,* 432–437.

Cappuccio, F. P., Taggart, F. M., Kandala, N.-B., Currie, A., Peile, E., Stranges, S. et al. (2008). Meta-analysis of short sleep duration and obesity in children and adults. *Sleep, 31,* 619–626.

Carmelli, D., Bliwise, D. L., Swan, G. E., & Reed, T. (2001). Genetic factors in self-reported snoring and excessive daytime sleepiness – A twin study. *American Journal of Respiratory and Critical Care Medicine, 164,* 949–952.

Carmelli, D., Colrain, I. M., Swan, G. E., & Bliwise, D. L. (2004). Genetic and environmental influences in sleep-disordered breathing in older male twins. *Sleep, 27,* 917–922.

Carskadon, M. A., Sharkey, K. M., Knopik, V. S., & McGeary, J. E. (2012). Short sleep as an environmental exposure: a preliminary study associating 5-HTTLPR genotype to self-reported sleep duration and depressed mood in first-year university students. *Sleep, 35,* 791–796.

Chabas, D., Taheri, S., Renier, C., & Mignot, E. (2003). The genetics of narcolepsy. *Annual Review of Genomics and Human Genetics, 4,* 459–483.

Chemelli, R. M., Willie, J. T., Sinton, C. M., Elmquist, J. K., Scammell, T., Lee, C. et al. (1999). Narcolepsy in orexin knockout mice: Molecular genetics of sleep regulation. *Cell, 98,* 437–451.

Chervin, R. D., Dillon, J. E., Archbold, K. H., & Ruzicka, D. L. (2003). Conduct problems and symptoms of sleep disorders in children. *Journal of the American Academy of Child and Adolescent Psychiatry, 42,* 201–208.

Chouvet, G., Blois, R., Debilly, G., & Jouvet, M. (1983). Temporal patterns of rapid eye-movements during paradoxical sleep are similar in human monozygotic twins. *Comptes Rendus de l Academie des Sciences Serie Iii-Sciences de la Vie-Life Sciences, 296,* 1063–1068.

Cirelli, C. (2009). The genetic and molecular regulation of sleep: from fruit flies to humans. *Nature Reviews Neuroscience, 10,* 549–560.

Dahl, R. E. (1996). The regulation of sleep and arousal: Development and psychopathology. *Development and Psychopathology, 8,* 3–27.

Dauvilliers, Y., Maret, S., & Tafti, M. (2005a). Genetics of normal and pathological sleep in humans. *Sleep Medicine Reviews, 9,* 91–100.

Dauvilliers, Y., Morin, C., Cervena, K., Carlander, B., Touchon, J., Besset, A. et al. (2005b). Family studies in insomnia. *Journal of Psychosomatic Research, 58,* 271–278.

De Gennaro, L., Ferrara, M., Vecchio, F., Curcio, G., & Bertini, M. (2005). An electroencephalographic fingerprint of human sleep. *Neuroimage, 26,* 114–122.

De Gennaro, L., Marzano, C., Fratello, F., Moroni, F., Pellicciari, M. C., Ferlazzo, F. et al. (2008). The electroencephalographic fingerprint of sleep is genetically determined: a twin study. *Annals of Neurology, 64,* 455–460.

Desai, A. V., Cherkas, L. F., Spector, T. D., & Williams, A. J. (2004). Genetic influences in self-reported symptoms of obstructive sleep apnoea and restless legs: A twin study. *Twin Research, 7,* 589–595.

Deuschle, M., Schredl, M., Schilling, C., Wust, S., Frank, J., Witt, S. H. et al. (2010). Association between a serotonin transporter length polymorphism and primary insomnia. *Sleep, 33,* 343–347.

Dick, D. M., Agrawal, A., Schuckit, M. A., Bierut, L., Hinrichs, A., Fox, L. et al. (2006). Marital status, alcohol dependence, and GABRA2: Evidence for gene-environment correlation and interaction. *Journal of Studies on Alcohol, 67,* 185–194.

Dijk, D.-J., & von Schantz, M. (2005). Timing and consolidation of human sleep, wakefulness, and performance by a symphony of oscillators. *Journal of Biological Rhythms, 20,* 279–290.

Drake, C. L., Friedman, N. P., Wright, K. P., & Roth, T. (2011). Sleep reactivity and insomnia: genetic and environmental influences. *Sleep, 34,* 1179–1188.

Driver, H. S., & Taylor, S. R. (2000). Exercise and sleep. *Sleep Medicine Reviews, 4,* 387–402.

Fenzl, T., Touma, C., Romanowski, C. P., Ruschel, J., Holsboer, F., Landgraf, R. et al. (2011). Sleep disturbances in highly stress reactive mice: Modeling endophenotypes of major depression. *BMC Neuroscience, 12.*

Ford, D. E., & Kamerow, D. B. (1989). Epidemiologic study of sleep disturbances and psychiatric disorders. An opportunity for prevention? *Journal of the American Medical Association, 262,* 1479–1484.

Gedda, L., & Brenci, G. (1979). Sleep and dream characteristics in twins. *Acta Geneticae Medicae et Gemellologiae, 28,* 237–239.

Gehrman, P. R., Meltzer, L. J., Moore, M., Pack, A. I., Perlis, M. L., Eaves, L. J. et al. (2011). Heritability of insomnia symptoms in youth and their relationship to depression and anxiety. *Sleep, 34,* 1641–1646.

Goodnight, J. A., Bates, J. E., Staples, A. D., Pettit, G. S., & Dodge, K. A. (2007). Temperamental resistance to control increases the association between sleep problems and externalizing behavior development. *Journal of Family Psychology, 21,* 39–48.

Gottlieb, D. J., O'Connor, G. T., & Wilk, J. B. (2007). Genome-wide association of sleep and circadian phenotypes. *BMC Medical Genetics, 8,* S9.

Grandner, M. A., Patel, N. P., Gehrman, P. R., Xie, D. W., Sha, D. H., Weaver, T. et al. (2010). Who gets the best sleep? Ethnic and socioeconomic factors related to sleep complaints. *Sleep Medicine, 11,* 470–478.

Gregory, A. M. (2008). A genetic decomposition of the association between parasomnias and dyssomnias in 8-year-old twins. *Archives of Pediatrics & Adolescent Medicine, 162,* 299–304.

Gregory, A. M., Cox, J., Crawford, M. R., Holland, J., & Harvey, A. G. (2009a). Dysfunctional beliefs and attitudes about sleep in children. *Journal of Sleep Research, 18,* 422–426.

Gregory, A. M., & Franken, P. (2009). Genetic approaches to the problem of sleep. In M. G. Francos (Ed.), *Current advances in sleep biology* (pp. 41–62). New York: Nova Biomedical Books.

Gregory, A. M., Moffitt, T. C., Ambler, A., Arseneault, L., Houts, R. M., & Caspi, A. (2012). Insomnia and children's family socialization environments. *Sleep, 35,* 579–582.

Gregory, A. M., Noone, D. M., Eley, T. C., & Harvey, A. G. (2010). Catastrophising and symptoms of sleep disturbances in children. *Journal of Sleep Research, 19,* 175–182.

Gregory, A. M., & O'Connor, T. G. (2002). Sleep problems in childhood: A longitudinal study of developmental change and association with behavioral problems. *Journal of the American Academy of Child and Adolescent Psychiatry, 41,* 964–971.

Gregory, A. M., Rijsdijk, F. V., Dahl, R. E., McGuffin, P., & Eley, T. C. (2006a). Associations between sleep problems, anxiety and depression in twins at 8 years of age. *Pediatrics, 118,* 1124–1132.

Gregory, A. M., Rijsdijk, F. V., & Eley, T. C. (2006b). A twin-study of sleep difficulties in school-aged children. *Child Development, 77,* 1668–1679.

Gregory, A. M., Rijsdijk, F. V., Lau, J. F., Dahl, R. E., & Eley, T. C. (2009b). The direction of longitudinal associations between sleep problems and depression symptoms: a study of twins aged 8 and 10 years. *Sleep, 32,* 189–199.

Gregory, A. M., Van den Ende, J., Willis, T. A., & Verhulst, F. C. (2008a). Parent-reported sleep problems during development predicts self-reported anxiety/depression, attention problems and aggression later in life. *Archives of Pediatrics & Adolescent Medicine, 162,* 330–335.

Gregory, A. M., Willis, T. A., Wiggs, L., & Harvey, A. G. (2008b). Pre-sleep arousal and sleep disturbances in children. *Sleep, 31,* 1745–1747.

Gregory, A. M., Buysse, D. J., Willis, T. A., Rijsdijk, F. V., Maughan, B., Rowe, R. et al. (2011). Associations between sleep quality and anxiety and depression symptoms in a sample of young adult twins and siblings. *Journal of Psychosomatic Research, 71,* 250–255.

Hale, L. (2005). Who has time to sleep? *Journal of Public Health, 27,* 205–211.

Hallmayer, J., Faraco, J., Lin, L., Hesselson, S., Winkelmann, J., Kawashima, M. et al. (2009). Narcolepsy is strongly associated with the T-cell receptor alpha locus. *Nature Genetics, 41,* 708–711.

Hariri, A. R., Mattay, V. S., Tessitore, A., Kolachana, B., Fera, F., Goldman, D. et al. (2002). Serotonin transporter genetic variation and the response of the human amygdala. *Science, 297,* 400–403.

Harold, G., Rice, F., Hay, D., Boivin, J., van den Bree, M., & Thapar, A. (2011). Familial transmission of depression and antisocial behavior symptoms: disentangling the contribution of inherited and environmental factors and testing the mediating role of parenting. *Psychological Medicine, 41,* 1175–1185.

Harvey, A. G. (2002). A cognitive model of insomnia. *Behaviour Research & Therapy, 40,* 869–893.

Harvey, A. G., & Greenall, E. (2003). Catastrophic worry in primary insomnia. *Journal of Behavior Therapy and Experimental Psychiatry, 34,* 11–23.

He, Y., Jones, C. R., Fujiki, N., Xu, Y., Guo, B., Holder, J. L. et al. (2009). The transcriptional repressor DEC2 regulates sleep length in mammals. *Science, 325,* 866–870.

Heath, A. C., Kendler, K. S., Eaves, L. J., & Martin, N. G. (1990). Evidence for genetic influences on sleep disturbances and sleep patterns in twins. *Sleep, 13,* 318–335.

Hublin, C., Kaprio, J., Partinen, M., Heikkila, K., & Koskenvuo, M. (1997). Prevalence and genetics of sleepwalking: A population-based twin study. *Neurology, 48,* 177–181.

Hublin, C., Kaprio, J., Partinen, M., & Koskenvuo, M. (1998a). Nocturnal enuresis in a nationwide twin cohort. *Sleep, 21,* 579–585.

Hublin, C., Kaprio, J., Partinen, M., & Koskenvuo, M. (1998b). Sleep bruxism based on self-report in a nationwide twin cohort. *Journal of Sleep Research, 7,* 61–67.

Hublin, C., Kaprio, J., Partinen, M., & Koskenvuo, M. (1998c). Sleeptalking in twins: Epidemiology and psychiatric comorbidity. *Behavior Genetics, 28,* 289–298.

Hublin, C., Partinen, M., Koskenvuo, M., & Kaprio, J. (2011). Heritability and mortality risk of insomnia-related symptoms: A genetic epidemiologic study in a population-based twin cohort. *Sleep, 34,* 957–964.

Hur, Y. M. (2007). Stability of genetic influence on morningness-eveningness: a cross-sectional examination of South Korean twins from preadolescence to young adulthood. *Journal of Sleep Research, 16,* 17–23.

Hur, Y. M., Bouchard, T. J., & Lykken, D. T. (1998). Genetic and environmental influence on morningness-eveningness. *Personality and Individual Differences, 25,* 917–925.

Imlah, N. W. (1961). Narcolepsy in identical twins. *Journal of Neurology Neurosurgery and Psychiatry, 24,* 158–160

Jaffee, S. R., & Price, T. S. (2007). Gene-environment correlations: a review of the evidence and implications for prevention of mental illness. *Molecular Psychiatry, 12,* 432–442.

Johnson, E. O., Chilcoat, H. D., & Breslau, N. (2000). Trouble sleeping and anxiety/depression in childhood. *Psychiatry Research, 94,* 93–102.

Jones, C. R., Campbell, S. S., Zone, S. E., Cooper, F., DeSano, A., Murphy, P. J. et al. (1999). Familial advanced sleep-phase syndrome: A short-period circadian rhythm variant in humans. *Nature Medicine, 5,* 1062–1065.

Jones, K. H. S., Ellis, J., von Schantz, M., Skene, D. J., Dijk, D. J., & Archer, S. N. (2007). Age-related change in the association between a polymorphism in the PER3 gene and preferred timing of sleep and waking activities. *Journal of Sleep Research, 16,* 12–16.

Jorm, A. F., Henderson, A. S., Jacomb, P. A., Christensen, H., Korten, A. E., Rodgers, B. et al. (2000). Association of a functional polymorphism of the monoamine oxidase A gene promoter with personality and psychiatric symptoms. *Psychiatric Genetics, 10,* 87–90.

Kaprio, J., Hublin, C., Partinen, M., Heikkila, K., & Koskenvuo, M. (1996). Narcolepsy-like symptoms among adult twins. *Journal of Sleep Research, 5,* 55–60.

Katzenberg, D., Young, T., Finn, L., Lin, L., King, D. P., Takahashi, J. S. et al. (1998). A CLOCK polymorphism associated with human diurnal preference. *Sleep, 21,* 569–576.

Kendler, K., & Baker, J. H. (2007). Genetic influences on measures of the environment: a systematic review. *Psychological Medicine, 37,* 615–626.

Koskenvuo, M., Hublin, C., Partinen, M., Heikkila, K., & Kaprio, J. (2007). Heritability of diurnal type: a nationwide study of 8753 adult twin pairs. *Journal of Sleep Research, 16,* 156–162.

Lesch, K. P., Bengel, D., Heils, A., Sabol, S. Z., Greenburg, B. D., Petri, S. et al. (1996). Association of anxiety-related traits with a polymorphism in the serotonin transporter gene regulatory region. *Science, 274,* 1527–1531.

Lewis, G., Rice, F., Harold, G. T., Collishaw, S., & Thapar, A. (2011). Investigating environmental links between parent depression and child depressive/anxiety symptoms using an assisted conception design. *Journal of the American Academy of Child and Adolescent Psychiatry, 50,* 451–459.

Lin, L., Faraco, J., Li, R., Kadotani, H., Rogers, W., Lin, X. Y. et al. (1999). The sleep disorder canine narcolepsy is caused by a mutation in the hypocretin (orexin) receptor 2 gene. *Cell, 98,* 365–376.

Linkowski, P., Kerkhofs, M., Hauspie, R., Susanne, C., & Mendlewicz, J. (1989). EEG sleep patterns in man – a twin study. *Electroencephalography and Clinical Neurophysiology, 73,* 279–284.

Manolio, T. A., Collins, F. S., Cox, N. J., Goldstein, D. B., Hindorff, L. A., Hunter, D. J. et al. (2009). Finding the missing heritability of complex diseases. *Nature, 461,* 747–753.

Marrus, S. B., Zeng, H. K., & Rosbash, M. (1996). Effect of constant light and circadian entrainment of per(S) flies: Evidence for light-mediated delay of the negative feedback loop in Drosophila. *Embo Journal, 15,* 6877–6886.

Mathur, R., & Douglas, N. J. (1995). Family studies in patients with the sleep-apnea hypopnea syndrome. *Annals of Internal Medicine, 122,* 174–178.

McCarren, M., Goldberg, J., Ramakrishnan, V., & Fabsitz, R. (1994). Insomnia in Vietnam era veteran twins: Influence of genes and combat experience. *Sleep, 17,* 456–461.

Meaney, M. J. (2010). Epigenetics and the biological definition of gene x environment interactions. *Child Development, 81,* 41–79.

Mignot, E. (1998). Genetic and familial aspects of narcolepsy. *Neurology, 50,* S16–S22.

Mignot, E., Lin, L., Rogers, W., Honda, Y., Qiu, X. H., Lin, X. Y. et al. (2001). Complex HLA-DR and -DQ interactions confer risk of narcolepsy-cataplexy in three ethnic groups. *American Journal of Human Genetics, 68,* 686–699.

Morin, C. M., Stone, J., Trinkle, D., Mercer, J., & Remsberg, S. (1993). Dysfunctional beliefs and attitudes about sleep among older adults with and without insomnia complaints. *Psychology and Aging, 8,* 463–467.

Munafo, M. R., Brown, S. M., & Hariri, A. R. (2008). Serotonin transporter (5-HTTLPR) genotype and amygdala activation: A meta-analysis. *Biological Psychiatry, 63,* 852–857.

Nicassio, P. M., Mendlowitz, D. R., Fussell, J. J., & Petras, L. (1985). The phenomenology of the pre-sleep state – the development of the pre-sleep arousal scale. *Behaviour Research and Therapy, 23,* 263–271.

Ohayon, M. M. (2002). Epidemiology of insomnia: what we know and what we still need to learn. *Sleep Medicine Reviews, 6,* 97–111.

Ondo, W. G., Vuong, K. D., & Wang, Q. (2000). Restless legs syndrome in monozygotic twins: Clinical correlates. *Neurology, 55,* 1404–1406.

Owens, J. A., Spirito, A., McGuinn, M., & Nobile, C. (2000). Sleep habits and sleep disturbance in elementary school-aged children. *Journal of Developmental and Behavioral Pediatrics, 21,* 27–36.

Partinen, M., Kaprio, J., Koskenvuo, M., Putkonen, P., & Langinvainio, H. (1983). Genetic and environmental determination of human sleep. *Sleep, 6,* 179–185.

Perlis, R. H., Mischoulon, D., Smoller, J. W., Wan, Y. J. Y., Lamon-Fava, S., Lin, K. M. et al. (2003). Serotonin transporter polymorphisms and adverse effects with fluoxetine treatment. *Biological Psychiatry, 54,* 879–883.

Plomin, R., DeFries, J. C., Knopik, V. S., & Neiderhiser, J. M. (2013). *Behavioral Genetics.* (6th ed.) New York: Worth Publishers.

Purcell, S. (2002). Variance components models for gene-environment interaction in twin analysis. *Twin Research, 5,* 554–571.

Retey, J. V., Adam, M., Honegger, E., Khatami, R., Luhmann, U. F. O., Jung, H. H. et al. (2005). A functional genetic variation of adenosine deaminase affects the duration and intensity of deep sleep in humans. *PNAS, 102,* 15676–15681.

Rice, F. (2012). Genetic influences on depression and anxiety in childhood and adolescence. In A.Ronald & S. H. Rhee (Eds), *Behavior genetics and psychopathology.* New York: Springer.

Saudino, K. J., & Plomin, R. (1997). Cognitive and temperamental mediators of genetic contributions to the home environment during infancy. *Merrill-Palmer Quarterly, 43,* 1–23.

Semiz, U. B., Alguel, A., Basoglu, C., Ates, M. A., Ebrinc, S., Cetin, M. et al. (2008). The relationship between subjective sleep quality and aggression in male subjects with antisocial personality disorder. *Turk Psikiyatri Dergisi, 19,* 373–381.

Serretti, A., Benedetti, F., Mandelli, L., Lorenzi, C., Pirovano, A., Colombo, C. et al. (2003). Genetic dissection of psychopathological symptoms: Insomnia in mood disorders and CLOCK gene polymorphism. *American Journal of Medical Genetics Part B-Neuropsychiatric Genetics, 121B,* 35–38.

Shaw, P. J., Tononi, G., Greenspan, R. J., & Robinson, D. F. (2002). Stress response genes protect against lethal effects of sleep deprivation in Drosophila. *Nature, 417,* 287–291.

Silberg, J. L., Maes, H., & Eaves, L. J. (2010). Genetic and environmental influences on the transmission of parental depression to children's depression and conduct disturbance: an extended Children of Twins study. *Journal of Child Psychology and Psychiatry, 51,* 734–744.

Spilsbury, J. C., Storfer-Isser, A., Drotar, D., Rosen, C. L., Kirchner, L., & Redline, S. (2005). Effects of the home environment on school-aged children's sleep. *Sleep, 28,* 1419–1427.

Toh, K. L., Jones, C. R., He, Y., Eide, E. J., Hinz, W. A., Virshup, D. M. et al. (2001). An hPer2 phosphorylation site mutation in familiar advanced sleep phase syndrome. *Science, 291,* 1040–1043.

Utge, S., Kronholm, E., Partonen, T., Soronen, P., Ollila, H., Loukola, A. et al. (2011). Shared genetic background for regulation of mood and sleep: association of GRIA3 with sleep duration in healthy Finnish women. *Sleep, 34,* 1309–1316.

Utge, S., Soronen, P., Partonen, T., Loukola, A., Kronholm, E., Pirkola, S. et al. (2010). A population-based association study of candidate genes for depression and sleep disturbance. *American Journal of Medical Genetics Part B, 153B,* 468–476.

Vahtera, J., Kivimaki, M., Hublin, C., Korkeila, K., Suominen, S., Paunio, T. et al. (2007). Liability of anxiety and severe life events as predictors of new-onset sleep disturbances. *Sleep, 30,* 1537–1546.

Van den Oord, E. J. C. G., Boomsma, D. I., & Verhulst, F. C. (2000). A study of genetic and environmental effects on the co-occurrence of problem behaviors in three-year-old twins. *Journal of Abnormal Psychology, 109,* 360–372.

Vink, J. M., Groot, A. S., Kerkhof, G. A., & Boomsma, D. I. (2001). Genetic analysis of morningness and eveningness. *Chronobiology International, 18,* 809–822.

Viola, A. U., Archer, S. N., James, L. M., Groeger, J. A., Lo, J. C. Y., Skene, D. J. et al. (2007). PER3 polymorphism predicts sleep structure and waking performance. *Current Biology, 17,* 613–618.

Watson, N. F., Goldberg, J., Arguelles, L., & Buchwald, D. (2006). Genetic and environmental influences on insomnia, daytime sleepiness, and obesity in twins. *Sleep, 29,* 645–649.

Watson, N. F., Buchwald, D., Vitiello, M. V., Noonan, C., & Goldberg, J. (2010). A twin study of sleep duration and body mass index. *Journal of Clinical Sleep Medicine, 6,* 11–17.

Williams, K., Chambers, M., Logan, S., & Robinson, D. (1996). Association of common health symptoms with bullying in primary school children. *British Medical Journal, 313,* 17–19.

Winkelmann, J., Wetter, T. C., Collado-Seidel, V., Gasser, T., Dichgans, M., Yassouridis, A. et al. (2000). Clinical characteristics and frequency of the hereditary restless legs syndrome in a population of 300 patients. *Sleep, 23,* 597–602.

Xu, Y., Padiath, Q. S., Shapiro, R. E., Jones, C. R., Wu, S. C., Saigoh, N. et al. (2005). Functional consequences of a CKI delta mutation causing familial advanced sleep phase syndrome. *Nature, 434,* 640–644.

Zhang, E. E., & Kay, S. A. (2010). Clocks not winding down: unravelling circadian networks. *Nature Reviews Molecular Cell Biology, 11,* 764–776.

11

The Role of Genetic and Environmental Risk Factors in Aetiology of Suicidal Behaviour

Aigul Zainullina, Adelina Valiullina, and Elza Khusnutdinova

Suicidal behaviour is an extreme manifestation of psychological problems. According to the model of suicidal continuum, this behaviour includes suicidal thoughts, intentions, suicide attempts and completed suicide (Wasserman, 2001). It is thought that suicidal tendencies (including suicidal attempts) and completed suicide are aetiologically distinct. On the other hand, suicidal behaviour may intensify over time (Voytsekh, 2007; Wasserman, 2001).

Research has shown that suicidal behaviour differs across the life span. Suicidal behaviour in childhood is related to situational and personal reactions, so that it is associated not with the desire to die, but rather with the desire to avoid stressful situations or punishment. Suicidal behaviour in adolescence may result from emotional instability, a temporary personality modification that is found in almost a quarter of healthy adolescents. The key feature of adolescent suicidal behaviour is that suicide attempts occur 8–10 times more often than completed suicide. Psychological problems at this age may stem from the mismatch between increasing demands of the environment (e.g., economic, social, ecological) and inadequate ability of adolescents to react to these demands. Risk factors for suicidal behaviour in adolescents include aggressive and antisocial behaviour; substance abuse (alcohol, nicotine, drugs); auto-aggressive actions; and affiliation to youth subcultures, sexual and other minorities. Moreover, family-related factors can also be important and include lack of parental warmth, communication and emotional responsiveness; and high levels of violence, disapproval, severe discipline, abuse and conflict in the family (Bridge, Goldstein, & Brent, 2006; Ivanova et al., 2011).

In explaining the aetiology of suicidal behaviour, it is necessary to consider a broad variety of factors involved in the diathesis (predisposition): genetic, physiological, neurobiological, psychological and physical characteristics. Such predisposition to suicidal behaviour determines whether suicidality will be manifested in the stressful conditions (e.g., traumatic events or chronic stress; Costanza et al., 2014; Mann, Brent, & Arango, 2001; Mann et al., 2009; van Heeringen, 2003; Wasserman, 2001).

Research has shown that stress in different forms and at different stages of life may lead to a variety of negative developmental outcomes, such as poor mental health in general, neurosis, psychopathy, addiction and other psychological disorders (Gunnar & Cheatham, 2003; Heim et al., 2008; Hochberg et al., 2010; Paus et al., 2008; Weinstock, 2008). The links between stress and these outcomes are complex involving mediating and moderating effects between different factors; directional and reciprocal causal paths; and pleiotropic effect of genes. Biological mechanisms underlying the long-term effects of stress are extensively studied in humans and animals. For example, prenatal stress has been linked to negative programming of the hippocampus, amygdala and frontal cortex functioning, which in turn may result in problems with cognitive functions, personality deviations, fears and anxiety (Weinstock, 2008). Stress in childhood may affect the hippocampus (e.g., volume reduction and dysfunction), and result in related cognitive deficits, school performance and other problems (Tarquis, 2005). Stress in adolescence has been found to be associated with the frontal and prefrontal cortex and the mesolimbic system dysfunction (Bremner, 2006; Heim & Nemeroff, 2001), which may partly explain the relationship problems, personality deviations, antisocial behaviour, addictions, anxiety, depression, low self-esteem and school failure (Charmandari et al., 2005; Lupien et al., 2009). In adulthood, stress may exacerbate inhibition of neurogenesis and attenuate neuronal survival that are associated with secondary cognitive deficits, affective disorders, aggression and suicidal behaviour (Bremner, 2006; Wolf, 2009).

The hypothalamic-pituitary-adrenal axis (HPA axis) mediates stress stimuli and creates a background for auto-aggressive and suicidal behaviour. Virtually all components of the HPA axis (corticotropin-releasing hormone (CRH), adrenocorticotropic hormone (ACTH) and cortisol) can be detected in blood, saliva or urine. Cortisol, together with its receptor proteins, regulates gene expression acting as a transcription factor. In addition to controlling the release of ACTH, CRH affects various neuronal structures acting as a neurotransmitter (Sher, 2008). Early childhood stress can change the subsequent physiological response to stressful situations later in life (Aguilera et al., 2007; Korosi & Baram, 2008). In adulthood, those who experienced stress early in life are at higher risk for anxiety, depression, aggressive and suicidal behaviour in response to adversity (Ladd et al., 2000; Rutter, 2002).

Behavioural genetic studies of suicidal behaviour

Behavioural genetic studies indicate genetic influences on suicidal behaviour (Gilger, 2000; Moffitt, 2005). It has been shown that in 6% of suicide cases one of the victim's parents also committed suicide. In children with suicidal behaviour, 44% had a close relative who attempted suicide, and 25% had a parent who attempted suicide. Children of suicide victims are approximately four times more likely to attempt or complete suicide than children from "non-suicidal" families. Probability of completed suicide in monozygotic twins is about 20 times higher than in dizygotic twins, but in dizygotic twins it is 3.5 times higher on average than in the population. Adopted children, who had a family history of suicide, are six times more likely to attempt suicide compared to the control group of adopted children whose biological parents did not make suicidal attempts (Currier & Mann, 2008).

Research suggests that multiple genetic factors, each with small effect, contribute to the heritability of suicidal behaviour. Recent studies suggest that a number of genes may contribute to suicidal behaviour, and their effects can be moderated by adverse environmental conditions: gene–environment interaction (see Chapter 2 for more detail).

Neurotransmitter systems genes in suicidal behaviour

The serotonergic system of the brain plays an important role in suicidal behaviour. Serotonin deficiency in the frontal regions of the brain can lead to behavioural dysregulation and disinhibition as well as aggression and depression, which in turn are associated with suicidal behaviour. Genetic factors contribute to regulation of the serotonergic system (Mann & Currier, 2007; Oquendo & Mann, 2001), with individual differences in this system being potentially associated with individual differences in suicidal behaviour (van Heeringen, 2003). Genes that are involved the functioning of the serotonergic system are candidates genes for susceptibility to suicidal behaviour. The research to date has focused on genes that control key stages in the biosynthesis of serotonin, its re-uptake into the cell, its reception and degradation.

Serotonin is synthesized from tryptophan in all mammals. This process includes two consecutive stages. The first stage is controlled by tryptophan hydroxylase (TPH), an enzyme with high substrate specificity and exclusively involved in the biosynthesis of serotonin (Mockus & Vrana, 1998). The second stage involves the activation of decarboxylase of aromatic L-amino acids. This enzyme is involved in many molecular processes affecting derivatives of tryptophan, phenylalanine and tyrosine (Zhu & Juorio, 1995). Because the first stage is rate-limiting for serotonin synthesis, molecular genetic studies have focused on genes encoding two isoforms of TPH – TPH1 and TPH2. These isoforms are 71% homologous in their amino acid composition (Hasegawa & Ichiyama, 1987; Kuhn et al., 1980; Walther et al., 2003). The results of studies of associations of the *TPH1* gene with suicidal behaviour, mood disorders and other psychiatric disorders are inconsistent (Brezo et al., 2010; Gaysina et al., 2008; Saetre et al., 2010). The most studied Single

Nucleotide Polymorphism (SNP) of the *TPH1* gene (*rs1800532*) is located in intron 7 and affects the processes of splicing, translation and mRNA stability (Hong et al., 2001). Another polymorphic locus, *rs4537731*, located in the promoter region of the gene, influences the translation of the *TPH1* gene (Rotondo et al., 1999). There is evidence for association of suicidal behaviour with haplotypes constructed from different polymorphic loci of the *TPH1* gene. For example, the study by Turecki and colleagues (2001) found an association of GTC haplotype (*–6526A>G, –5806G>T, 218A>C*) with completed suicides. Liu and colleagues (2006) showed an association of TCAA haplotype of polymorphic loci *–7680G> T, –7065C>T, –6526A> G; 218A> C, 779A> C* with suicide in a Chinese sample. Zaboli and colleagues (2006) examined six polymorphic loci of *TPH1* gene in women with borderline personality disorders who had at least two suicide attempts. They found that TCAAA haplotype (*–7680G> T, –7065C>T, –6526A>G; 218A>C, 779A>C*) was associated with suicidal behaviour.

Any neurotransmitter system activity can be measured by the state of its transmembrane transporters. A number of studies suggest the involvement of the serotonin transporter in suicidal behaviour (Mann et al., 2001). The serotonin transporter gene (*SLC6A4*) is located on chromosome 17 (17q11.2–12). The insertion-deletion of the 44 base pair fragment in the promoter region of the gene (*5-HTTLPR*) is represented by a long form (allele *L*) or a short form (allele *S*). Studies in cell cultures and human post mortem brain tissues have showed that the *S*-allele is associated with a lower level of expression of the gene (Hranilovic et al., 2003; Lesch et al., 1996). Moreover, in the long form of *5-HTTLPR*, another polymorphism with two alleles A and G (*rs25531*) is present (Nakamura et al., 2000), creating the binding site for the transcription factor AP-2 which affects the level of mRNA expression of the gene. The presence of allele L_G leads to a lower level of mRNA expression of the gene that roughly corresponds to the level of gene expression in the presence of the *S*-allele (Hu et al., 2006). Another polymorphism in this gene, the varying number tandem repeats (*VNTR*) in an intron 2, is represented by the alleles with 9, 10 or 12 repeats, and can also affect the transcription of the *SLC6A4* gene (Fiskerstrand et al., 1999; Heils et al., 1997).

Polymorphic loci of the *SLC6A4* gene have been intensively investigated for association with suicidal behaviour (Brezo et al., 2010; Božina et al., 2012; Gaysina et al., 2006; Linkowska et al., 2009; Mann et al., 2009; Zalsman et al., 2011; Zupanc et al., 2010). This research indicates an association of allele *S* (or L_G) with depression (Zalsman et al., 2006) and suicidal behaviour (Kraft et al., 2008; Neves et al., 2008, 2010). However, in the sample of Russian women, the long allele (*L*) and L-10 haplotype (*5-HTTLPR-VNTR*) were associated with suicidal behaviour (Gaysina et al., 2006). The contradictory results may be due to ethnic differences between the investigated samples (Bennett et al., 2002; Chiao & Blizinsky, 2010; Manuck et al., 2004; Noskova et al., 2008).

The excess of serotonin accumulating in the presynaptic terminal undergoes sufficiently rapid degradation. This process is jointly served by several enzymes, the most important of which is monoamine oxidase (MAO). This enzyme is involved in the degradation of not only serotonin but also dopamine, adrenaline and noradrenaline (Ma et al., 2004) and plays a very important role in the regulation of many synaptic contacts; as a result, it may affect a number of behavioural traits. There are two forms of monoamine oxidase enzyme – MAOA and MAOB. The genes encoding these proteins are located on the X-chromosome. A number of studies of polymorphic loci of the *MAOA* and *MAOB* genes showed associations with suicidal behaviour and various mental disorders (Brummett et al., 2007; De Luca et al., 2008; Dlugos et al., 2009; Hung et al., 2012; Linkowska et al., 2009; Lung et al., 2011; Zalsman et al., 2011).

In humans, there are fifteen specific serotonin receptor subtypes, each of which is encoded by a separate gene. These that are of particular interest in studies of suicidal behaviour are genes encoding type 1 receptors (i.e., *HTR1A, HTR1B*) and type 2 receptors (i.e., *HTR2A, HTR2C*). Serotonin receptor 1A (HTR1A) is a target for the action of antidepressants. The *HTR1A* gene is localized on chromosome 5 (5q11.2-q13). A functional polymorphism in the promoter region of this gene (*G1019C*) controls the transcription activity of the gene and is associated with suicidal behaviour, unipolar depression and anxiety disorder, as well as anxiety- and depression-related personality traits (Lemonde et al., 2003; Samadi Rad et al., 2012; Strobel et al., 2003). However, other studies did not confirm the association of this polymorphism with suicidal behaviour or personality traits (Koller et al., 2006; Serretti et al., 2009; Videtič et al., 2009b; Wasserman et al., 2007).

Serotonin 1B receptor gene (*HTR1B*) is located on chromosome 6 (6q13). A number of studies in transgenic mice lacking the *HTR1B* gene confirmed its involvement in the aetiology of aggressiveness and alcoholism (Brunner et al., 1999; Saudou et al., 1994). A number of human studies reported the association of the *HTR1B* polymorphic variants with suicidal behaviour (Gaysina et al., 2003; Lappalainen et al., 1998; New et al., 2001), whereas in some other studies this association was not found (Murphy et al., 2011; Wang et al., 2009; Wrzosek et al., 2011; Zupanc et al., 2010).

In humans, the serotonin 2A receptor gene (*HTR2A*) is located on the long arm of chromosome 13 (13q14-q21). Three important polymorphisms have been identified in this gene: *T102C* and *A1438G* in the promoter region and *His452Tyr* in the coding region. Studies on brain autopsy material showed reduced binding of serotonin, and increased expression of the HTR2A receptors in the brain tissues of suicide victims as compared with the control group (Turecki et al., 2014). It was also reported that signal transduction through the HTR2A receptors may be damaged as a result of alternative splicing and altered protein structure (Arango et al., 2003). The results of association studies are inconsistent, with some, but not all, studies suggesting the association of the *HTR2A* gene with suicidal behaviour (Brezo et al., 2010; Gaysina et al., 2003; Gaysina et al., 2008; Wrzosek et al., 2011; Zalsman et al., 2011; Zhang et al., 2008; Zupanc et al., 2010).

The serotonin 2C receptor gene (*HTR2C*) is not only a functional, but also a positional, candidate gene for suicidal behaviour. The results of association candidate gene analysis (Videtič et al., 2009a) and genome-wide analysis (Zubenko et al., 2004) indicated a functional importance of the Xq25–26.1 region, in which the *HTR2C* gene is located, for the risk of suicide attempts in people with unipolar depression. However, some other studies did not confirm the association of this gene with suicidal behaviour (De Luca et al., 2008; Serretti et al., 2009; Videtič et al., 2009a).

Dopaminergic and noradrenergic systems are also important neurotransmitter systems associated with suicidal behaviour. It is believed that reduction of dopamine in the mesolimbic brain structures may cause anhedonia (inability to enjoy daily activities) that is associated with depression. Reduced levels of norepinephrine in the *locus coeruleus* neurons in the brain may be involved in anxiety, restlessness and inadequate responses to stressful situations (Aguilera et al., 2007; Korosi & Baram, 2008). Disturbances in these systems can occur under stress. Repeated or chronic stress undermines the ability of dopaminergic mesolimbic structures to compensate for unpleasant emotions; and the ability of the noradrenergic system to assess the real danger coming from environmental signals (Nemeroff & Vale, 2005).

Studies of brain tissues of suicide victims revealed a reduced amount of noradrenergic neurons in the *locus coeruleus*, reduced concentration of norepinephrine, increased tyrosine hydroxylase activity, and increased number of α1-adrenoceptors (Mann et al., 2001). According to several studies, pathological aggression and suicidal behaviour is accompanied by increased levels of 3-methoxy-4-hydroxy-phenylglycol – one of the end products of the metabolism of norepinephrine in the cerebrospinal fluid. Norepinephrine transporter (NET) is responsible for the process of norepinephrine reuptake in the presynaptic nerve endings. Its level was found to be reduced in the *locus coeruleus* in patients with unipolar depression (Klimek et al., 1997). The norepinephrine transporter gene (NET) was shown to be associated with depressive disorders (Inoue et al., 2004; Ryu et al., 2004).

Pathological aggression and suicidal behaviour may be associated with dysfunction of the dopaminergic system (Chen et al., 2005; Sher, 2006). Low levels of homovanillic acid were found in patients with depression and a history of suicidal behaviour (Sher et al., 2006). It was also shown that dopaminergic dysfunction was associated with impulsivity, mood disorders and alcohol dependence (Sher et al., 2006; Tupala & Tiihonen, 2004). The associations of polymorphic loci of the dopamine D2 receptor gene (*DRD2*) and dopamine transporter gene (*SLC6A3*) with suicidal behaviour and aggressive behaviour in adolescents and adults have been reported (Chen et al., 2005; Gaysina et al., 2004; Suda et al., 2009).

Several studies examined the role of catechol-O-methyltransferase (COMT), an enzyme involved in the biosynthesis and metabolism of dopamine and norepinephrine. In humans, the *COMT* gene is located on the chromosome 22 (22q11.1-q11.2). Functional polymorphism *Val158Met* of this gene is associated with different enzyme activity (Lachman et al., 1996). A number of association studies, as well as a meta-analysis of these studies, indicated the role of the *COMT* gene in suicidal behaviour (Kia-Keating et al., 2007; Lee & Kim, 2011; Nedic et al., 2011; Pivac et al., 2011).

Other neurotransmitter brain systems have also been implicated in suicidal behaviour and depression, for example, the cholinergic, glutamatergic and gamma-aminobutyric acid-ergic(GABAergic) systems (Cannon et al., 2011; Cohen-Woods et al., 2009; Mann et al., 2009; Sequeira et al., 2009). Glutamatergic and GABAergic systems of the hippocampus are involved in cognitive processes often impaired in people with suicidal behaviour, such as autobiographical memory and a decreased ability to overcome life's difficulties (Müller & Schwarz, 2007). These disturbances can be associated with the death of hippocampal neurons due to hyperactivity of glutamatergic processes or neurogenesis impairment in the hippocampus (Campbell et al., 2014) as a result of the toxic effect of stress hormones.

It is known that glutamate is a major excitatory neurotransmitter in the CNS (about 70% of excitatory neurons have glutamatergic neurotransmission). There are several types of glutamate receptors. Molecular biological studies identified 15 genes that fall into three groups, corresponding to functional groups of ionotropic glutamate receptors – NMDA, AMPA and kainate; and eight genes constituting three groups of metabolotropic glutamate receptors (Nakanishi, 1992). The *GRIN2B* gene encodes NR2B-subunit NMDA-glutamate receptor. Associations of polymorphic variants of this gene with schizophrenia (Jia et al., 2012) and suicidal behaviour (Sokolowski et al., 2013) were found.

GABA receptors belong to a group of cellular receptors, for which γ-aminobutyric acid is an endogenous agonist (a main inhibitory neurotransmitter in the vertebrate nervous system). Mutations in the genes encoding subunits of GABAA receptor cause different types of epilepsy (Jentsch et al., 2002; Planells-Cases & Jentsch, 2009), as well as depression and other affective disorders (Cascio, 2006). Several studies showed an association of the of gamma-amino-butyric acid type 6 receptor gene (*GABRA6*) and b1-subunit of GABAA receptor gene (*GABRB1*) with an increased risk of alcohol dependence (Buckley et al., 2006).

Stress is often associated with the dysfunction of the cholinergic system in the brain, such as decreased acetylcholinesterase activity in neurons; decreased the biosynthesis of acetylcholine and choline reuptake; destruction of cholinergic neurons of the central nervous system (cerebral cortex, reticular formation, etc.). Animal studies showed significant differences in behaviour between rats with lesions of the cholinergic system of the brain and control animals: rats with cholinergic deficit were more fearful, stressed and aggressive (Steimer, 2002). The cholinergic neurotransmitter system is involved in the regulation of memory, sleep, mood, emotions, states of consciousness, functions of internal organs, muscles, blood vessels. Cholinergic neurons are involved in functionally distinct neuronal networks, where they have a mainly excitatory function, activating emotiogenic structures (amygdala, hippocampus) and striatum. The neurotransmitter acetylcholine binds to cholinergic receptors of the postsynaptic membrane that belong to the family of G-protein linked receptors. There is a group of cholinergic receptors that react to muscarine (fly-agaric poison) – M-choline receptors (muscarin-sensitive) (Comings et al., 2002). Corticosteroids may increase expression of the M2 muscarinic receptor. In addition, injection of muscarinic agonists resulted in a significant increase in corticosterone in normal mice, and did not lead to an increase in corticosterone levels in knockout mice lacking the gene of the M2 muscarinic receptor. Cholinergic hypersensitivity was also detected in patients with depression and bipolar disorder, after the injection of muscarinic receptor agonists or acetylcholinesterase inhibitors. Polymorphic loci located in the 5′-UTR region of the gene *CHRM2* may affect gene transcription and expression of this gene (Cannon et al., 2011; Luo et al., 2005). Studies of the role of this gene in depressive disorders and suicidal behaviour have not consistently found significant associations (Cannon et al., 2011; Cohen-Woods et al., 2009; Zainullina et al., 2011).

The hypothalamic-pituitary-adrenocortical axis genes

Stress activates the hypothalamic-pituitary-adrenal (HPA) axis. This system plays an important role in the pathogenesis of impulsivity, aggression, suicidality and affective disorders via feedback-based activation of glucocorticoid hormone receptors. It is believed that, in the case of affective disorders, the negative feedback mechanism – by which high cortisol levels affect glucocorticoid receptors (mainly in the hippocampus) – is damaged, thereby reducing the synthesis of corticotropin-releasing hormones. The decrease in free receptors in the hippocampus weakens inhibition and improves the reactivity of the HPA axis. Elevated levels of cortisol in the blood have been associated with depressive symptoms (Habib et al., 2001). There is convincing evidence that aggression in patients with personality disorders is not only negatively correlated with the level of cortisol and β-endorphin in plasma, but also with the level of free cortisol in urine. Reduced levels of cortisol in urine have been found in people with suicide attempts against a background of alcoholism (Mann & Currier, 2007).

The HPA axis functioning is regulated with corticotropin-releasing hormone (CRH), which activates the corticotropin-releasing hormone receptor type 1 (CRHR1). These receptors are located in the limbic system and in the anterior lobe of the pituitary gland. The *CRHR1* gene is localized on chromosome 17 (17q21–22), encompassing 20 kb of genomic DNA and containing 14 exons. There is evidence of the *CRHR1* gene being involved in the aetiology of suicidal behaviour (Wasserman et al., 2009a), depression (Papiol et al., 2007; Wasserman et al., 2009b) and panic disorder (Keck et al., 2008). Polymorphic locus *rs878886* in exon 13 of the gene is functionally significant as it affects the efficiency of the translation process (Keck et al., 2008). Polymorphic locus *rs12936511* in exon 2 regulates alternative splicing and has been associated with the risk of depressive episodes and suicide attempts (Wasserman et al., 2009b). Polymorphic loci *rs4792887, rs110402, rs1876828, rs242939, rs242941, rs7209436, rs16940665*, in intronic regions of the gene have been associated with the increased risk of depression and suicidal behaviour (De Luca et al., 2008; Papiol et al., 2007; Wasserman et al., 2009b).

The glucocorticoid receptor (GR, or NR3C1) plays an important role in the regulation of the HPA axis in response to stress. This receptor regulates the effect of cortisol on target tissues by direct binding to a specific DNA sequence responsive to the hormone effect and, through interaction with other transcription factors, leads to a change in gene transcription. A number of functional polymorphisms in this gene, such as *A3669G (rs6198)* in exon 9 *rs41423247* (BclI) in an intron region and *rs6195 (N363S)* in exon 2, have been associated with suicidal behaviour and depressive disorders (Mann, 2003; Schreiner et al., 2009).

Other candidate genes of suicidal behaviour

FKBP5. Immunophilins are intracellular proteins that serve as major receptors of immunosuppressive drugs such as cyclosporin A (CsA), FK506 and rapamycin. Immunophilin–drug complexes are involved in a number of intracellular signal transduction systems, especially in the immune and nervous systems. In neural tissues, immunophilins affect the neuronal growth process and the synthesis of nitric oxide, as well as the release of neurotransmitters. The protein FKBP5 modulates the function of the glucocorticoid receptor via heat shock protein 90, which plays a key role in signal transduction of steroid hormones (Willour et al., 2009). The *FKBP5* gene is located on the chromosome 6 (6p21.31). Several studies showed the association of the polymorphic loci of this gene with post-traumatic stress disorder (Boscarino et al., 2012), bipolar disorder (Willour et al., 2009), anxiety (Binder, 2009) and unipolar depression (Tatro et al., 2010). Studies of depressive disorders in patients with a history of suicide attempts reported the association of the *FKBP5* gene with suicidal behaviour (Brent et al., 2010). There is evidence for the interaction between the *FKBP5* gene and child trauma in the development of suicidal behaviour (Roy et al., 2012).

MTHFR. The enzyme 5,10-methylenetetrahydrofolate reductase (MTHFR) is one of the key enzymes of the folate cycle. It catalyses the conversion of 5,10-methylenetetrahydrofolate to 5-methylenetetrahydrofolate – the main form of folate in plasma. The latter is a donor of methyl groups for remethylation of homocysteine in the synthesis of methionine. In turn, methionine is converted to S-adenosylmethionine, which plays an important role in the methylation of proteins, neurotransmitters, phospholipids and DNA, and therefore in the regulation of gene expression (Födinger et al., 1999). A polymorphic locus *rs1801133 (Ala222Val)* in the *MTHFR* gene results in a decrease of enzyme activity to 35–70% (Weisberg et al., 1998). The biological mechanisms behind the role of the *MTHFR* gene might involve aberrant methylation, disruption of DNA synthesis and metabolism of neurotransmitters (Zeisel, 2009). This polymorphism has been widely investigated for association with psychopathology, including schizophrenia (Allen et al., 2008; Ellingrod et al., 2008; Feng et al., 2009; Peerbooms et al., 2011; Roffman et al., 2011; Saetre et al., 2011; Shi et al., 2008), autism (Goin-Kochel et al., 2009; Mohammad et al., 2009), alcohol addiction (Benyamina et al., 2009) and depression (Gaysina et al., 2008; Peerbooms et al., 2011; Słopien et al., 2008).

14–3-3 genes. 14–3-3 proteins are highly effective regulators of apoptosis, cell cycle, division, transcription, replication and functioning of ion channels and transporters, as well as the organization of the cytoskeleton (Morrison, 2009). 14–3-3 protein is found in neural tissue, where its concentration exceeds 1% of the proteome (Boston et al., 1982, Muslin et al., 1996). 14–3-3 proteins contain at least two sites that provide binding of phosphorylated ligands with particular primary structures. This allows 14–3-3 proteins to bind target proteins, leading to the increase of the activity of enzymes and controlling their subcellular localization (Berg et al., 2003). It was found that 14–3-3 proteins activate key enzymes for the synthesis of serotonin and catecholamines – tryptophan- and tyrosine-monooxygenase (Ichimura et al., 1987). 14–3-3 proteins are involved in the development of a number of neurodegenerative (Di Fede et al., 2007; Mackie & Aitken, 2005) and mental illnesses (Ikeda et al., 2008; Sekiguchi et al., 2011; Yanagi et al., 2005).

In human tissues 7 isoforms of 14–3-3 protein were found: β, γ, ε, η, σ, τ / θ and ζ (Ichimura et al., 1987). Each isoform is encoded by a separate gene (Aitken, 2002). In the literature, there is evidence for the association of 14–3-3 genes with Creutzfeldt-Jakob disease, depression and schizophrenia (Cumbler et al., 2009; Di Fede et al., 2007; Fountoulakis et al., 1999; Grover et al., 2009; Jia et al., 2004; Martins-de-Souza et al., 2009; Middleton et al., 2005; Nagamani et al., 2009; Sánchez-Valle et al., 2002; Sugimori et al., 2007; Wong et al., 2005; Yanagi et al., 2005).

BDNF. A number of studies have investigated the mechanisms of cellular interactions in the brain that may underlie suicidal behaviour. In particular, attention is paid to neurotrophins (e.g., brain-derived neurotrophic factor, BDNF) and other regulatory neuropeptides (Neves et al., 2011; Pregelj et al., 2011). It is possible that stress undermines neuronal viability and plasticity of nervous tissue due to inhibition of BDNF. There is evidence that the concentration of BDNF is decreased in the brain and other tissues (e.g., blood) in suicide victims (Deveci et al., 2007). A recent meta-analysis of 12 studies showed an association of the *BDNF* gene with suicidal behaviour (Paska et al., 2013).

X-chromosome genes. A number of genes located on the X-chromosome have been associated with suicide (Fiori et al., 2011; Zubenko et al., 2004). The functions of these genes (e.g., *SAT1, RPS6KA3, AP1S2, ARHGAP6* and *GPM6B*) are diverse. For example, the *SAT1* gene is involved in polyamine catabolism, the *ARHGAP6* gene activates neurogenesis and the *GPM6B* gene is involved in neuronal differentiation and myelination.

Genes identified in GWAS studies. In recent years, a number of genome-wide association studies (GWASs) of suicidal behaviour have been conducted (Perlis et al., 2010; Schosser et al., 2011; Willour et al., 2012). These GWASs of suicidal behaviour found no associations significant at the genome-wide level ($p < 5 \times 10^{-8}$) and any suggestive associations have not been replicated in independent samples. Polygenic score analysis suggested pleiotropy between psychiatric disorders and suicidal ideation, whereas the tendency to act on such thoughts was shown to have a partially independent genetic diathesis (Mullins et al., 2014).

Gene-environment interaction and epigenetics

As mentioned earlier in this chapter, genetic predisposition to suicidal behaviour is more likely to lead to suicide in adverse environments – a phenomenon known as gene–environment interaction (see Chapter 2 for more detail). For example, people homozygous for the *L*-allele of a functional polymorphism of the serotonin-transporter-linked polymorphic region (5-*HTTLPR*) are more likely to develop depression (Caspi et al., 2003) and suicidal behaviour due to accumulation of stressful life events (Roy et al., 2007). Other examples of interactions of specific genes (e.g., BDNF, HPA-axis genes) with environmental risk in depression and, to a lesser extent, in suicidal behaviour are known; however only a few reliable results have been obtained (Mandelli & Serretti, 2013).

Gene–environment interaction may involve epigenetic mechanisms, whereby environment affects gene expression. Epigenetic processes lead to activation or suppression of gene expression as a result of changes in the levels of DNA methylation and changes in the protein components of chromatin (acetylation, etc.). While the structure of DNA remains largely unchanged throughout life, DNA expression is a dynamic process and depends on both genetic and environmental factors. Epigenome is a dynamic interface between the inherited genome, environmental factors and behaviour (Champagne & Curley, 2009; Meaney & Szyf, 2005; Szyf, 2009).

The role of epigenetic mechanisms of behaviour has been shown in both humans and animals (Suomi, 2006; Szyf et al., 2010; Tsankova et al., 2007). For example, early isolation of infant rats from their mother leads to hypersecretion of corticosterone in adulthood, due to a decrease of methylation of the gene encoding the release of vasopressin. The increase in secretion of vasopressin in the hypothalamus stimulates secretion of the adrenocorticotropic hormone (ACTH) (Murgatroyd et al., 2009). In humans, suicide victims with experience of childhood violence had increased levels of methylation of the glucocorticoid receptor gene on the brains, which entails the inhibition of the synthesis of this protein and leads to hyperactivity of the HPA axis (McGowan et al., 2009). In other words, stressful events in early childhood program the activity of HPA axis through epigenetic processes.

Conclusion

The search for genetic and environmental risk factors for suicidal behaviour is complicated because each factor makes only a small contribution to this trait, and multiple factors interact with each other. Despite the large number of behavioural genetic studies of suicidal behaviour, the results are inconsistent. This inconsistency can be attributed to a number of reasons including differences across studies in gender, age and ethnicity of the samples. For example, the frequencies of alleles in different populations may vary. Moreover, geographical, economic and social conditions also vary across populations and, through interaction with genetic factors, may contribute differently to the development of suicidal behaviour. It is therefore necessary to carry out research into the interplay of environmental and genetic factors at different developmental stages in different populations. Modern behavioural genetics, genomics and epigenomics provide new strategies for early diagnostics and prevention of suicidal behaviour.

References

Aguilera, G., Kiss, A., Liu, Y., & Kamitakahara, A. (2007). Negative regulation of corticotropin releasing factor expression and limitation of stress response. *Stress, 10*(2), 153–161.

Aitken, A. (2002). Functional specificity in 14–3-3 isoform interactions through dimer formation and phosphorylation. Chromosome location of mammalian isoforms and variants. *Plant Molecular Biology, 50*(6), 993–1010.

Allen, N. C., Bagade, S., McQueen, M. B., Ioannidis, J. P., Kavvoura, F. K., Khoury, M. J., ... & Bertram, L. (2008). Systematic meta-analyses and field synopsis of genetic association studies in schizophrenia: the SzGene database. *Nature Genetics, 40*(7), 827–834.

Almeida, O. P., McCaul, K., Hankey, G. J., Norman, P., Jamrozik, K., & Flicker, L. (2008). Homocysteine and depression in later life. *Archives of General Psychiatry, 65*(11), 1286–1294.

Arango, V., Huang, Y. Y., Underwood, M. D., & Mann, J. J. (2003). Genetics of the serotonergic system in suicidal behavior. *Journal of Psychiatric Research, 37*(5), 375–386.

Bardi, M., & Huffman, M. A. (2002). Effects of maternal style on infant behavior in Japanese macaques (Macaca fuscata). *Developmental Psychobiology, 41*(4), 364–372.

Barrett, J. C., Fry, B., Maller, J. D. M. J., & Daly, M. J. (2005). Haploview: analysis and visualization of LD and haplotype maps. *Bioinformatics, 21*(2), 263–265.

Bennett, A. J., Lesch, K. P., Heils, A., Long, J. C., Lorenz, J. G., Shoaf, S. E., ... & Higley, J. D. (2002). Early experience and serotonin transporter gene variation interact to influence primate CNS function. *Molecular Psychiatry, 7*(1), 118–122.

Benyamina, A., Saffroy, R., Blecha, L., Pham, P., Karila, L., Debuire, B., ... & Reynaud, M. (2009). Brief report: Association between MTHFR 677C-T polymorphism and alcohol dependence according to Lesch and Babor typology. *Addiction Biology, 14*(4), 503–505.

Berg, D., Holzmann, C., & Riess, O. (2003). 14-3-3 proteins in the nervous system. *Nature Reviews Neuroscience, 4*(9), 752–762.

Bet, P. M., Penninx, B. W., Bochdanovits, Z., Uitterlinden, A. G., Beekman, A. T., van Schoor, N. M., ... & Hoogendijk, W. J. (2009). Glucocorticoid receptor gene polymorphisms and childhood adversity are associated with depression: new evidence for a gene–environment interaction. *American Journal of Medical Genetics Part B: Neuropsychiatric Genetics, 150*(5), 660–669.

Binder, E. B. (2009). The role of FKBP5, a co-chaperone of the glucocorticoid receptor in the pathogenesis and therapy of affective and anxiety disorders. *Psychoneuroendocrinology, 34*, S186–S195.

Binder, E. B., Bradley, R. G., Liu, W., Epstein, M. P., Deveau, T. C., Mercer, K. B., ... & Ressler, K. J. (2008). Association of FKBP5 polymorphisms and childhood abuse with risk of posttraumatic stress disorder symptoms in adults. *JAMA, 299*(11), 1291–1305.

Bondy, B., Buettner, A., & Zill, P. (2006). Genetics of suicide. *Molecular Psychiatry, 11*(4), 336–351.

Boscarino, J. A., Erlich, P. M., Hoffman, S. N., & Zhang, X. (2012). Higher FKBP5, COMT, CHRNA5, and CRHR1 allele burdens are associated with PTSD and interact with trauma exposure: implications for neuropsychiatric research and treatment. *Neuropsychiatric Disease and Treatment, 8*, 131.

Boston, P. F., Jackson, P., Kynoch, P. A., & Thompson, R. J. (1982). Purification, Properties, and Immunohistochemical Localisation of Human Brain 14-3-3 Protein. *Journal of Neurochemistry, 38*(5), 1466–1474.

Božina, N., Jovanović, N., Podlesek, A., Kuzman, M. R., Slijepčević, M. K., Roguljić, A., ... & Medved, V. (2012). Suicide ideators and attempters with schizophrenia–the role of 5-HTTLPR, rs25531, and 5-HTT VNTR intron 2 variants. *Journal of Psychiatric Research, 46*(6), 767–773.

Bremner, J. D. (2006). Traumatic stress: effects on the brain. *Dialogues in Clinical Neuroscience, 8*(4), 445–461.

Brent, D., Melhem, N., Ferrell, R., Emslie, G., Wagner, K. D., Ryan, N., ... & Keller, M. (2010). Association of FKBP5 polymorphisms with suicidal events in the Treatment of Resistant Depression in Adolescents (TORDIA) study. *American Journal of Psychiatry*, 167(2), 190–197.

Brezo, J., Bureau, A., Mérette, C., Jomphe, V., Barker, E. D., Vitaro, F., ... & Turecki, G. (2010). Differences and similarities in the serotonergic diathesis for suicide attempts and mood disorders: a 22-year longitudinal gene–environment study. *Molecular Psychiatry, 15*(8), 831–843.

Brezun, J. M., & Daszuta, A. (1999). Depletion in serotonin decreases neurogenesis in the dentate gyrus and the subventricular zone of adult rats. *Neuroscience, 89*(4), 999–1002.

Bridge, J. A., Goldstein, T. R., & Brent, D. A. (2006). Adolescent suicide and suicidal behavior. *Journal of Child Psychology and Psychiatry, 47*(3–4), 372–394.

Brummett, B. H., Krystal, A. D., Siegler, I. C., Kuhn, C., Surwit, R. S., Züchner, S., ... & Williams, R. B. (2007). Associations of a regulatory polymorphism of monoamine oxidase-A gene promoter (MAOA-uVNTR) with symptoms of depression and sleep quality. *Psychosomatic Medicine, 69*(5), 396–401.

Brunner, D., Buhot, M. C., Hen, R., & Hofer, M. (1999). Anxiety, motor activation, and maternal–infant interactions in 5HT 1B knockout mice. *Behavioral Neuroscience, 113*(3), 587–601.

Buckley, S. T., Foley, P. F., Innes, D. J., Loh, E. W., Shen, Y., Williams, S. M., ... & Dodd, P. R. (2006). GABA A receptor β isoform protein expression in human alcoholic brain: interaction with genotype. *Neurochemistry International, 49*(6), 557–567.

Butler, A. W., Breen, G., Tozzi, F., Craddock, N., Gill, M., Korszun, A., ... & McGuffin, P. (2010). A genome wide linkage study on suicidality in major depressive disorder confirms evidence for linkage to 2p12. *American Journal of Medical Genetics Part B: Neuropsychiatric Genetics, 153*(8), 1465–1473.

Campbell, S., Marriott, M., Nahmias, C., & MacQueen, G. M. (2014). Lower hippocampal volume in patients suffering from depression: a meta-analysis. *American Journal of Psychiatry*, 161(4), 598–607.

Cannon, D. M., Klaver, J. K., Gandhi, S. K., Solorio, G., Peck, S. A., Erickson, K., ... & Drevets, W. C. (2011). Genetic variation in cholinergic muscarinic-2 receptor gene modulates M2 receptor binding in vivo and accounts for reduced binding in bipolar disorder. *Molecular Psychiatry, 16*(4), 407–418.

Cascio, M. (2006). Modulating inhibitory ligand-gated ion channels. *AAPS Journal, 8*(2), E353–E361.

Caspi, A., McClay, J., Moffitt, T. E., Mill, J., Martin, J., Craig, I. W., ... & Poulton, R. (2002). Role of genotype in the cycle of violence in maltreated children. *Science, 297*(5582), 851–854.

Caspi, A., Sugden, K., Moffitt, T. E., Taylor, A., Craig, I. W., Harrington, H., ... & Poulton, R. (2003). Influence of life stress on depression: moderation by a polymorphism in the 5-HTT gene. *Science, 301*(5631), 386–389.

Champagne, F. A., & Curley, J. P. (2009). Epigenetic mechanisms mediating the long-term effects of maternal care on development. *Neuroscience & Biobehavioral Reviews, 33*(4), 593–600.

Charmandari, E., Tsigos, C., & Chrousos, G. (2005). Endocrinology of the stress response 1. *Annual Review of Physiology, 67*, 259–284.

Chen, T. J., Blum, K., Mathews, D., Fisher, L., Schnautz, N., Braverman, E. R., ... & Comings, D. E. (2005). Are dopaminergic genes involved in a predisposition to pathological aggression?: Hypothesizing the importance of "super normal controls" in psychiatric genetic research of complex behavioral disorders. *Medical Hypotheses, 65*(4), 703–707.

Chiao, J. Y., & Blizinsky, K. D. (2010). Culture–gene coevolution of individualism–collectivism and the serotonin transporter gene. *Proceedings of the Royal Society of London B: Biological Sciences, 277*(1681), 529–537.

Cohen-Woods, S., Gaysina, D., Craddock, N., Farmer, A., Gray, J., Gunasinghe, C., ... & McGuffin, P. (2009). Depression Case Control (DeCC) Study fails to support involvement of the muscarinic acetylcholine receptor M2 (CHRM2) gene in recurrent major depressive disorder. *Human Molecular Genetics, 18*(8), 1504–1509.

Comings, D. E., Wu, S., Rostamkhani, M., McGue, M., Iacono, W. G., & MacMurray, J. P. (2002). Association of the muscarinic cholinergic 2 receptor (CHRM2) gene with major depression in women. *American Journal of Medical Genetics, 114*(5), 527–529.

Costanza, A., D'Orta, I., Perroud, N., Burkhardt, S., Malafosse, A., Mangin, P., & La Harpe, R. (2014). Neurobiology of suicide: do biomarkers exist? *International Journal of Legal Medicine, 128*(1), 73–82.

Cumbler, E., Furfari, K., & Guerrasio, J. (2009). Creutzfeldt-Jacob disease presenting as severe depression: a case report. *Cases Journal, 2*(1), 122.

Currier, D., & Mann, J. J. (2008). Stress, genes and the biology of suicidal behavior. *Psychiatric Clinics of North America, 31*(2), 247–269.

De Luca, V., Strauss, J., & Kennedy, J. L. (2008). Power based association analysis (PBAT) of serotonergic and noradrenergic polymorphisms in bipolar patients with suicidal behaviour. *Progress in Neuro-Psychopharmacology and Biological Psychiatry, 32*(1), 197–203.

De Luca, V., Tharmaligam, S., Strauss, J., & Kennedy, J. L. (2008). 5-HT2C receptor and MAO-A interaction analysis: no association with suicidal behaviour in bipolar patients. *European Archives of Psychiatry and Clinical Neuroscience, 258*(7), 428–433.

De Luca, V., Tharmalingam, S., Zai, C., Potapova, N., Strauss, J., Vincent, J., & Kennedy, J. L. (2008). Association of HPA axis genes with suicidal behaviour in schizophrenia. *Journal of Psychopharmacology*, 24(5), 677–682.

Deveci, A., Aydemir, O., Taskin, O., Taneli, F., & Esen-Danaci, A. (2007). Serum BDNF levels in suicide attempters related to psychosocial stressors: a comparative study with depression. *Neuropsychobiology, 56*(2–3), 93–97.

Di Fede, G., Giaccone, G., Limido, L., Mangieri, M., Suardi, S., Puoti, G., ... & Tagliavini, F. (2007). The ε isoform of 14-3-3 protein is a component of the prion protein amyloid deposits of Gerstmann-Sträussler-Scheinker disease. *Journal of Neuropathology & Experimental Neurology, 66*(2), 124–130.

Dlugos, A. M., Palmer, A. A., & de Wit, H. (2009). Negative emotionality: monoamine oxidase B gene variants modulate personality traits in healthy humans. *Journal of Neural Transmission, 116*(10), 1323–1334.

Ellingrod, V. L., Miller, D. D., Taylor, S. F., Moline, J., Holman, T., & Kerr, J. (2008). Metabolic syndrome and insulin resistance in schizophrenia patients receiving antipsychotics genotyped for the methylenetetrahydrofolate reductase (MTHFR) 677C/T and 1298A/C variants. *Schizophrenia Research, 98*(1), 47–54.

Feng, L. G., Song, Z. W., Xin, F., & Hu, J. (2009). Association of plasma homocysteine and methylenetetrahydrofolate reductase C677T gene variant with schizophrenia: a Chinese Han population-based case–control study. *Psychiatry Research, 168*(3), 205–208.

Fiori, L. M., Zouk, H., Himmelman, C., & Turecki, G. (2011). X chromosome and suicide. *Molecular Psychiatry, 16*(2), 216–226.

Fiskerstrand, C. E., Lovejoy, E. A., & Quinn, J. P. (1999). An intronic polymorphic domain often associated with susceptibility to affective disorders has allele dependent differential enhancer activity in embryonic stem cells. *FEBS Letters, 458*(2), 171–174.

Födinger, M., Hörl, W. H., & Sunder-Plassmann, G. (1999). Molecular biology of 5, 10-methylenetetrahydrofolate reductase. *Journal of Nephrology, 13*(1), 20–33.

Fountoulakis, M., Cairns, N., & Lubec, G. (1999). *Increased levels of 14-3-3 gamma and epsilon proteins in brain of patients with Alzheimer's disease and Down syndrome* (pp. 323–335). Springer Vienna.

Galfalvy, H., Zalsman, G., Huang, Y. Y., Murphy, L., Rosoklija, G., Dwork, A. J., ... & Mann, J. J. (2013). A pilot genome wide association and gene expression array study of suicide with and without major depression. *World Journal of Biological Psychiatry, 14*(8), 574–582.

Gaysina, D., Cohen, S., Craddock, N., Farmer, A., Hoda, F., Korszun, A., ... & McGuffin, P. (2008). No association with the 5, 10-methylenetetrahydrofolate reductase gene and major depressive disorder: Results of the depression case control (DeCC) study and a meta-analysis. *American Journal of Medical Genetics Part B: Neuropsychiatric Genetics*, 147(6), 699–706.

Gaysina, D., Jurjev, E., Gabdulhatov, R., & Khusnutdinova, E. (2003). Association Study of the Serotonin 2A and 1B Receptor Genes with Suicidal Behavior in Two Different Populations from Russia. *Balkan Journal of Medical Genetics, 6*, 51–56.

Gaysina, D. A., Jurjev, E. B., Gumerova, R. B., Gabdulhatov, R. M., Khusnutdinova, E. K. (2004). Associaciya polimorfnyh DNK – lokusov gena D2 receptora dofamina i gena perenoschika dofamina s agressivnym povedeniem [Association of the dopamine D2 receptor and transporter genes with human aggressive behaviour]. *Medicinskaya Genetika, 3*(3), 145–148.

Gaysina, D. A., Jurjev, E. B., Khusnutdinova, E. K. (2003). Izuchenie associacii ryada polimorfnyh DNK – lokusov serotoninergicheskoj sistemy s suicidal'nym povedeniem [Association study of polymorphic loci of serotonergic system with suicidal attempts]. *Medicinskaya Genetika*, 2(6), 278–281.

Gaysina, D. A., Khallilova, Z. L., Khusnutdinova, E. K. (2008). Geneticheskie faktory riska suicidal'nogo povedeniya [Genetic risk factors of suicidal behaviour]. *Zhurnal Nevrologii I Psikhiatrii Imeni S.S. Korsakova, 108*(1), 87–91.

Gaysina, D., Zainullina, A., Gabdulhakov, R., & Khusnutdinova, E. (2006). The serotonin transporter gene: polymorphism and haplotype analysis in Russian suicide attempters. *Neuropsychobiology, 54*(1), 70–74.

Gaysina, D. A., Zainullina, A. G., Kazantseva, A. V., Jurjev, E. B, & Khusnutdinova, E. K. (2007). The serotonin transporter gene: is it a gender-specific candidate gene for suicidal behavior? In S. B. Malykh & A. M. Torgersen (Eds), *Genes, Brain, Behavior* (pp. 193–203). Oslo, Moscow.

Gaysina, D., Zainullina, A., & Khusnutdinova, E. (2012). Inherited neuropsychiatric disorders in Russia. In D. Kumar (Ed.), *Genomics and Health in the Developing World.* Oxford University Press.

Gilbody, S., Lewis, S., & Lightfoot, T. (2007). Methylenetetrahydrofolate reductase (MTHFR) genetic polymorphisms and psychiatric disorders: a HuGE review. *American Journal of Epidemiology, 165*(1), 1–13.

Gilger, J. W. (2000). Contributions and promise of human behavioral genetics. *Human Biology*, 229–255.

Goin-Kochel, R. P., Porter, A. E., Peters, S. U., Shinawi, M., Sahoo, T., & Beaudet, A. L. (2009). The MTHFR 677C→ T polymorphism and behaviors in children with autism: exploratory genotype–phenotype correlations. *Autism Research, 2*(2), 98–108.

Grover, D., Verma, R., Goes, F. S., Mahon, P. L. B., Gershon, E. S., McMahon, F. J., & Potash, J. B. (2009). Family-based association of YWHAH in psychotic bipolar disorder. *American Journal of Medical Genetics Part B: Neuropsychiatric Genetics, 150*(7), 977–983.

Gunnar, M. R., & Cheatham, C. L. (2003). Brain and behavior interface: Stress and the developing brain. *Infant Mental Health Journal, 24*(3), 195–211.

Habib, K. E., Gold, P. W., & Chrousos, G. P. (2001). Neuroendocrinology of stress. *Endocrinology and Metabolism Clinics of North America, 30*(3), 695–728.

Hasegawa, H., & Ichiyama, A. (1987). Tryptophan 5-monooxygenase from mouse mastocytoma: High-performance liquid chromatography assay. *Methods in Enzymology, 142*, 88–92.

Heils, A., Mössner, R., & Lesch, K. P. (1997). The human serotonin transporter gene polymorphism-basic research and clinical implications. *Journal of Neural Transmission, 104*(10), 1005–1014.

Heim, C., & Nemeroff, C. B. (2001). The role of childhood trauma in the neurobiology of mood and anxiety disorders: preclinical and clinical studies. *Biological Psychiatry, 49*(12), 1023–1039.

Heim, C., Newport, D. J., Mletzko, T., Miller, A. H., & Nemeroff, C. B. (2008). The link between childhood trauma and depression: insights from HPA axis studies in humans. *Psychoneuroendocrinology, 33*(6), 693–710.

Hochberg, Z. E., Feil, R., Constancia, M., Fraga, M., Junien, C., Carel, J. C., ... & Albertsson-Wikland, K. (2010). Child health, developmental plasticity, and epigenetic programming. *Endocrine Reviews, 32*(2), 159–224.

Hong, C. J., Tsai, S. J., & Wang, Y. C. (2001). Association between tryptophan hydroxylase gene polymorphism (A218C) and schizophrenic disorders. *Schizophrenia Research, 49*(1), 59–63.

Hranilovic, D., Stefulj, J., Furac, I., Kubat, M., Balija, M., & Jernej, B. (2003). Serotonin transporter gene promoter (5-HTTLPR) and intron 2 (VNTR) polymorphisms in Croatian suicide victims. *Biological Psychiatry, 54*(9), 884–889.

Hu, X. Z., Lipsky, R. H., Zhu, G., Akhtar, L. A., Taubman, J., Greenberg, B. D., ... & Goldman, D. (2006). Serotonin transporter promoter gain-of-function genotypes are linked to obsessive-compulsive disorder. *American Journal of Human Genetics, 78*(5), 815–826.

Hung, C. F., Lung, F. W., Hung, T. H., Chong, M. Y., Wu, C. K., Wen, J. K., & Lin, P. Y. (2012). Monoamine oxidase A gene polymorphism and suicide: an association study and meta-analysis. *Journal of Affective Disorders, 136*(3), 643–649.

Ichimura, T., Isobe, T., Okuyama, T., Yamauchi, T., & Fujisawa, H. (1987). Brain 14–3–3 protein is an activator protein that activates tryptophan 5-monooxygenase and tyrosine 3-monooxygenase in the presence of Ca 2+, calmodulin-dependent protein kinase II. *FEBS Letters, 219*(1), 79–82.

Ikeda, M., Hikita, T., Taya, S., Uraguchi-Asaki, J., Toyo-oka, K., Wynshaw-Boris, A., ... & Iwata, N. (2008). Identification of YWHAE, a gene encoding 14–3-3 epsilon, as a possible susceptibility gene for schizophrenia. *Human Molecular Genetics, 17*(20), 3212–3222.

Inoue, K., Itoh, K., Yoshida, K., Shimizu, T., & Suzuki, T. (2004). Positive association between T-182C polymorphism in the norepinephrine transporter gene and susceptibility to major depressive disorder in a Japanese population. *Neuropsychobiology, 50*(4), 301–304.

Ivanova, A. E., Sabgajda, T. P., Semenova, V. G., Antonova, O. I., Nikitina, S. Y., Evdokushkina, G. N., Chernobavskij, M. V. (2011). Smertnost' rossijskih podrostkov ot samoubijstv [Death as a result of suicide among Russian adolescents]. UNISEF, pp. 133.

Jacobs, B. L., Van Praag, H., & Gage, F. H. (2000). Adult brain neurogenesis and psychiatry: a novel theory of depression. *Molecular Psychiatry, 5*(3), 262–269.

Jentsch, T. J., Stein, V., Weinreich, F., & Zdebik, A. A. (2002). Molecular structure and physiological function of chloride channels. *Physiological Reviews, 82*(2), 503–568.

Jia, P., Wang, L., Fanous, A. H., Pato, C. N., Edwards, T. L., & Zhao, Z. (2012). Network-assisted investigation of combined causal signals from genome-wide association studies in schizophrenia. *PLoS Computational Biology, 8*(7), e1002587.

Jia, Y., Yu, X., Zhang, B., Yuan, Y., Xu, Q., Shen, Y., & Shen, Y. (2004). An association study between polymorphisms in three genes of 14–3-3 (tyrosine 3-monooxygenase/tryptophan 5-monooxygenase activation protein) family and paranoid schizophrenia in northern Chinese population. *European Psychiatry, 19*(6), 377–379.

Keck, M. E., Kern, N., Erhardt, A., Unschuld, P. G., Ising, M., Salyakina, D., ... & Binder, E. B. (2008). Combined effects of exonic polymorphisms in CRHR1 and AVPR1B genes in a case/control study for panic disorder. *American Journal of Medical Genetics Part B: Neuropsychiatric Genetics, 147*(7), 1196–1204.

Kia-Keating, B. M., Glatt, S. J., & Tsuang, M. T. (2007). Meta-analyses suggest association between COMT, but not HTR1B, alleles, and suicidal behavior. *American Journal of Medical Genetics Part B: Neuropsychiatric Genetics, 144*(8), 1048–1053.

Klimek, V., Stockmeier, C., Overholser, J., Meltzer, H. Y., Kalka, S., Dilley, G., & Ordway, G. A. (1997). Reduced levels of norepinephrine transporters in the locus coeruleus in major depression. *Journal of Neuroscience, 17*(21), 8451–8458.

Koller, G., Bondy, B., Preuss, U. W., Zill, P., & Soyka, M. (2006). The C (-1019) G 5-HT1A promoter polymorphism and personality traits: no evidence for significant association in alcoholic patients. *Behavioral and Brain Functions, 2*(7).

Korosi, A., & Baram, T. Z. (2008). The central corticotropin releasing factor system during development and adulthood. *European Journal of Pharmacology, 583*(2), 204–214.

Kraft, J. B., Slager, S. L., McGrath, P. J., & Hamilton, S. P. (2005). Sequence analysis of the serotonin transporter and associations with antidepressant response. *Biological Psychology, 58*(5), 374–381.

Kuhn, D. M., O'Callaghan, J. P., Juskevich, J., & Lovenberg, W. (1980). Activation of brain tryptophan hydroxylase by ATP-Mg2+: dependence on calmodulin. *Proceedings of the National Academy of Sciences, 77*(8), 4688–4691.

Lachman, H. M., Papolos, D. F., Saito, T., Yu, Y. M., Szumlanski, C. L., & Weinshilboum, R. M. (1996). Human catechol-O-methyltransferase pharmacogenetics: description of a functional polymorphism and its potential application to neuropsychiatric disorders. *Pharmacogenetics and Genomics, 6*(3), 243–250.

Ladd, C. O., Huot, R. L., Thrivikraman, K. V., Nemeroff, C. B., Meaney, M. J., Plotsky, P., ... & Saper, C. B. (2000). Long-term behavioral and neuroendocrine adaptations to adverse early experience. In E. A. Mayer & C. B. Saper (Eds), The biological basis for mind body interactions. *Progress in Brain Research* (Vol. 122, pp. 81–104). Amsterdam: Elsevier.

Laje, G., Allen, A. S., Akula, N., Manji, H., Rush, A. J., & McMahon, F. J. (2009). Genome-wide association study of suicidal ideation emerging during citalopram treatment of depressed outpatients. *Pharmacogenetics and Genomics, 19*(9), 666–674.

Lappalainen, J., Long, J. C., Eggert, M., Ozaki, N., Robin, R. W., Brown, G. L., ... & Goldman, D. (1998). Linkage of antisocial alcoholism to the serotonin 5-HT1B receptor gene in 2 populations. *Archives of General Psychiatry, 55*(11), 989–994.

Lee, H. Y., & Kim, Y. K. (2011). Gender effect of catechol-O-methyltransferase Val158Met polymorphism on suicidal behavior. *Neuropsychobiology, 63*(3), 177–182.

Lemonde, S., Turecki, G., Bakish, D., Du, L., Hrdina, P. D., Bown, C. D., ... & Albert, P. R. (2003). Impaired repression at a 5-hydroxytryptamine 1A receptor gene polymorphism associated with major depression and suicide. *Journal of Neuroscience, 23*(25), 8788–8799.

Lesch, K. P., Bengel, D., Heils, A., Sabol, S. Z., Greenberg, B. D., Petri, S., ... & Murphy, D. L. (1996). Association of anxiety-related traits with a polymorphism in the serotonin transporter gene regulatory region. *Science, 274*(5292), 1527–1531.

Linkowska, K., Daca, P., Sykutera, M., Pufal, E., Bloch-Bogusławska, E., & Grzybowski, T. (2009). Search for association between suicide and 5-HTT, MAOA and DAT polymorphism in Polish males. *Archiwum Medycyny Sadowej i Kryminologii, 60*(2–3), 112–117.

Liu, D., Diorio, J., Tannenbaum, B., Caldji, C., Francis, D., Freedman, A., ... & Meaney, M. J. (1997). Maternal care, hippocampal glucocorticoid receptors, and hypothalamic-pituitary-adrenal responses to stress. *Science, 277*(5332), 1659–1662.

Liu, D., Diorio, J., Day, J. C., Francis, D. D., & Meaney, M. J. (2000). Maternal care, hippocampal synaptogenesis and cognitive development in rats. *Nature Neuroscience, 3*(8), 799–806.

Liu, X., Li, H., Qin, W., He, G., Li, D., Shen, Y., ... & He, L. (2006). Association of TPH1 with suicidal behaviour and psychiatric disorders in the Chinese population. *Journal of Medical Genetics, 43*(2), 4–10.

Lopez-Leon, S., Janssens, A. C. J. W., Ladd, A. G. Z., Del-Favero, J., Claes, S. J., Oostra, B. A., & Van Duijn, C. M. (2008). Meta-analyses of genetic studies on major depressive disorder. *Molecular Psychiatry, 13*(8), 772–785.

Lung, F. W., Tzeng, D. S., Huang, M. F., & Lee, M. B. (2011). Association of the MAOA promoter uVNTR polymorphism with suicide attempts in patients with major depressive disorder. *BMC Medical Genetics, 12*(1), 74.

Luo, X., Kranzler, H. R., Zuo, L., Wang, S., Blumberg, H. P., & Gelernter, J. (2005). CHRM2 gene predisposes to alcohol dependence, drug dependence and affective disorders: results from an extended case–control structured association study. *Human Molecular Genetics, 14*(16), 2421–2434.

Lupien, S. J., McEwen, B. S., Gunnar, M. R., & Heim, C. (2009). Effects of stress throughout the lifespan on the brain, behaviour and cognition. *Nature Reviews Neuroscience, 10*(6), 434–445.

Ma, J., Yoshimura, M., Yamashita, E., Nakagawa, A., Ito, A., & Tsukihara, T. (2004). Structure of rat monoamine oxidase A and its specific recognitions for substrates and inhibitors. *Journal of Molecular Biology, 338*(1), 103–114.

MacKay, R. K., Colson, N. J., Dodd, P. R., & Lewohl, J. M. (2011). Differential Expression of 14-3-3 Isoforms in Human Alcoholic Brain. *Alcoholism: Clinical and Experimental Research, 35*(6), 1041–1049.

Mackie, S., & Aitken, A. (2005). Novel brain 14-3-3 interacting proteins involved in neurodegenerative disease. *FEBS Journal, 272*(16), 4202–4210.

Mandelli, L., & Serretti, A. (2013). Gene environment interaction studies in depression and suicidal behavior: an update. *Neuroscience & Biobehavioral Reviews,* 37(10), 2375–2397.

Mann, J. J. (2003). Neurobiology of suicidal behaviour. *Nature Reviews Neuroscience, 4*(10), 819–828.

Mann, J. J., Arango, V. A., Avenevoli, S., Brent, D. A., Champagne, F. A., Clayton, P., ... & Wenzel, A. (2009). Candidate endophenotypes for genetic studies of suicidal behavior. *Biological Psychiatry, 65*(7), 556–563.

Mann, J. J., Brent, D. A., & Arango, V. (2001). The neurobiology and genetics of suicide and attempted suicide: a focus on the serotonergic system. *Neuropsychopharmacology, 24*(5), 467–477.

Mann, J. J., & Currier, D. (2007). A review of prospective studies of biologic predictors of suicidal behavior in mood disorders. *Archives of Suicide Research, 11*(1), 3–16.

Manuck, S. B., Flory, J. D., Ferrell, R. E., & Muldoon, M. F. (2004). Socio-economic status covaries with central nervous system serotonergic responsivity as a function of allelic variation in the serotonin transporter gene-linked polymorphic region. *Psychoneuroendocrinology, 29*(5), 651–668.

Martins-de-Souza, D., Gattaz, W. F., Schmitt, A., Rewerts, C., Maccarrone, G., Dias-Neto, E., & Turck, C. W. (2009). Prefrontal cortex shotgun proteome analysis reveals altered calcium homeostasis and immune system imbalance in schizophrenia. *European Archives of Psychiatry and Clinical Neuroscience, 259*(3), 151–163.

Mathew, C. G. P. (1984). The isolation of high molecular weight eukaryotic DNA. In *Nucleic Acids* (pp. 31–34). Humana Press.

McGowan, P. O., Sasaki, A., D'Alessio, A. C., Dymov, S., Labonté, B., Szyf, M., ... & Meaney, M. J. (2009). Epigenetic regulation of the glucocorticoid receptor in human brain associates with childhood abuse. *Nature Neuroscience, 12*(3), 342–348.

Meaney, M. J., & Szyf, M. (2005). Maternal care as a model for experience-dependent chromatin plasticity? *Trends in Neurosciences, 28*(9), 456–463.

Middleton, F. A., Peng, L., Lewis, D. A., Levitt, P., & Mirnics, K. (2005). Altered expression of 14–3–3 genes in the prefrontal cortex of subjects with schizophrenia. *Neuropsychopharmacology, 30*(5), 974–983.

Mockus, S. M., & Vrana, K. E. (1998). Advances in the molecular characterization of tryptophan hydroxylase. *Journal of Molecular Neuroscience, 10*(3), 163–179.

Moffitt, T. E. (2005). Genetic and environmental influences on antisocial behaviors: evidence from behavioral–genetic research. *Advances in Genetics, 55*, 41–104.

Mohammad, N. S., Jain, J. M. N., Chintakindi, K. P., Singh, R. P., Naik, U., & Akella, R. R. D. (2009). Aberrations in folate metabolic pathway and altered susceptibility to autism. *Psychiatric Genetics, 19*(4), 171–176.

Morrison, D. K. (2009). The 14–3-3 proteins: integrators of diverse signaling cues that impact cell fate and cancer development. *Trends in Cell Biology, 19*(1), 16–23.

Müller, N., & Schwarz, M. J. (2007). The immune-mediated alteration of serotonin and glutamate: towards an integrated view of depression. *Molecular Psychiatry, 12*(11), 988–1000.

Murgatroyd, C., Patchev, A. V., Wu, Y., Micale, V., Bockmühl, Y., Fischer, D., ... & Spengler, D. (2009). Dynamic DNA methylation programs persistent adverse effects of early-life stress. *Nature Neuroscience, 12*(12), 1559–1566.

Murphy, T. M., Ryan, M., Foster, T., Kelly, C., McClelland, R., O'Grady, J., ... & Malone, K. M. (2011). Risk and protective genetic variants in suicidal behaviour: association with SLC1A2, SLC1A3, 5-HTR1B &NTRK2 polymorphisms. *Behavioral and Brain Functions*, 28, *7–22*.

Muslin, A. J., Tanner, J. W., Allen, P. M., & Shaw, A. S. (1996). Interaction of 14–3-3 with signaling proteins is mediated by the recognition of phosphoserine. *Cell, 84*(6), 889–897.

Nagamani, S. S., Zhang, F., Shchelochkov, O. A., Bi, W., Ou, Z., Scaglia, F., ... & Stankiewicz, P. (2009). Microdeletions including YWHAE in the Miller–Dieker syndrome region on chromosome 17p13. 3 result in facial dysmorphisms, growth restriction, and cognitive impairment. *Journal of Medical Genetics, 46*(12), 825–833.

Nakamura, M., Ueno, S., Sano, A., & Tanabe, H. (2000). The human serotonin transporter gene linked polymorphism (5-HTTLPR) shows ten novel allelic variants. *Molecular Psychiatry, 5*(1), 32–38.

Nakanishi, S. (1992). Molecular diversity of glutamate receptors and implications for brain function. *Science*, 258(5082), 597–603.

Nedic, G., Nikolac, M., Sviglin, K. N., Muck-Seler, D., Borovecki, F., & Pivac, N. (2011). Association study of a functional catechol-O-methyltransferase (COMT) Val108/158Met polymorphism and suicide attempts in patients with alcohol dependence. *International Journal of Neuropsychopharmacology, 14*(3), 377–388.

Nemeroff, C. B., & Vale, W. W. (2005). The neurobiology of depression: inroads to treatment and new drug discovery. *Journal of Clinical Psychiatry, 66*, 5–13.

Neves, F. S., Malloy-Diniz, L. F., Romano-Silva, M. A., Aguiar, G. C., de Matos, L. O., & Correa, H. (2010). Is the serotonin transporter polymorphism (5-HTTLPR) a potential marker for suicidal behavior in bipolar disorder patients? *Journal of Affective Disorders, 125*(1), 98–102.

Neves, F. S., Malloy-Diniz, L., Romano-Silva, M. A., Campos, S. B., Miranda, D. M., De Marco, L., ... & Correa, H. (2011). The role of BDNF genetic polymorphisms in bipolar disorder with psychiatric comorbidities. *Journal of Affective Disorders, 131*(1), 307–311.

Neves, F. S., Silveira, G., Romano-Silva, M. A., Malloy-Diniz, L., Ferreira, A. A., De Marco, L., & Correa, H. (2008). Is the 5-HTTLPR polymorphism associated with bipolar disorder or with suicidal behavior of bipolar disorder patients? *American Journal of Medical Genetics Part B: Neuropsychiatric Genetics, 147*(1), 114–116.

New, A. S., Gelernter, J., Goodman, M., Mitropoulou, V., Koenigsberg, H., Silverman, J., & Siever, L. J. (2001). Suicide, impulsive aggression, and HTR1B genotype. *Biological Psychiatry, 50*(1), 62–65.

Nielsen, D. A., Goldman, D., Virkkunen, M., Tokola, R., Rawlings, R., & Linnoila, M. (1994). Suicidality and 5-hydroxyindoleacetic acid concentration associated with a tryptophan hydroxylase polymorphism. *Archives of General Psychiatry, 51*(1), 34–38.

Nielsen, D. A., Virkkunen, M., Lappalainen, J., Eggert, M., Brown, G. L., Long, J. C., ... & Linnoila, M. (1998). A tryptophan hydroxylase gene marker for suicidality and alcoholism. *Archives of General Psychiatry, 55*(7), 593–602.

Noskova, T., Pivac, N., Nedic, G., Kazantseva, A., Gaysina, D., Faskhutdinova, G., ... & Seler, D. M. (2008). Ethnic differences in the serotonin transporter polymorphism (5-HTTLPR) in several European populations. *Progress in Neuro-Psychopharmacology and Biological Psychiatry, 32*(7), 1735–1739.

Oquendo, M. A., & Mann, J. J. (2001). Neuroimaging findings in major depression, suicidal behavior and aggression. *Clinical Neuroscience Research, 1*(5), 377–380.

Ozbek, Z., Kucukali, C. I., Ozkok, E., Orhan, N., Aydin, M., Kilic, G., ... & Kara, I. (2008). Effect of the methylenetetrahydrofolate reductase gene polymorphisms on homocysteine, folate and vitamin B12 in patients with bipolar disorder and relatives. *Progress in Neuro-Psychopharmacology and Biological Psychiatry, 32*(5), 1331–1337.

Paska, A. V., Zupanc, T., & Pregelj, P. (2013). The role of brain-derived neurotrophic factor in the pathophysiology of suicidal behavior. *Psychiatr Danub, 25*(Suppl 2), S341–4.

Papiol, S., Arias, B., Gastó, C., Gutiérrez, B., Catalán, R., & Fañanás, L. (2007). Genetic variability at HPA axis in major depression and clinical response to antidepressant treatment. *Journal of Affective Disorders, 104*(1), 83–90.

Paus, T., Keshavan, M., & Giedd, J. N. (2008). Why do many psychiatric disorders emerge during adolescence? *Nature Reviews Neuroscience, 9*(12), 947–957.

Peerbooms, O. L., van Os, J., Drukker, M., Kenis, G., Hoogveld, L., De Hert, M., ... & Rutten, B. P. (2011). Meta-analysis of MTHFR gene variants in schizophrenia, bipolar disorder and unipolar depressive disorder: evidence for a common genetic vulnerability? *Brain, Behavior, and Immunity, 25*(8), 1530–1543.

Perlis, R. H., Huang, J., Purcell, S., Fava, M., Rush, A. J., Sullivan, P. F., ... & Wellcome Trust Case Control Consortium Bipolar Disorder Group. (2010). Genome-wide association study of suicide attempts in mood disorder patients. *American Journal of Psychiatry*, 167, 1499–1507.

Pivac, N., Pregelj, P., Nikolac, M., Zupanc, T., Nedic, G., Muck Seler, D., & Videtic Paska, A. (2011). The association between catechol-O-methyl-transferase Val108/158Met polymorphism and suicide. *Genes, Brain and Behavior, 10*(5), 565–569.

Planells-Cases, R., & Jentsch, T. J. (2009). Chloride channelopathies. *Biochimica et Biophysica Acta (BBA)-Molecular Basis of Disease, 1792*(3), 173–189.

Podolsky, A. I., Karabanova, O. A., Idobaeva, O. A., Heymans, P. (2011). Psihoemocional'noe blagopoluchie sovremennyh podrostkov: opyt mezhdunarodnogo issledovaniya [Psychoemotional wellbeing of contemporary adolescents: international study experience]. *Vestnik Moskovskogo Universiteta*, 14, Psychology, 2, 9–20.

Pregelj, P., Nedic, G., Paska, A. V., Zupanc, T., Nikolac, M., Balažic, J., ..., & Pivac, N. (2011). The association between brain-derived neurotrophic factor polymorphism (BDNF Val66Met) and suicide. *Journal of Affective Disorders, 128*(3), 287–290.

Roffman, J. L., Brohawn, D. G., Nitenson, A. Z., Macklin, E. A., Smoller, J. W., & Goff, D. C. (2011). Genetic variation throughout the folate metabolic pathway influences negative symptom severity in schizophrenia. *Schizophrenia Bulletin, 39*(2), 330–338.

Rot, Maan het., Mathew, S. J., & Charney, D. S. (2009). Neurobiological mechanisms in major depressive disorder. *Canadian Medical Association Journal, 180*(3), 305–313.

Rotondo, A., Schuebel, K., Bergen, A., Aragon, R., Virkkunen, M., Linnoila, M., ..., & Nielsen, D. (1999). Identification of four variants in the tryptophan hydroxylase promoter and association to behavior. *Molecular Psychiatry, 4*(4), 360–368.

Roy, A., Hu, X. Z., Janal, M. N., & Goldman, D. (2007). Interaction between childhood trauma and serotonin transporter gene variation in suicide. *Neuropsychopharmacology, 32*(9), 2046–2052.

Roy, A., Hodgkinson, C. A., DeLuca, V., Goldman, D., & Enoch, M. A. (2012). Two HPA axis genes, CRHBP and FKBP5, interact with childhood trauma to increase the risk for suicidal behavior. *Journal of Psychiatric Research, 46*(1), 72–79.

Rutter, M. (2002). The interplay of nature, nurture, and developmental influences: the challenge ahead for mental health. *Archives of General Psychiatry, 59*(11), 996–1000.

Ryu, S. H., Lee, S. H., Lee, H. J., Cha, J. H., Ham, B. J., Han, C. S., ... & Lee, M. S. (2004). Association between norepinephrine transporter gene polymorphism and major depression. *Neuropsychobiology, 49*(4), 174–177.

Saetre, P., Lundmark, P., Wang, A., Hansen, T., Rasmussen, H. B., Djurovic, S., ..., & Jönsson, E. G. (2010). The tryptophan hydroxylase 1 (TPH1) gene, schizophrenia susceptibility, and suicidal behavior: A multi-centre case-control study and meta-analysis. *American Journal of Medical Genetics Part B: Neuropsychiatric Genetics, 153*(2), 387–396.

Saetre, P., Vares, M., Werge, T., Andreassen, O. A., Arinami, T., Ishiguro, H., ..., & Jönsson, E. G. (2011). Methylenetetrahydrofolate reductase (MTHFR) C677T and A1298C polymorphisms and age of onset in schizophrenia: a combined analysis of independent samples. *American Journal of Medical Genetics Part B: Neuropsychiatric Genetics, 156*(2), 215–224.

Samadi Rad, B., Ghasemi, A., Seifi, M., Samadikuchaksaraei, A., Baybordi, F., & Danaei, N. (2012). Serotonin 1A receptor genetic variations, suicide, and life events in the Iranian population. *Psychiatry and Clinical Neurosciences, 66*(4), 337–343.

Sánchez-Valle, R., Saiz, A., & Graus, F. (2002). 14-3-3 Protein isoforms and atypical patterns of the 14-3-3 assay in the diagnosis of Creutzfeldt-Jakob disease. *Neuroscience Letters, 320*(1), 69–72.

Saudou, F., Amara, D. A., Dierich, A., LeMeur, M., Ramboz, S., Segu, L., ..., & Hen, R. (1994). Enhanced aggressive behavior in mice lacking 5-HT1B receptor. *Science, 265*(5180), 1875–1878.

Schlesselman, J. J., & Schneiderman, M. A. (1982). Case control studies: design, conduct, analysis. *Journal of Occupational and Environmental Medicine, 24*(11), 879.

Schosser, A., Butler, A. W., Ising, M., Perroud, N., Uher, R., Ng, M. Y., ..., & Lewis, C. M. (2011). Genomewide association scan of suicidal thoughts and behaviour in major depression. *PLoS One, 6*(7), e20690.

Schreiner, F., Tozakidou, M., Maslak, R., Holtkamp, U., Peter, M., Gohlke, B., & Woelfle, J. (2009). Functional glucocorticoid receptor gene variants do not underlie the high variability of 17-hydroxyprogesterone screening values in healthy newborns. *European Journal of Endocrinology, 160*(4), 667–673.

Sekiguchi, H., Iritani, S., Habuchi, C., Torii, Y., Kuroda, K., Kaibuchi, K., & Ozaki, N. (2011). Impairment of the tyrosine hydroxylase neuronal network in the orbitofrontal cortex of a genetically modified mouse model of schizophrenia. *Brain Research, 1392*, 47–53.

Sequeira, A., Mamdani, F., Ernst, C., Vawter, M. P., Bunney, W. E., Lebel, V., ..., & Turecki, G. (2009). Global brain gene expression analysis links glutamatergic and GABAergic alterations to suicide and major depression. *PLoS One*, 4(8), e6585.

Serretti, A., Calati, R., Giegling, I., Hartmann, A. M., Möller, H. J., & Rujescu, D. (2009). Serotonin receptor HTR1A and HTR2C variants and personality traits in suicide attempters and controls. *Journal of Psychiatric Research, 43*(5), 519–525.

Sher, L. (2006). Alcoholism and suicidal behavior: a clinical overview. *Acta Psychiatrica Scandinavica, 113*(1), 13–22.

Sher, L. (2006). Dopamine and the pathophysiology of suicidal behavior in alcoholism. *Medical Hypotheses, 66*(5), 1043–1044.

Sher, L. (2008). The concept of post-traumatic mood disorder and its implications for adolescent suicidal behavior. *Minerva Pediatrica, 60*(6), 1393–1399.

Sher, L., Mann, J. J., Traskman-Bendz, L., Winchel, R., Huang, Y. Y., Fertuck, E., & Stanley, B. H. (2006). Lower cerebrospinal fluid homovanillic acid levels in depressed suicide attempters. *Journal of Affective Disorders, 90*(1), 83–89.

Shi, Y. Y., He, G., Zhang, Z., Tang, W., Zhang, J., Zhao, Q., ..., & He, L. (2008). A study of rare structural variants in schizophrenia patients and normal controls from Chinese Han population. *Molecular Psychiatry, 13*(10), 911–913.

Słopien, R., Jasniewicz, K., Meczekalski, B., Warenik-Szymankiewicz, A., Lianeri, M., & Jagodziński, P. P. (2008). Polymorphic variants of genes encoding MTHFR, MTR, and MTHFD1 and the risk of depression in postmenopausal women in Poland. *Maturitas, 61*(3), 252–255.

Sokolowski, M., Ben-Efraim, Y. J., Wasserman, J., & Wasserman, D. (2013). Glutamatergic GRIN2B and polyaminergic ODC1 genes in suicide attempts: associations and gene–environment interactions with childhood/adolescent physical assault. *Molecular Psychiatry, 18*(9), 985–992.

Stanley, B., Molcho, A., Stanley, M., Winchel, R., Gameroff, M. J., Parsons, B., & Mann, J. J. (2000). Association of aggressive behavior with altered serotonergic function in patients who are not suicidal. *American Journal of Psychiatry, 157*(4), 609–614.

Steimer, T. (2002). The biology of fear-and anxiety-related behaviors. *Dialogues in Clinical Neuroscience*, 4(3), 231–249.

Strobel, A., Gutknecht, L., Rothe, C., Reif, A., Mössner, R., Zeng, Y., ..., & Lesch, K. P. (2003). Allelic variation in 5-HT1A receptor expression is associated with anxiety-and depression-related personality traits. *Journal of Neural Transmission, 110*(12), 1445–1453.

Suda, A., Kawanishi, C., Kishida, I., Sato, R., Yamada, T., Nakagawa, M., ..., & Hirayasu, Y. (2009). Dopamine D2 receptor gene polymorphisms are associated with suicide attempt in the Japanese population. *Neuropsychobiology, 59*(2), 130–134.

Sugimori, K., Kobayashi, K., Kitamura, T., Sudo, S., & Koshino, Y. (2007). 14-3-3 protein beta isoform is associated with 3-repeat tau neurofibrillary tangles in Alzheimer's disease. *Psychiatry and Clinical Neurosciences, 61*(2), 159–167.

Suomi, S. J. (2006). Risk, resilience, and gene× environment interactions in rhesus monkeys. *Annals of the New York Academy of Sciences, 1094*(1), 52–62.

Supriyanto, I., Sasada, T., Fukutake, M., Asano, M., Ueno, Y., Nagasaki, Y., ..., & Hishimoto, A. (2011). Association of FKBP5 gene haplotypes with completed suicide in the Japanese population. *Progress in Neuro-Psychopharmacology and Biological Psychiatry, 35*(1), 252–256.

Szyf, M. (2009). The early life environment and the epigenome. *Biochimica et Biophysica Acta (BBA)-General Subjects, 1790*(9), 878–885.

Szyf, M., McGowan, P., & Meaney, M. J. (2010). The social environment and the epigenome. In C. M. Worthman, P. M. Plotsky, D. S. Schechter & C. A. Cummings (Eds), *Formative experiences: The interaction of caregiving, culture, and developmental psychobiology* (pp. 53–81). Cambridge University Press.

Tabery, J. (2007). Biometric and developmental gene–environment interactions: Looking back, moving forward. *Development and Psychopathology, 19*(04), 961–976.

Tarquis, N. (2005). Hypothèses neurobiologiques concernant les liens entre psychopathie et maltraitance infantile [Neurobiological hypothesis relating to connections between psychopathy and childhood maltreatment]. *L'Encephale, 32*(3 Pt 1), 377–384.

Tatro, E. T., Nguyen, T. B., Bousman, C. A., Masliah, E., Grant, I., Atkinson, J. H., & Everall, I. P. (2010). Correlation of major depressive disorder symptoms with FKBP5 but not FKBP4 expression in human immunodeficiency virus-infected individuals. *Journal of Neurovirology, 16*(5), 399–404.

Tommerup, N., & Leffers, H. (1996). Assignment of the Human Genes Encoding 14–3-3 Eta (YWHAH) to 22q12, 14–3-3 Zeta (YWHAZ) to 2p25. 1–p25. 2, and 14–3-3 Beta (YWHAB) to 20q13. 1 byin SituHybridization. *Genomics, 33*(1), 149–150.

Tsankova, N., Renthal, W., Kumar, A., & Nestler, E. J. (2007). Epigenetic regulation in psychiatric disorders. *Nature Reviews Neuroscience, 8*(5), 355–367.

Tupala, E., & Tiihonen, J. (2004). Dopamine and alcoholism: neurobiological basis of ethanol abuse. *Progress in Neuro-Psychopharmacology and Biological -Psychiatry, 28*(8), 1221–1247.

Turecki, G., Brière, R., Dewar, K., Antonetti, T., Lesage, A. D., Séguin, M., ..., & Rouleau, G. A. (2014). Prediction of level of serotonin 2A receptor binding by serotonin receptor 2A genetic variation in postmortem brain samples from subjects who did or did not commit suicide. *American Journal of Psychiatry*, 156, 1456–1458.

Turecki, G., Zhu, Z., Tzenova, J., Lesage, A., Seguin, M., Tousignant, M., ..., & Rouleau, G. A. (2001). TPH and suicidal behavior: a study in suicide completers. *Molecular Psychiatry, 6*(1), 98–102.

van Heeringen, K. (2003). The neurobiology of suicide and suicidality. *Canadian Journal of Psychiatry/La Revue canadienne de psychiatrie*, 48(5), 292–300.

Videtič, A., Peternelj, T. T., Zupanc, T., Balažic, J., & Komel, R. (2009a). Promoter and functional polymorphisms of HTR2C and suicide victims. *Genes, Brain and Behavior, 8*(5), 541–545.

Videtič, A., Zupanc, T., Pregelj, P., Balažic, J., Tomori, M., & Komel, R. (2009b). Suicide, stress and serotonin receptor 1A promoter polymorphism-1019C> G in Slovenian suicide victims. *European Archives of Psychiatry and Clinical Neuroscience, 259*(4), 234–238.

Voytsekh, V. F. (2007). Klinicheskaya suicidologiya [Clinical suicidology]. Miklosh: Moscow,.

Walther, D. J., Peter, J. U., Bashammakh, S., Hörtnagl, H., Voits, M., Fink, H., & Bader, M. (2003). Synthesis of serotonin by a second tryptophan hydroxylase isoform. *Science, 299*(5603), 76–76.

Wang, S., Zhang, K., Xu, Y., Sun, N., Shen, Y., & Xu, Q. (2009). An association study of the serotonin transporter and receptor genes with the suicidal ideation of major depression in a Chinese Han population. *Psychiatry Research, 170*(2), 204–207.

Wasserman, D. (Ed.) (2001). *Suicide. An unnecessary death.* London: Martin Dunitz.

Wasserman, D., Geijer, T., Sokolowski, M., Frisch, A., Michaelovsky, E., Weizman, A.,..., & Wasserman, J. (2007). Association of the serotonin transporter promotor polymorphism with suicide attempters with a high medical damage. *European Neuropsychopharmacology, 17*(3), 230–233.

Wasserman, D., Sokolowski, M., Wasserman, J., & Rujesku, D. (2009a) Neurobiology and genetics of suicide. In D. Wasserman & C. Wasserman (Eds), *Oxford Textbook on Suicidology and Suicide Prevention* (pp. 165–182). New York: Oxford University Press.

Wasserman, D., Wasserman, J., Rozanov, V., & Sokolowski, M. (2009b). Depression in suicidal males: genetic risk variants in the CRHR1 gene. *Genes, Brain and Behavior, 8*(1), 72–79.

Wasserman, D., Wasserman, J., & Sokolowski, M. (2010). Genetics of HPA-axis, depression and suicidality. *European Psychiatry, 25*(5), 278–280.

Weinstock, M. (2008). The long-term behavioural consequences of prenatal stress. *Neuroscience & Biobehavioral Reviews, 32*(6), 1073–1086.

Weisberg, I., Tran, P., Christensen, B., Sibani, S., & Rozen, R. (1998). A second genetic polymorphism in methylenetetrahydrofolate reductase (MTHFR) associated with decreased enzyme activity. *Molecular Genetics and Metabolism, 64*(3), 169–172.

Wendland, J. R., Martin, B. J., Kruse, M. R., Lesch, K. P., & Murphy, D. L. (2006). Simultaneous genotyping of four functional loci of human SLC6A4, with a reappraisal of 5-HTTLPR and rs25531. *Molecular Psychiatry,* 11, 224–226.

Willour, V. L., Chen, H., Toolan, J., Belmonte, P., Cutler, D. J., Goes, F. S.,..., & Potash, J. B. (2009). Family-based association of FKBP5 in bipolar disorder. *Molecular Psychiatry, 14*(3), 261–268.

Willour, V. L., Seifuddin, F., Mahon, P. B., Jancic, D., Pirooznia, M., Steele, J.,..., & Potash, J. B. (2012). A genome-wide association study of attempted suicide. *Molecular Psychiatry, 17*(4), 433–444.

Wolf, O. T. (2009). Stress and memory in humans: twelve years of progress? *Brain Research, 1293,* 142–154.

Wong, A. H., Likhodi, O., Trakalo, J., Yusuf, M., Sinha, A., Pato, C. N.,..., & Kennedy, J. L. (2005). Genetic and post-mortem mRNA analysis of the 14–3–3 genes that encode phosphoserine/threonine-binding regulatory proteins in schizophrenia and bipolar disorder. *Schizophrenia Research, 78*(2), 137–146.

Wrzosek, M., Łukaszkiewicz, J., Wrzosek, M., Serafin, P., Jakubczyk, A., Klimkiewicz, A.,..., & Wojnar, M. (2011). Association of polymorphisms in HTR2A, HTR1A and

TPH2 genes with suicide attempts in alcohol dependence: a preliminary report. *Psychiatry Research, 190*(1), 149–151.

Yanagi, M., Shirakawa, O., Kitamura, N., Okamura, K., Sakurai, K., Nishiguchi, N., ..., & Maeda, K. (2005). Association of 14–3–3 ε gene haplotype with completed suicide in Japanese. *Journal of Human Genetics, 50*(4), 210–216.

Zaboli, G., Gizatullin, R., Nilsonne, Å., Wilczek, A., Jönsson, E. G., Ahnemark, E., ..., & Leopardi, R. (2006). Tryptophan hydroxylase-1 gene variants associate with a group of suicidal borderline women. *Neuropsychopharmacology, 31*(9), 1982–1990.

Zainullina, A., Gaysina, D., Akhmerova, I., Valinurov, R., Cohen-Woods, S., Craig, I., ..., & Khusnutdinova, E. (2011). P. 1. a. 018 Association analysis of glucocorticoid receptor gene and the muscarinic cholinergic 2 receptor gene in suicidal behaviour. *European Neuropsychopharmacology, 21* (Supplement 3).

Zalsman, G., Huang, Y. Y., Oquendo, M. A., Burke, A. K., Hu, X. Z., Brent, D. A., ..., & J John Mann, M. D. (2006). Association of a triallelic serotonin transporter gene promoter region (5-HTTLPR) polymorphism with stressful life events and severity of depression. *American Journal of Psychiatry, 163*(9).

Zalsman, G. I. L., Patya, M., Frisch, A., Ofek, H., Schapir, L., Blum, I., ..., & Tyano, S. (2011). Association of polymorphisms of the serotonergic pathways with clinical traits of impulsive-aggression and suicidality in adolescents: a multi-center study. *World Journal of Biological Psychiatry, 12*(1), 33–41.

Zeisel, S. H. (2009). Epigenetic mechanisms for nutrition determinants of later health outcomes. *American Journal of Clinical Nutrition, 89*(5), 1488S-1493S.

Zhang, J., Shen, Y., He, G., Li, X., Meng, J., Guo, S., ..., & He, L. (2008). Lack of association between three serotonin genes and suicidal behavior in Chinese psychiatric patients. *Progress in Neuro-Psychopharmacology and Biological Psychiatry, 32*(2), 467–471.

Zhu, M. Y., & Juorio, A. V. (1995). Aromatic L-amino acid decarboxylase: biological characterization and functional role. *General Pharmacology: The Vascular System, 26*(4), 681–696.

Zubenko, G. S., Maher, B. S., Hughes, H. B., Zubenko, W. N., Scott Stiffler, J., & Marazita, M. L. (2004). Genome-wide linkage survey for genetic loci that affect the risk of suicide attempts in families with recurrent, early-onset, major depression. *American Journal of Medical Genetics Part B: Neuropsychiatric Genetics, 129*(1), 47–54.

Zupanc, T., Pregelj, P., Tomori, M., Komel, R., & Videtič Paska, A. (2010). No association between polymorphisms in four serotonin receptor genes, serotonin transporter gene and alcohol-related suicide. *Psychiatria Danubina, 22*(4.), 522–527.

12
Genetically Informative Investigations of Neurophysiological Traits

Sergey Malykh

Genetic psychophysiology is a fairly new area of research that has emerged at the intersection of behavioural genetics and psychophysiology. It is obvious that the specifics of the structure and functions of the nervous system are the mediating link in the "genes-behaviour" chain. The effects of human genes on behaviour are mediated at a number of different levels – cellular (physical and chemical processes within the cell), intercellular (physical and chemical processes between cells), the organs (their structure and functions) and the systems of organs. To understand mechanisms of genetic influences on behaviour, it is necessary to investigate genetic variability underlying differences in the structure and function of the nervous system at all levels – from biochemical processes within a cell to electrochemical processes of the whole brain.

Genetically informative studies of brain activity are the key to understanding the nature of the neural mechanisms that underlie individual differences in human behaviour. However, due to high complexity and high cost of such research, there have been only a limited number of such studies to date. Recent advances in digital data analysis and new technologies of measuring brain activity have led to increased interest (especially in the last decade) in this area of research. Neuroimaging methods, such as computerized tomography (CT), diffuse optical tomography, magnetic resonance imaging (MRI), functional MRI (fMRI) and positron emission tomography (PET), are actively used in genetically informative studies of the structure and functions of the brain.

The purpose of this chapter is to analyse the results of previous studies in genetic psychophysiology over the past 15 years (results from earlier research have been reviewed elsewhere, e.g., Malykh, 1998; van Beijsterveldt & Boomsma, 1994; van Beijsterveldt & van Baal, 1998).

Genetically informative studies of the brain structure

One of the best known methods for studying the brain structure is magnetic resonance imaging (MRI; see Box 2). This method is widely used in the study of the aetiology of individual differences in the morphology of the human brain. A meta-analysis of 62 genetically informative studies, using neuroimaging techniques, showed that genetic factors have a significant influence on general parameters of the brain structure, including the intracranial volume, the total brain volume, the volume of the cerebral hemispheres, and general and local volumes of grey matter and white matter. The results of this meta-analysis showed considerable variability in heritability of the volume of different structures, including the cerebellum, subcortical structures and corpus callosum (Blokland, de Zubicaray, McMahon, & Wright, 2012). A number of studies, using the high-resolution mode and a new method of three-dimensional voxel data transformation, showed a high level of genetic influence (up to 83%) on the density of grey matter in the frontal area, the Broca's and Wernicke's areas, the Geshl's gyrus, the left occipital and left posterior cingular region, the amygdala and the medial temporal cortex (Cannon et al., 2006; Hulshoff Pol et al., 2006a; Peper et al., 2009; Thompson et al., 2001). High heritability estimates (up to 93%) were also shown for density of the white matter of the corpus callosum, corticospinal tract, upper occipital frontal and longitudinal fascicles (Hulshoff Pol et al., 2006a; Peper et al., 2009). A number of studies found significant genetic influences on cortical thickness, especially in the frontal and parietal areas (Joshi et al., 2011; Lenroot et al., 2009; Rimol et al., 2010; Yoon, Fahim, Perusse, & Evans, 2010). Studies, using the method of diffusion-tensor morphometry, showed a strong genetic influence on the corpus callosum (Brun et al., 2009; Yoon et al., 2011), occipital lobe (Brun et al., 2009), lateral fronto-orbital gyrus, cerebellum, brain stem, hippocampal gyrus, right temporal white matter and the upper frontal gyrus (Yoon et al., 2011).

In the first quantitative study of the structure of the brain in healthy MZ and DZ twins computerized tomography was used (Reveley et al., 1984). It was shown that individual differences in the volume of the lateral ventricles are largely attributed to genetic factors. Further studies, using MRI, revealed high heritability of general parameters of the brain, including the intracranial volume (>81%; Baaré et al., 2001; Carmelli et al., 1998; Pfefferbaum et al., 2000), as well as total brain volume (66–97%; Baaré et al., 2001; Bartley et al., 1997; Pennington et al., 2000; Wright et al., 2002). The first family (twins and non-twin siblings) study of individual characteristics of the total volume of white and grey matter revealed the following estimates of heritability: 82% for grey matter and 88% for white matter (Baaré et al., 2001). Genetic factors explained around 65% of the phenotypic variance in hemisphere volume (Geschwind et al., 2002); 88% of the cerebellum volume (Posthuma et al., 2000); and 79%–94% of the corpus callosum volume (Pfefferbaum et al., 2000; Scamvougeras et al., 2003). In a twin study of children, genetic influences explained a significant proportion of the variance in almost all areas of the brain, with the exception of the cerebellum (Wallace et al., 2006). In a study of 139 twin pairs of older adults (69–80 years) significant genetic influences on the brain volume in certain areas were found. For some regions (e.g., right frontal, right temporal, and left and right lateral ventricular) genetic influences accounted for about 70% of their variance. For other regions (e.g., the temporal region), these estimates were lower (60%). Since most of the participants lived separately, this may indicate stable genetic control of the volume of certain brain areas in old age (Carmelli et al., 2002).

Genetically informative studies have also been conducted on the measures of three-dimensional fibre tractography, using diffusion tensor imaging (DTI) methods (see box 2) (Chiang et al., 2009; Kochunov et al., 2010; Lee et al., 2009). This method allows one to visualize bundles of nerve fibres that connect different areas of the brain, as well as communication between axons of the white matter of the brain. Diffusion tensor imaging is used to evaluate anisotropy (direction) in the diffusion of water molecules in the brain tissue (Basser et al., 1994).

One twin study using the DTI method examined the aetiology of brain asymmetry in a sample of 269 twin and non-twin sibling pairs. The results showed that genetic factors accounted for 33% of the asymmetry variability in the lower frontoparietal fascicle, 37% of the variability of the anterior thalamic radiations and 20% of the variability for the uncinate fasciculus and the major forceps. The contribution of shared environmental factors was 15% in the variation in asymmetry for the corticospinal tract, and 10% for the minor forceps (Jahanshad et al., 2010).

Other studies using the DTI method showed significant genetic influences on the corpus callosum (Brun et al., 2009; Yoon et al., 2011), the occipital lobes (Brun et al., 2009), the lateral fronto-orbital gyrus and the cerebellum, as well as several subcortical structures, the brain stem and the uncus, the right temporal white matter and the upper frontal gyrus (Yoon et al., 2011).

The DTI method was also used to study the aetiology of individual differences in integrity of the white matter of the brain in a large twin sample (705 adolescent and adult twin and non-twin siblings) (Chiang et al., 2011). The integrity of the white matter was evaluated with fractional anisotropy, a measure of "directed" brain structure, which depends on the number and orientation of conductive paths of the white matter of the brain. Heritability estimates of parameters of the white matter were higher in adolescents as compared with adults, and in men as compared with women. Socio-economic status (SES) and IQ interacted with genetic influences on the integrity of fibres. Heritability estimates were higher in participants with higher SES.

In participants with higher IQ, genetic factors explained more than 80% of the observed variability in fractional anisotropy in the thalamus, the genu of the corpus callosum, the posterior limb of the internal capsule and the upper radiate crown; whereas in participants with lower IQ, genetic factors explained 40% of the variance on average. There are several potential explanations of why SES and intelligence can interact with heritability of the brain structures. For example, it is known that education and learning can influence gene expression in the brain, such as the expression of the brain derived neurotrophic factor (BDNF) involved in neuronal growth and cognitive functions (Kesslak et al., 1998). It is possible that genetic factors play an active role in the transformation of physical and social environment – gene–environment correlation (Kendler & Baker, 2007). In particular, the genes affecting white matter can accelerate axonal conduction in the thalamus and the cortico-spinal tract, which may indirectly contribute to the development of higher IQ. Thus, genetic factors influence individual differences in the integrity of fibres, but their effects depend on sex, age, SES and IQ (Chiang et al., 2011).

The links between brain structures and intelligence have also been investigated for more global parameters, such as total brain volume, and grey and white matter volume have been shown to correlate with IQ. Such studies found that common genetic factors underlie differences in brain volume and in verbal and nonverbal intelligence (Hulshoff Pol et al., 2006b; Posthuma, 2002; Thompson et al., 2001). For example, one study showed a positive correlation between the volume of grey matter in the frontal region and intelligence (Thompson et al., 2001). Since the magnitude of the correlation was greater in MZ than in DZ twins, the authors concluded that there are common genetic factors for the covariance of the volume of grey matter of the frontal lobe and intelligence (Thompson et al., 2001; Toga and Thompson, 2005). Another twin study revealed the influence of genetic factors on covariance of verbal and nonverbal intelligence with the anatomical neural network involving grey matter of the frontal, occipital and parahippocampal areas, as well as white matter connecting the upper occipital-frontal fasciculus and the corpus callosum (Hulshoff Pol et al., 2006b). In contrast, a study of older twins showed that the relationship between the volume of the frontal-temporal regions and executive functions was explained by common environmental influences. It is possible that individuals with high level of cognitive functioning select environments that contribute to greater brain volume (Carmelli et al., 2002).

Genetically informative studies have also been conducted to investigate the links between brain structure and psychopathology. For example, significant decreases over time in whole brain and frontal and temporal lobe volumes were found in patients with schizophrenia and their unaffected co-twins compared with control twins. Significant additive genetic influences have been found for correlations between schizophrenia liability and progressive whole brain (66%), frontal lobe (76%) and temporal lobe (79%) volume change (Brans et al., 2008). Moreover, it was shown that common genetic factors explain a large proportion of the phenotypic relationship between psychotic symptoms and grey matter concentration in the left posterior and right dorsal frontal cingulate area (Rijsdijk et al., 2010). Reduced volume in white matter was associated with a substantial genetic risk for bipolar disorder (van der Schot et al., 2009).

Thus, a number of twin and family studies showed that genetic factors affect many of the structural characteristics of the brain. Despite the high heritability of many characteristics of the brain, specific genetic variants that contribute to heritability remain largely unknown (Cohen et al., 2001; den Heijer et al., 2002; Lu et al., 2011; Richter-Schmidinger et al., 2011).

Genetically informative studies of the brain function

Unlike behavioural genetic studies of brain structure, much less is known about the contribution of genetic and environmental factors to brain functioning. The focus of this section is on genetically informative studies that use neuroimaging techniques to investigate brain activity.

Behavioural genetic research using electroencephalogram

Electroencephalogram (EEG) is sensitive to a variety of external influences and to changes in the functional state of the cerebral cortex responsible for sensory, motor, cognitive and emotional processes. EEG characteristics are often analysed as biological markers (or endophenotypes) of complex psychological symptoms, including psychological disorders. EEG characteristics are stable over time (test-retest reliability of about 0,8; Pollock et al., 1991; Salinsky et al., 1991) and have been shown to be heritable (e.g., Almasy et al., 1999; Anokhin et al., 2001; de Geus, 2010; Smit, Posthuma, Boomsma, & de Geus, 2005; Smit et al., 2008, 2010; van Baal et al., 1996; van Beijsterveldt et al., 1996, 1998a; Zietsch et al., 2007).

A number of studies have shown associations of several EEG parameters with different psychological characteristics: cognitive abilities (Anokhin & Vogel, 1996; Giannitrapani, 1985; Jaušovec & Jaušovec, 2000; Klimesch et al., 1993; Onton et al., 2005; Schmid et al., 2002; Tesche & Karhu, 2000; Thatcher et al., 2008); personality traits (Tran et al., 2006); emotions (Davidson, 1998); temperament (Gale, 1988; Mundy-Castle, 1957); disorders, such as dyslexia, learning disabilities, ADHD, depression, alcoholism and autism (Barry et al., 2003; Cameron et al., 2003; Duffy & McAnulty, 1990; John et al., 1980; Kemner et al., 1999; Knott, 2001; Lazzaro et al., 1998; Linkenkaer-Hansen et al., 2005; Pogarell et al., 2006; Porjesz & Begleiter, 1996; Porjesz et al., 1998).

Research has demonstrated that genetic factors are largely responsible for individual differences in many EEG parameters (See Box 2), such as EEG spectral power, estimates of coherence and synchronization of EEG signals, network efficiency, dynamic complexity of the signal and others (Almasy et al., 1999; Anokhin et al., 2001; Posthuma et al., 2005; Smit et al., 2005; Smit et al., 2008, 2010; van Baal et al., 1996; van Beijsterveldt et al., 1996; van Beijsterveldt et al., 1998a, b; Zietsch et al., 2007).

EEG spectral power. Spectral power estimates reflect the total contribution of a large number of processes occurring in the brain. The individual pattern of spectral power reflects specific psychological states. Research has shown that resting EEG spectral power is one of the most heritable human traits (e.g., de Geus, 2010; Posthuma et al., 2001; Smit et al., 2005; Smit et al., 2010; Zietsch et al., 2007).

The earliest behavioural genetic studies in adults have shown that individual differences in spectral power of the alpha rhythm are under strong genetic influences (van Beijsterveldt & Boomsma, 1994). The results from different twin and other family studies are quite consistent (Anokhin, 1989; Anokhin et al., 2006a). The heritability of spectral characteristics of the EEG has been shown to be as high as that for such features as height and weight. The few twin studies that have been conducted in children and adolescents also showed high rates of heritability of spectral characteristics of the EEG. One study of twins aged 8–11 months (21 MZ and 20 DZ pairs) showed that individual differences in EEG characteristics are influenced by both genetic factors and shared environment. For example, the frequency of the alpha rhythm in infants was largely explained by genetic factors (73%). In contrast, individual differences in the spectral amplitude were largely influenced by environmental factors (Orekhova et al., 2003). In a study of the same children at the age of 5–6 years, individual differences in both alpha rhythm parameters (spectral amplitude and frequency) were largely explained by genetic factors (68–90% and about 80%, respectively) (Stroganova et al., 2009).

In a study using a larger sample of 5-year old children (71 MZ and 96 DZ pairs) high heritabilities have been found for: the alpha-1 and alpha-2 subbands (81% and 78%); beta-1 and beta-2 subbands (73% and 64%); and theta band (81%) (Baal et al., 1996). In another study, similar results have been obtained for 16-year old twins (van Beijsterveldt et al., 1996). The averaged heritabilities for the alpha, beta, delta and theta, frequencies was 89%, 86%, 76%, and 89% respectively. Multivariate analyses suggested that the same genes for EEG alpha rhythm were expressed in different brain areas in the left and right hemisphere (van Beijsterveldt et al., 1996).

Other research has also shown that largely the same genes contribute to the variability of all spectral powers of the delta, theta, alpha and beta bands. Genetic correlations for all the four bands ranged from 55% to 75% (Smith et al., 2005). A study on the largest sample to date (239 MZ and 280 DZ twin pairs, average age 16.24 years) supports the results of previous studies (Zietsch et al., 2007). This study analysed spectral powers of the four bands in the frontal and occipital regions of both brain hemispheres. The results showed that heritabilities of the spectral power band in the occipital regions were higher than in the frontal areas. Multivariate genetic analysis (see box 1) revealed both general and specific (for each band and brain area) genetic influences on the variability of the spectral powers of all the frequency bands. One common factor was responsible for a large part of genetic variability for all frequency bands in the frontal (0.41–0.68) and occipital (0.52–0.87) areas. This genetic factor may affect structural features such as the thickness of the skull, which determines the distance from the electrode to the pulse generator (a pyramidal cell) and thus affects the EEG power in the entire frequency spectrum.

One of the first molecular genetic studies of EEG spectral power was conducted in a sample of people selected for having low voltage EEG (LVEEG). This type of bioelectric activity was found to be associated with the chromosomal region 20q. The results also suggested that this feature may follow the Mendelian pattern of inheritance, with clear differences between relatives (Anokhin et al., 1992). Genetic association studies have recently been undertaken for the power of different frequency bands. For example, Ducci and colleagues (2009) found an association of the power of the alpha rhythm with the serotonin 3B receptor gene (*HTR3B*). However, it should be noted that this association was shown only for two of the three samples analysed. A genome-wide study (Hodgkinson et al., 2010) on a sample of 322 Native Americans showed an association of SNPs in the gene *SGIP1* (SH3-domain GRB2-like endophilin-interacting protein 1) with the EEG theta-band power.

Frontal asymmetry. Another EEG indicator is a decrease in the power of the rhythm (e.g., alpha) in one of the cerebral hemispheres – functional hemispheric asymmetry (Gao et al., 2009). A number of studies have investigated frontal asymmetry. Frontal asymmetry pattern is formed in early life. Atypical frontal asymmetry was found in children with ADHD, with oppositional defiant disorder and affective disorders, such as anxiety and depression (Baving, Laucht, & Schmidt, 1999, 2002; Forbes et al., 2006). Research has also found the link between the frontal hemispheric asymmetry of the alpha rhythm and different aspects of emotionality, both trait and situational, as well as basic emotional reactions such as approach-avoidance. Negative affect was found to be associated with increased activation in the frontal right hemisphere, whereas positive emotional state was associated with increased activation in the frontal left hemisphere (Davidson, 2003; Tomarken et al., 1990; Wheeler et al., 1993).

One study, using a sample of 264 female twins (with average age of 21), investigated the aetiology of asymmetry as a neural correlate of emotions. Significant moderate heritability (27%) of the frontal asymmetry was found for the mid-frontal region, whereas no significant genetic influence was found for the lateral-frontal area. The relatively low heritability of functional asymmetry may indicate the plasticity of neural bases of emotions in the course of development (Anokhin et al., 2006). In another twin study of both adult men and women (under the age of 35) moderate heritability of frontal asymmetry parameters was found: 37% for women and 32% for men (Smit et al., 2007). In a large-scale twin study of frontal asymmetry in children (951 twins, boys and girls, age 9–10 years) similar results to those in adults were found: genetic factors explained 11–28% of the variability in frontal asymmetry (Gao et al., 2009).

A number of molecular genetic studies found associations of the frontal asymmetry with specific genetic factors, for example, with polymorphisms in the serotonin 1A receptor gene (*HTR1A*) (Bismark et al., 2010).

EEG coherence. This EEG indicator is used to assess the degree of functional connectivity between the brain regions. EEG coherence is a quadratic cross-correlation between the signals from two areas (leads) of the brain for each component of the frequency domain. There are two types of EEG coherence: longer-distance coherence and short-distance coherence. There is evidence for association between EEG coherence and different psychological characteristics, such as intelligence (Anokhin et al., 1999c; Thatcher et al., 1983), emotional response (Reiser et al., 2012) and obsessive-compulsive disorder (Olbrich & Arns, 2013).

A twin study of the role of genetic and environmental factors of individual variance of interhemispheric coherence was conducted in a sample of 167 5-year old twin pairs. The coherence of the theta rhythms (4.0 to 7.5 Hz) was analysed between the prefrontal, frontal, central, parietal and occipital regions. Heritability of EEG coherence indices ranged from 37% to 71%, depending on the location of electrodes and distance between them, with distant coherence being more heritable than proximal coherence (van Baal, De Geus, & Boomsma et al., 1998a). The same twin sample was examined at the age of 7 years (van Baal et al., 2001). This age is characterized by significant changes in cortico-cortical connectivity reflected in greater EEG coherence. Interestingly, these periods of coherence spikes coincide with transition to a new stage of cognitive development by Piaget. It is assumed that after an early period of development, which is characterized by an excess of synaptic contacts, synaptic "pruning" (removing the non-functional part of these contacts) takes place (Thatcher, Walker, & Guidice, 1987; Thatcher, 1991; Thatcher, 1994a; Thatcher et al., 1994b). A longitudinal analysis of asymmetry found an increase between ages 5 and 7 years in coherence in the right prefrontal area, no change in coherence in the central prefrontal area and reduced coherence in the posterior areas. Genetic analysis showed that at the age of 5, heritability of EEG coherence ranged from 44% to 76% in the left hemisphere (61% on average) and from 38% to 64% in the right hemisphere (49% on average), whereas at the age of 7, heritability ranged from 37% to 74% (59% on average) in the left hemisphere, and from 25% to 85% (62% on average) in the right hemisphere. The influence of genetic factors on EEG coherence correlated with the distance between the electrodes in both age groups (the greater the distance, the higher the heritability). In addition, at the age of 7, new genetic influences were identified than were not present at age 5. It has been suggested that the new additive genetic factors identified at the age of 7 are related to the development of the visual-spatial loop that connects the dorsal stream and the dorsolateral prefrontal cortex. The maturation of this stream may underlie a qualitative change in the brain architecture that supports the development of conservation and other cognitive skills proposed by Piaget (van Baal et al., 2001).

Another study, using a sample of adolescent twins, found significant additive genetic effects on proximal and distal interhemispheric coherence (60% on average) in four frequency bands, with similar heritability across 18 combinations of electrodes (van Beijsterveldt, 1996). In a study of 690 siblings (age range: 6–65 years) intra-hemispheric and inter-hemispheric bipolar coherences were analysed (unlike previous studies that used monopolar coherence). The heritability of bipolar coherence indicators ranged from 22% to 63% (Chorlian et al., 2007).

EEG complexity and other quantitative characteristics of EEG. The complexity of EEG is another EEG parameter that reflects a large number of independent oscillations of neural networks. EEG curve has a complex structure that reflects the complexity of the underlying generators. In recent years, cortical oscillations were considered in terms of the theory of nonlinear dynamic systems, focusing on the study of complex aperiodic systems capable of self-organization. The behaviour of complex systems can be calculated by measuring the degree of dynamic complexity (Pritchard and Duke, 1995). Typically, EEG complexity is higher for tasks that require greater diversity of neural representations, that is, simultaneous activation of associative neural networks with fluctuations at different frequencies. On the contrary, tasks that require selective and focused attention result in low EEG complexity, possibly reflecting the suppression of excessive brain activity.

Individual differences in EEG complexity have been shown to be associated with individual differences in intelligence. It is assumed that EEG complexity may be an indicator of significant individual differences in brain dynamics (Anokhin et al., 1999). A twin study of the aetiology of EEG complexity (PD2, pointwise correlation dimension) in a sample of 214 female twins (age range: 18–28) found that PD2 heritability estimates ranged from 62% to 68% with closed eyes, and from 46% to 60% with open eyes, with the highest heritability in the central region (Cz) (Anokhin et al., 2006a).

Traditionally, EEG coherence has been used to assess the mutual influences of different areas of the brain. However, this approach is limited to measuring only processes that are based on linear dynamics. An alternative approach is based on EEG synchronization likelihood (SL) that allows analysis of non-linear influences between different areas of the brain using the graph theory. Clustering coefficient (CC) calculated on the basis of SL is the ratio of interconnected local nodes, and ranges from 0 to 1. In respect to brain dynamics, each node is understood as activation in the local area, as measured by EEG. Another indicator of EEG SL is the average path length (L), that is, the average number of steps from one network node to another. The CC and the average path length have been associated with information processing (Micheloyannis et al., 2006).

In a study of functional brain connectivity of 760 twins and siblings, it was shown that the contribution of genetic factors to CC was 37–62% and contribution to L was 46–89% (Smit et al., 2008). The results were replicated in another sample of 16-year old twins (n=879) and their siblings (Schutte et al., 2013).

Another study, using a sample of 1,438 twins and their siblings from four different age groups (16, 18, 25 and 50 years), found significant additive genetic effects on functional brain connectivity. Individual variation in SL and L were moderately to highly heritable at each age (SL: 40–82%; L: 29–63%). Genetic factors underlying these phenotypes overlapped. CC was also heritable (25–49%) but showed no systematic overlap with SL and L (Smit et al., 2010).

EEG during sleep. A number of genetically informative studies of EEG during sleep reported the influence of genetic factors on the characteristics of sleep (Gould et al., 1978; Webb & Campbell, 1983). When a person falls asleep, several alternating stages of sleep occur, each characterized by a specific EEG pattern. One small-scale twin study of EEG during sleep revealed significant heritability of the second (82%) and fourth (62%) stages of sleep (Linkowskiet al., 1991). Another twin study using spectral analysis showed a significant genetic influence on the duration of the third phase of sleep and REM-sleep (Ambrosius et al., 2008).

Molecular genetic research of EEG during sleep has revealed that several genes, such as adenosine deaminase (*ADA*), Period 3 *(PER3)*, catechol-O-methyltransferase *(COMT)*, 5-HT3R and metabotropic glutamate receptor 8 *(GRM8)*, are associated with of EEG characteristics during sleep (Watson, 2011).

Behavioural genetic research of event-related potentials

Event-related potentials (ERPs) or evoked potentials represent changes in electrical activity of the brain in response to external or internal stimuli. Event-related potentials include two types of responses: responses to external stimuli (sensory ERP) and internal stimuli (e.g., brain potentials associated with movement, or conditional negative wave). Sensory evoked potentials are classified according to the modality of the stimulus: visual, somatosensory and auditory ERP. Each type of ERP has a certain shape. The following characteristics are used to describe ERP components: peak latency and amplitude of the corresponding oscillations. For description of ERP, a letter is used indicating polarity (P – positive oscillation, N – negative oscillation) and a number is used that is equal to the latent period, that is, the time from the start to the peak of an oscillation. For example, a component designated as P300 is a positive waveform with a 300 ms latency period. Sometimes, a sequence number of a component is used instead (e.g., P3).

P300 component. The P300 component of event-related potentials has been widely used in studies of human cognitive functions in normal and clinical groups (Smit et al., 2007). It was first described in studies that used the "oddball paradigm": a situation of a rarely occurring yet relevant event (the oddball) appearing randomly among many irrelevant stimuli to which the subject must respond (count or press the button; Desmedt et al., 1965; Sutton et al., 1965). The target stimulus leads to an increase of the P300 amplitude in healthy individuals, and it is associated with attention resources during working memory tasks, or with cognitive processing required for encoding a new stimulus or renewal of stimulus representation in working memory. The classical component P300 is observed 300–600 ms after the relevant stimulus and is associated with cognitive processes of update (corrections) of a context and categorization of an event (Dien et al., 2004; Donchin & Coles, 1988; Kok, 2001; Verleger, 1988). P300 amplitude may reflect the processes of working memory related to detection of changes in external environment, whereas P300 latency is associated with stimulus evaluation time: shorter latency corresponds to better cognitive functioning (Hall et al., 2006). Psychological disorders, such as autism, schizophrenia and alcoholism, all of which have a significant cognitive component, have been associated with deviant P300 amplitude (e.g., Courchesne, 1987; Friedman et al., 1986; Polich et al., 1994).

Behavioural genetic studies have reported moderate to high heritability estimates of P300 amplitude: from 39% to 79% (Almasy et al., 1999; Katsanis et al., 1997; O'Connor et al., 1994; Polich & Burns, 1987; Rogers & Deary, 1991; Surwillo, 1980; van Beijsterveldt et al., 1998b). Heritability estimates of P300 latency are inconsistent, however (Almasy et al., 1999; Eischen & Polich, 1994; Katsanis et al., 1997; Polich & Burns, 1987; Rogers & Deary, 1991; Surwillo, 1980).

In one of the first twin studies of P300 latency of auditory ERP (6 MZ and 6 DZ pairs) higher similarity of MZ twins as compared to DZ twins was observed (Surwillo, 1980). Greater similarity in P300 amplitude and latency of auditory ERP was observed in MZ twins than in unrelated participants (Polich & Burns, 1987) and greater similarity in P300 latency (but not amplitude) was shown in MZ twins than in DZ twins (Rogers & Deary, 1991). In a study of ten families with biological parents and two children, P300 amplitude and latency were investigated using visual and auditory oddball paradigms. The amplitude and latency of P300 among biologically related family members were more similar than those among unrelated individuals (Eischen & Polich, 1994).

In a larger twin study, negligible genetic influences on auditory P300 latency and strong genetic influences (i.e., heritability of 60%) on the amplitude in the caudal area were found (O'Connor et al., 1994). A higher heritability estimate (80%) of the amplitude of visual P300 was found in a sample of 30 MZ and 34 DZ twin pairs (age: 17–18 years; Katsanis et al., 1997). In two Dutch twin cohorts (209 pairs of 5-year old twins, and 213 pairs of 16-year old twins), visual P300 was investigated using the oddball paradigm at two time points with a 2-year follow up (at ages 7 and 18). The results of this study showed that the P300 latency was under genetic influence at age 5 (34%) and age 7 (70%), whereas no significant genetic influences were detected in the adolescent cohort (at ages 16 and 18). This difference in heritability between children and adolescents may be explained by the fact that the same tasks were used for both children and adolescents, and it is likely that the tasks were difficult for young children and too easy for adolescents (van Baal et al., 1998b; van Beijsterveldt et al., 2001). The results of another twin study of P300 in 18-year olds using both simple and complex conditions found significantly greater genetic heritability in more difficult conditions (Katsanis et al., 1997). At the same time, heritability of the amplitude and latency of P300 did not change significantly in the follow-up study; however, new genetic influences on P300 amplitude and latency were identified in some regions (van Baal et al., 1998b). Sex differences in the contribution of genetic and environmental factors to the amplitude of P300 were found in adolescents. In adolescent boys, heritability of P300 amplitude was substantial (62%) and did not differ between two time points. In adolescent girls, in contrast, shared environment explained most of the variance (60%), and this effect did not change between two ages (van Beijsterveldt et al., 2001).

The aetiology of individual differences in amplitude and latency of visual P300 was investigated in a study of 306 twin pairs (140 MZ and 166 DZ, aged 15–17 years) and siblings from 354 families (Wright et al., 2001). In this study, a working memory task (i.e., a delayed response situation) was used: participants were asked to memorize the location of the target stimulus for a short time and then perform motor response by pointing to the location of the stimulus. The authors suggested that the cognitive complexity of this task was higher than that of the oddball task, which only required a minimal effort to achieve 100% accuracy. Greater complexity of cognitive operations may be associated with greater genetic influence and family resemblance in P300. The results of this study showed that a large proportion of P300 amplitude variability was explained with genetic factors; P300 latency also showed significant family resemblance. Multivariate genetic analysis showed that common genetic factors contributed to individual differences in a number of brain areas. The extent of genetic influence on P300 amplitude ranged from 48% to 61%, which is similar to the data from a study by van Beijsterveldt et al. (1998b) where average heritability estimates were

42% for relevant and 62% for irrelevant stimuli. Higher rates of heritability (79%) were obtained in another study (Katsanis et al., 1997), which can be explained by differences in the stimulus materials, age of participants and sample size.

The analysis of temporal characteristics of P300 showed significant family resemblance in the frontal, central and parietal regions. Wright and colleagues (2001) found 44% to 50% of the variance of P300 latency was explained by genetic influences. These results are consistent with several previous twin studies of temporal characteristics of P300 (Almasy et al., 1999; Eischen & Polich, 1994; Katsanis et al., 1997; Polich & Burns, 1987; Rogers & Deary, 1991; Surwillo, 1980). However, in two other twin studies, using the oddball paradigm, genetic influences on temporal characteristics of P300 were negligible (O'Connor et al., 1994; van Beijsterveldt et al., 1998).

Several studies reported a decrease in heritability estimates of visual P300 amplitude with age (e.g., Courchesne, 1978; Johnson & Newport, 1989; Mullis, Holcomb, Diner, & Dykman, 1985; Taylor, 1993; Katsanis, Iacono, & McGue, 1996). For example, in a longitudinal twin study (161 MZ and 80 DZ, all men), visual P300 was examined at ages 17, 20 and 24 (Carlson & Iacono, 2006). Participants performed the rotated-heads visual oddball task (Begleiter et al., 1984). Heritability estimates for P300 amplitude obtained in the study were 64% at age 17, 68% at age 20 years and 53% at age 24 (Carlson & Iacono, 2006), similar to the results of a meta-analysis (van Beijsterveldt & van Baal, 2002).

In a study of two groups of adult twins and their siblings (760 family members of 390 families with twins), a younger (mean age = 26.5 years) and an older group (mean age = 48.8 years), heritability estimates were ~50% for amplitudes and ~45% for latencies of P300 across different brain regions (Smit et al., 2007). Despite the differences in means and variances of P300 amplitude and latency between the two age groups, genetic factors account for similar proportion of phenotypic variances in these groups.

Only a few molecular genetic studies of P300 component have been conducted to date. For example, one study found an association between a Taq I polymorphism of the dopamine D2 receptor (*DRD2*) gene and P300 amplitude in children of alcoholics: carriers of the A1 allele had a decrease in the P300 amplitude as compared with A2/A2 carriers (Hill et al., 1998). Genes potentially involved in the regulation of brain activity associated with the generation of P300 were located on chromosomes 2 and 6 (Begleiter et al., 1998). Ionotropic glutamate receptor (*GRIK2*) gene located in 6q21–23.2 was suggested as a candidate gene for P300. Association between the *GRIK2* gene and P300 amplitude in the frontal-parietal regions of the right hemisphere was found. Other genes, such as the chains of D and G acetylcholine receptor genes (*CHRND*; *CHRNG*) located in 2q32, may also be involved in P300 amplitude (Porjesz et al., 2002).

Differences in heritability estimates, as well as associations with potential genetic markers may be associated with functional heterogeneity of the P300 component. It is assumed that the basis for the generation of P300 activity is distributed networks in the brain, among which are the temporoparietal junction, the medial temporal, medial frontal and lateral prefrontal cortices, the inferior parietal lobule and the thalamus, as well as the frontal-temporoparietal interaction and the cortico-limbic interaction. Functioning of each of these networks can be based on different genetic basis, which should be considered when analysing the contribution of genetic factors to the variability of P300.

P50 component. In recent years, the P50 component of evoked potential has attracted special interest. In the 1990s, it was described as a biological marker for schizophrenia (Freedman et al., 2000). Twin studies have shown a significant influence of genetic factors on this component (Myles-Worsley et al., 1996; Young et al., 1996). For example, it was shown that 44% of the variability of P50 sensory gating (i.e., filtering of sensory signals) was accounted for by genetic factors (Young et al., 1996). In another study, the impaired ability to filter sensory information, which is considered a key dysfunction in schizophrenia and schizophrenia spectrum disorders, sensory filtering was analysed using the P50 component in a sample of 48 MZ and 40 DZ twin pairs (all women aged 18–29 years) (Anokhin et al., 2007). Another study examined multiple ERPs – P300, P50 and mismatch negativity – in a sample of 16 MZ twin pairs with schizophrenia, 9 MZ twin pairs discordant for schizophrenia and 77 twin pairs (45 MZ and 32 DZ twin pairs) without schizophrenia (controls). It was shown that 67% of the variability of P300 amplitude, 57% of P50 suppression ratio, 48% of mismatch negativity amplitude and 21% of P300 latency was accounted for by genetic factors. Significant phenotypic correlations were found between schizophrenia and ERP indicators. Genetic factors were shown to be the main source of these phenotypic correlations: P50 suppression ratio (rg = 0.59), P300 amplitude (rg = –0.48), P300 latency (0.44) and mismatch negativity amplitude (rg = 0.39) (Hall et al., 2007).

It was shown that between 36% and 68% of the variability in P50 characteristics are explained by genetic factors (Anokhin, 2014). This variation in heritability estimates may be due to significant intra-individual variability in P50. A number of studies with repeated measures reported low test-retest reliability for P50. The characteristics of this component vary substantially even within a single ERP study (Boutros et al., 1991; Cardenas et al., 1993; Clementz et al., 1997; Smith et al., 1994). Thus, it is possible that at least some differences between the heritability coefficients obtained in different studies may be related to the instability of P50.

To date, there are limited data on associations between genetic markers and P50 variability. For example, Freedman and colleagues (1997) found an association between P50 inhibition and the alpha7 neuronal nicotinic receptor subunit (*CHRNA7*) gene (Freedman et al., 1997). This association was also replicated in another study (Leonard et al., 2002).

N1 component. N1 is an endogenous ERP component recorded in the range 75–250 ms after stimulus presentation (Smit et al., 2007). In cognitive psychophysiology, it is related to sensory processing (orientation response), as well as to the process of active information gating. It was found that an attended stimulus causes a more pronounced N1 (Eason et al., 1964). This finding was replicated in many studies of attention. Behavioural genetic studies of N1 component are rare, and have provided inconsistent results. For example, some, but not all, studies of auditory modality have shown significant heritability of N1 amplitude (O'Connor et al., 1994; Polich & Burns, 1987; Surwillo, 1980). Two studies of visual modality have provided opposite results. One study found stronger genetic influences on the variation of N1 amplitude in the frontal (19–31%) and parietal-occipital regions (45–54%), and weaker genetic influences on the variation of N1 latency in these regions (10–16% and 3–12%, respectively), in response to irrelevant stimuli (Almasy et al.,1999). Another study found negligible genetic effects on the variation of N1 in response to the relevant stimulus in three parietal leads (Katsanis et al., 1997). A more recent study of 651 twins and siblings from 292 families assessed the contribution of genetic and environmental factors in the variance of the amplitude and latency of N1 using the visual oddball paradigm (Smit et al., 2007a). In this study, two visual N1 components were identified: an anterior N1 peaking from 88 to 168 ms after stimulus presentation, and a posterior N1 peaking from 132 to 220 ms. Heritability estimates of the anterior N1 amplitude was 22%, and 50% for the posterior N1 amplitude. For the latency, heritability estimates were significantly higher: 45% for the anterior N1 and 43% for the posterior N1. A possible explanation for the differences in findings across these three studies is the differences in characteristics of the samples. For example, in a study by Almasy and colleagues, families with a history of alcoholism were included, whereas a study by Smit and colleagues included a sample of healthy participants.

Topography of N1 component is sensitive to stimulus modality (Altenmüller & Gerloff, 1999). In the auditory modality, anterior and posterior N1 components are not observed, however in the central region, a component with a similar time of occurrence and similar morphology is found. For this auditory N1 in the central region, heritability estimates were 60% for amplitude and 56% for latency (O'Connor et al., 1994).

Findings of high heritability of N1 component suggest that it can serve as a possible endophenotype of some psychopathologies. For example, individual differences in the N1 amplitude were associated with attention deficit hyperactivity disorder (ADHD; Johnstone et al., 2001; Satterfield et al., 1984) and alcoholism (Cohen et al., 1996; Patterson et al., 1987).

N2 component. Electrophysiological studies, using the Go/No-Go paradigm, found a negative component (N2) with latency of 200 to 350 ms, with the maximum amplitude in the mid-frontal region (Anokhin et al., 2004). The frontal N2 component increases sharply in No-Go trials and decreases or is absent during Go trials. A similar pattern of N2 component was observed in tasks without explicit motor response, for example in tasks involving different responses to different stimuli in a high-conflict situation. It is assumed that N2 represents not only the motor response inhibition but the overall process of conflict monitoring. The anterior cingulate cortex (ACC) is considered the most likely potential generator of N2. In addition, a correlation of the N2 component with activation of the lateral prefrontal cortex areas was found. The ACC and the dorsolateral prefrontal network are involved in inhibition processes. Inhibitory control deficit is a central dysfunction in a number of neuropsychiatric disorders such as ADHD, conduct disorder and substance abuse. These inhibitory control disorders are highly heritable, however little is known about neurocognitive mechanisms that mediate genetic influences.

In a twin study of 194 female twins (52 MZ and 45 DZ twin pairs) aged 18 to 28 years, aetiology of N2 component was investigated using the Go/No-Go paradigm (Anokhin et al., 2004). EEG was recorded in 19 leads. Three indicators of N2 amplitude were recorded: the peak amplitude in comparison to the pre-stimulus level; the difference between the N2 peak and the amplitude of the next positive peak; as well as the difference between the peaks for the amplitude of a positive peak using the partial regression analysis (it excludes a part of the variability in the between-peak amplitude of N2, caused by the variability of the preceding positive peak). In addition, the difference between the No-Go amplitude, and the maximum amplitude of N2 for the Go condition was calculated. Significant heritability of No-Go ERP components was found. Approximately 60% of phenotypic variance in event-related activation of the frontal brain regions (N2) was explained by genetic factors. Moreover, all three types of N2 measures showed similar results. No-Go ERP components may indicate genetically determined individual differences in brain processes associated with cognitive control (Anokhin et al., 2004).

Error-related negativity (ERN). Adaptive goal-directed behaviour requires tracking of the correspondence between planned and implemented actions, as well as adjustment in case of discrepancy. Numerous neuropsychological studies have shown association between mechanisms of self-regulation and the prefrontal cortex. Contemporary research using ERP measures and neuroimaging can define the role of the anterior cingulate and the lateral prefrontal cortex in processing errors and adjusting actions. When making errors in complex cognitive tasks, for example, tasks measuring choice reaction time, there is a sharp deviation of EEG in the fronto-central region of approximately 50–150 ms after an incorrect answer. This component of ERP is referred to as error negativity (Falkenstein et al., 1991). In the following studies, using different experimental paradigms, a similar phenomenon was identified, often called error-related negativity (ERN). Negativity associated with the correct answer is called CRN (correct-related negativity). ERN is an important tool in experimental studies of neurophysiological mechanisms of behaviour control in humans.

ERN is seen as a reflection of inconsistencies between neural presentations of current (incorrect) and desired (correct) response (Falkenstein et al., 1991). It has been suggested that ERN is not specific to an error situation and reflects the conflict between representations of responses or inconsistent information flows (Botvinick et al., 1999; Botvinick et al., 2001; Carter et al., 1998; Gehring & Fencsik, 2001; van Veen & Carter, 2002). This assumption is confirmed by the presence of a similar wave of smaller amplitude, so-called "correct negativity", for trials with correct answers. ERN is followed by a slow positive wave – error positivity, reaching a peak in about 400ms after a wrong answer. It may reflect a process of error realization, error recognition (Falkenstein, Hoormann, Christ, & Hohnsbein, 2000) and processes associated with response adjustment and response strategy adaptation, as well as emotional processing associated with the error.

A number of studies have found associations between ERN and behaviour, personality and psychopathology. This may indicate a possible role of ERN as an indicator of functionally significant differences in neurophysiological mechanisms of action regulation (Ruchsow et al., 2005; Potts, George, Martin, & Barratt, 2006; Dikman & Allen, 2000; Santesso, Segalowitz, & Schmidt, 2005 and others).

A genetically informative study of ERN, CRN and error positivity (Pe) of 99 MZ and 175 DZ adolescent twins showed significant heritability (40–60%) for all three components of ERP. In men and women, 47% of the phenotypic variance of the ERN amplitude was explained by genetic factors. CRN amplitude heritability estimates were 61% for men and 60% for women, whereas heritability estimates for Pe amplitude were slightly lower: 52% for men and 39% for women. The results of a multivariate genetic analysis indicated that there were common genetic factors for different ERP components.

A number of molecular genetic studies of ERN have been conducted. For example, it was shown that carriers of a short allele (*S*) of the serotonin transporter gene (5-*HTTLPR*) had on average higher ERN amplitude than homozygotes *L/L* (Fallgatter et al., 2004). However, in another study with a larger sample size (n=89), this finding was not replicated (Olvet et al., 2010). Significant association of ERN amplitude with the *BDNF* gene was found in another study (Beste et al., 2010).

Slow wave (SW). In delayed-response tasks, in which a stimulus needs to be kept in mind after its disappearance, a slow wave appears 1.4s after the stimulus disappearance. This slow wave is longer than the alpha wave, that is, more than 1/8 of a second. This long ERP component reflects the processes associated with working memory (Ruchkin et al., 1995).

Hansel and colleagues (2001) studied the aetiology of individual differences in the SW recorded during memory tests specific to the prefrontal region and/or SW as a whole. In this study, 391 pairs of 16-year old twins were asked to solve a visual spatial task. The results showed that 35–37% and 51–52% of the variability of SW in the prefrontal and parietal area, respectively, were explained by genetic factors. Multivariate results suggested that different genetic factors influenced SW activity in different brain areas (i.e., prefrontal executive processing and parietal spatial processing). These brain regions are reported to have different functional roles. The prefrontal cortex is associated with general cognitive control, while the parietal cortex is responsible for spatial information processing. The results of this study did not support the hypothesis that ERP recorded during working memory tasks may be under stronger genetic influence than ERP during less demanding tasks (Boomsma & Somsen, 1991; Vernon & Mori, 1989).

Another type of slow wave is the low-frequency electrophysiological wave, referred to as slow cortical potential (SCP). Generated by a warning stimulus (cue) before the target stimulus, SCP occurs in the upper layers of the cortex and has a minimum period of 200–300ms. Two of the most notable examples of SCP are readiness potential and contingent negative variation (Smit et al., 2009). Working memory load in inter-stimulus interval greatly increases the SCP. Motivational factors, both positive (reward) and negative (avoidance), also have this effect. Tasks with delayed response, which cause SCP, also produce a small but significant synchronization of the upper range of the alpha rhythm (Bastiaansen et al., 2002; Jensen et al., 2002; Klimesch et al., 1999; Sauseng et al., 2005a,b). In addition to SCP and alpha synchronization, tasks with delayed response also generate theta desynchronization in the interval between the warning and the imperative stimulus (Bastiaansen et al., 2002). Research has found associations of individual differences in SCP and alpha and theta synchronization with cognitive abilities (e.g., Basile et al., 2007; Doppelmayr et al., 2005; Hansell et al., 2005; Jausovec & Jausovec, 2004; Klimesch, 1999; Perez-Edgar et al.,2006).

Thus, with an increase in the amplitude of SCP in a working memory experiment, changes in alpha synchronization and theta desynchronization can also be expected. Support for this was found in a study by Smit and colleagues (2009). The sample included 760 twins and their siblings from 309 families split into two age groups: a younger group (mean age 26.2 years) and an older group (mean age 49.4 years). Consistent with previous results from studies using the working memory delayed-response task (Filipovic et al., 2001; Hansell et al., 2001); alpha synchronization (Bastiaansen et al., 2002; Filipovic et al., 2001; Klimesch et al., 1999; Klimesch et al., 2007; Krause et al., 1996), distinct SCP, alpha synchronization and theta desynchronization were revealed within the expectation period. Not only SCP, but also alpha synchronization and theta desynchronization increased significantly during the delay in working memory tasks. Significant heritability estimates were identified for the three parameters: SCP (25% to 43%), alpha synchronization (35% to 65%) and theta desynchronization (31% to 50%). The genetic contribution to SCP concerned mainly the right frontal, the left parietal-temporal and the central regions, whereas for alpha synchronization and theta desynchronization no distinct topographic patterns of heritability were revealed. In general, heritability was higher for higher memory load. For theta desynchronization some increase in heritability in the frontal areas was obtained. It is possible that, to some extent, SCP is directly caused by changes in alpha synchronization (Klimesch et al., 2007; Min et al., 2007; Nikulin et al., 2007; Sauseng et al., 2005) and individual differences in both parameters can by a common set of genes.

Event Related Spectral Perturbation (ERSP) (oscillations). Until recently, ERPs were the basic electrophysiological indices of cognition that provided valuable insights into human brain processes. It has been shown that some ERSP features may arise from changes in the dynamics of ongoing EEG rhythms/oscillations of different frequency bands that reflect ongoing sensory and/or cognitive processes (Basar, 1999). In other words, the EEG oscillations that are measured in a resting state become organized, amplified and/or coupled during mental activity. In addition, the network activity, induced by an event or stimulus, may trigger specific oscillatory responses, thus giving rise to an "evoked" (strongly locked to the stimulus/event) or "induced" (weakly associated with the stimulus/event) rhythmicity (Basar, 1999; Makeig et al., 2002). ERSP is a measure of the average changes over time in the amplitude of the EEG frequency (EEG amplitude spectrum) that is related to a certain event (see Box 2).

In a genetically informative study, using a sample of 51 MZ and 48 DZ twin pairs from the Minnesota Twin Registry, EEG spectral power in response to presentation of a stimulus during the visual oddball task was investigated (Ethridge et al., 2013). In this study, Principal Component Analysis (PCA) was used for identification of neuronal activity with maximum change in response to the stimulus presentation. The components obtained with the PCA underwent the wavelet transform (a type of spectral analysis) for frequency analysis of characteristics of the signal. Wavelet analysis is used for a detailed investigation of inner components and structure of inhomogeneous signals and functions. Results of this study showed that heritability estimates of ERSP on the target stimulus varied from 21% to 68% depending on the frequency band. Similar data were obtained for a spectral power in response to presentation of the standard stimulus (Ethridge et al., 2013).

Analysis of ERSP allows one to find more specific characteristics of neural activity of the brain that can be used for molecular genetic investigations. To date, it has been shown that ERSP in the delta and theta bands, upon presentation of the target stimulus, were associated with the muscarinic receptor (*CHRM2*) gene (Jones et al., 2004), the serotonin receptor 7 (*HTR7*) gene (Zlojutro et al., 2011), the glutamate receptor (*GRM8*) gene (Chen et al.,

2009), the corticotropin-releasing hormone receptor (*CRHR1*) gene (Chen et al., 2010) and inwardly-rectifying-K-channel (*KCNJ6*) gene (Kang et al., 2012).

Another approach to the analysis of rhythmic oscillations is based on EEG source localization using Independent Component Analysis (ICA; McLoughlin et al., 2014b). ICA uses the entire EEG dataset, not just the characteristics of the signal at certain time intervals and/or for separate leads. Modern computational capacity allows implementing algorithms that lead to high spatial resolution (5–10 mm) in localization of dipole sources based on identification of independent components in the total EEG activity.

In a twin study using a sample of 34 MZ and 33 DZ twin pairs, concordant and discordant for ADHD, the aetiology of the phenotypic relationship between frontal midline theta activity, reaction time variability (RTV) and ADHD was investigated (McLoughlin et al., 2014a). It was shown that variability in stimulus event-related theta phase from the frontal midline cortex was strongly related to both RTV and ADHD, both phenotypically and genetically. This suggests that variability in the timing of the theta signal in ADHD may be part of a dysfunctional brain network that impairs regulation of task-relevant responses in this disorder.

Thus, ICA-based analysis of electrical activity of the brain allows for a more detailed investigation of underlying brain mechanisms of various psychopathologies. In addition, this approach can identify new stable EEG markers of certain syndromes.

Imaging genetics

Brain imaging methods, such as MRI, MEG, PET, provide a unique opportunity for detailed analysis of structure and functions of the brain, as well as its role in human behaviour. These neuroimaging techniques have been used in genetically informative studies with the aim of understanding the aetiology of relationships between neurophysiological and behavioural phenotypes (imaging genetics). Imaging genetic studies have been used to study intelligence (Deary et al., 2006), working memory (Blokland et al., 2011; Koten et al., 2009) and emotional states (Côté et al., 2007). However, imaging genetic studies of brain functioning are not as common as imaging genetic studies of the brain structure. The results of existing studies are inconsistent (Blokland et al., 2008; Blokland et al., 2011; Côté et al., 2007; Koten et al., 2009; Matthews et al., 2007; Park et al., 2012; Polk et al., 2007), which may be explained by the use of samples of small size (for a meta-analysis, see Blokland, 2012). Nevertheless, overall, these studies suggest the influence of genetic factors on brain functioning.

The most commonly used method in imaging genetic studies of brain function is functional MRI (fMRI). This technique is based not on the analysis of static images, but on changes in properties of the local blood flow in the brain. When certain areas of the brain are actively involved in information processing, they need more oxygen. Oxygen is delivered to nerve cells as a part of haemoglobin in red blood cells. Due to different properties of the haemoglobin and deoxyhaemoglobin (i.e., haemoglobin without oxygen), fMRI can show changes in the concentration of oxygen in blood during brain activity.

A study of brain activity during working memory tasks using the methods of voxel analysis was conducted in a sample of 319 MZ and DZ twin pairs (Blokland et al., 2011). The participants were presented with numerical stimuli and they had to determine whether the presented stimuli appeared before (1 or 2 slides ago). This task (n-back) is used to assess basic working memory processes. Significant genetic influences on brain activity associated with working memory processes were found: the highest heritability estimates (40–65%) were observed in the lower, middle and superior frontal gyrus, the left supplementary motor area, the precentral and postcentral gyrus, the middle cingulate cortex, the angular gyrus, the superior parietal lobule (including the precuneus) and the upper occipital gyrus. As in other studies of cognitive abilities that revealed low levels of total environmental variability in adults (Plomin & Spinath, 2004), this study of working memory showed no shared environmental influences on brain activity and a significant contribution of non-shared environment (Blokland et al., 2011).

As shown in this study, MZ twins have more similar patterns of brain activity during working memory tasks than DZ twins. In addition to the effect of common genetic influences, it was hypothesized that MZ twins could use more similar strategies when performing tasks, and this similarity could be reflected in more similar patterns of brain activity (Kirchhoff & Buckner, 2006). However, this hypothesis was not supported by the findings of the present study (Blokland et al., 2011): the distribution of different strategies (48% with numerical strategy; 19% with a spatial strategy; and 67% with a combination of these two strategies, or none of them) did not differ between MZ and DZ twins; and MZ twins were not more similar in their strategies than DZ twins. The study showed that, although the type of strategy was not under genetic influences, accuracy and time spent on solving the problem were under genetic influences.

One of the first studies of the role of genetic and environmental factors in the brain reward system was conducted by Silverman and colleagues (2014). In a sample of 48 MZ twin pairs, 18 brain regions associated with the expectation of reward using the monetary incentive delay task were investigated. Significant within-pair correlations, suggesting genetic influences on the reward system function, were obtained for the accumbens nucleus, the caudate nucleus, the insula and the anterior cingulate cortex.

Default mode of brain function (DMBF) is considered as activity in a number of brain regions, observed when a person is not focused on a specific task and is awake. DMBF activity is characterized by slow coherent oscillations at a frequency of 0.1 Hz. Deviations in DMBF have been associated with a wide range of psychopathology, such as autism (Kennedy et al., 2006), schizophrenia (Whitfield-Gabrieli et al., 2009) and Alzheimer's disease (Mawuenyega et al., 2010). A number of genetically informative studies of DMBF have been conducted. In one study of 16 MZ and 13 DZ twin pairs, it was shown that at the level of interneuronal connections 60% of variability in DMBF was explained by additive genetic factors (Fornito et al., 2011). Similar results were obtained in another study of 22 MZ and 21 DZ adolescent twin pairs (van den Heuvel et al., 2013).

A new initiative in the area of imaging genetics is the development of ENIGMA Network (The Enhancing NeuroImaging Genetics through Meta-Analysis) – the network of research groups with the aim of finding new genetic variants that can explain the variability in the structure and functions of the brain, and their associations with behavioural traits (Thompson et al., 2014).

Conclusions

In recent years, a number of genetically informative studies of the structure and functions of the brain have been conducted. Twin studies have shown a significant contribution of genetic factors to individual differences in various neurophysiological parameters, such as EEG spectral power, coherence and synchronization of EEG signals, network efficiency and dynamic complexity of the signal. The influence of genetic factors on the individual differences in evoked related brain potentials was shown to be somewhat lower. Molecular genetic studies have started to reveal the relationships among psychophysiological characteristics, specific DNA markers and behavioural phenotypes.

Continuously advancing neuroimaging techniques and the use of genetically informative approaches in brain research open up new possibilities for understanding neurophysiological processes underlying learning. Results obtained with different imaging methods complement each other and give a more accurate picture of brain functioning. For example, the method of evoked related potentials is used to evaluate how information is processed in real time, whereas the fMRI method more accurately localizes areas of the brain where information is processed. Neuroimaging measures are sensitive to external influences, as well as to changes in the functional state of the cerebral cortex that reflect motor, cognitive and emotional processes. A number of studies have demonstrated associations of psychophysiological measures with both normal psychological characteristics (e.g., cognitive abilities, personality traits, emotions and temperament), and psychological disorders (e.g., dyslexia and other learning disabilities, depression, ADHD and autism). The paths from genes to psychological phenotypes are mediated by structural and functional features of the brain. Genetically informative studies of psychophysiological measures allow us to understand the nature of neural mechanisms that underlie individual differences in psychological traits that are essential for learning.

References

Almasy, L., Porjesz, B., Blangero, J., Chorlian, D. B., O'connor, S. J., Kuperman, S., ... & Begleiter, H. (1999). Heritability of event-related brain potentials in families with a history of alcoholism. *American Journal of Medical Genetics, 88*(4), 383–390.

Altenmüller, E. O., & Gerloff, C. (1999). Psychophysiology and the EEG. In E. Niedermeyer & F. Lopes da Silva (Eds), *Electroencephalography* (4th ed.). Philadelphia: Lippincott Williams & Wilkins.

Ambrosius, U., Lietzenmaier, S., Wehrle, R., Wichniak, A., Kalus, S., Winkelmann, J., ... & Friess, E. (2008). Heritability of sleep electroencephalogram. *Biological psychiatry, 64*(4), 344–348.

Anokhin, A. P. (1989). The genetic-aspects of human-brain functioning evidence from EEG multivariate-analysis. *International Journal of Psychophysiology,* 7, 120–121.

Anokhin, A. (2014). Genetic psychophysiology: Advances, problems, and future directions. *International Journal of Psychophysiology, 93,* 173–197

Anokhin, A. P., Golosheykin, S. A., & Myers, E. (2005). Error-related negativity in twins: Preliminary findings. *Psychophysiology, 42*(Suppl. 1), S32.

Anokhin, A. P., Heath, A. C., Myers, E. (2004). Genetics, prefrontal cortex, and cognitive control: a twin study of event-related potentials in a response inhibition task. *Neuroscience Letters, 368,* 314–318.

Anokhin, A. P., Heath, A. C., & Myers, E. (2006a). Genetic and environmental influences on frontal EEG asymmetry: A twin study. *Biological psychology, 71*(3), 289–295.

Anokhin, A. P, Lutzenberger, W., &c Birbaumer, N. (1999a). Spatiotemporal organization of brain dynamics and intelligence: An EEG study in adolescents. *International Journal of Psychophysiology, 33,* 259–273.

Anokhin, A. P., Müller, V., Lindenberger, U., Heath, A. C., & Myers, E. (2006b). Genetic influences on dynamic complexity of brain oscillations. *Neuroscience letters, 397*(1), 93–98.

Anokhin, A. P., Rohrbaugh, J. W., Todorov, A. A., Vedeniapin, A. B. (1999b. The P300 event-related brain potential in neuropsychiatric disorders: a moderator of genetic risk? *Behavior Genetics,* 29, 349.

Anokhin, A., Steinlein, O., Fischer, C., Mao, Y. P., Vogt, P., Schalt, E., Vogel, F. (1992). A genetic-study of the human low-voltage electroencephalogram. *Human Genetics, 90,* 99–112.

Anokhin, A. P, Lutzenberger, W., & Birbaumer, N. (1999c) Spatiotemporal organization of brain dynamics and intelligence: An EEG study in adolescents. *International Journal of Psychophysiology,* 33, 259–273.

Anokhin, A. P., Todorov, A. A., Madden, P. A. F., Grant, J. D., & Heath, A. C. (1999b). Brain event-related potentials, dopamine D2 receptor gene polymorphism, and smoking. *Genetic Epidemiology, 17*(S1), S37–S42.

Anokhin, A. P., van Baal, G. C., van Beijsterveldt, C. E., de Geus, E. J., Grant, J., Boomsma, D. I. (2001). Genetic correlation between the P300 event-related brain potential and the EEG power spectrum. *Behavior Genetics,* 31, 545–554.

Anokhin, A. P., Vedeniapin, A. B., Heath, A. C., Korzyukov, O., Boutros, N. N. (2007). Genetic and environmental influences on sensory gating of mid-latency auditory evoked responses: a twin study. *Schizophrenia Research*, 89, 312–319.

Anokhin, A., Vogel, F. (1996). EEG alpha rhythm frequency and intelligence in normal adults. *Intelligence*, 23, 1–14.

Baaré, W. F., Hulshoff, Pol, H. E., Boomsma, D. I., Posthuma, D., de Geus, E. J., Schnack, H. G., van Haren, N. E., van Oel, C. J., & Kahn, R. S. (2001). Quantitative genetic modeling of variation in human brain morphology. *Cerebral Cortex*, 11, 816–824.

Barry, R. J., Johnstone, S. J., & Clarke, A. R. (2003). A review of electrophysiology in attention-deficit/hyperactivity disorder: II. Event-related potentials. *Clinical Neurophysiology*, 114, 184–198.

Bartley, A. J., Jones, D. W., & Weinberger, D. R. (1997). Genetic variability of human brain size and cortical gyral patterns. *Brain*, 120, 257–269.

Başar, E., Başar-Eroglu, C., Karakaş, S., & Schurmann, M. (1999). Are cognitive processes manifested in event-related gamma, alpha, theta and delta oscillations in the EEG? *Neuroscience Letters, 259,* 165–168.

Basile, L. F., Anghinah, R., Ribeiro, P., Ramos, R. T., Piedade, R., Ballester, G., et al. (2007). Interindividual variability in EEG correlates of attention and limits of functional mapping. *International Journal of Psychophysiology*, 65, 238–251.

Basser, P. J., Mattiello, J., & LeBihan, D. (1994). Estimation of the effective self-diffusion tensor from the NMR spin echo. *Journal of Magnetic Resonance, Series B, 103*(3), 247–254.

Bastiaansen, M., & Hagoort, P. (2003). Event-induced theta responses as a window on the dynamics of memory. *Cortex*, 39, 967–992.

Bastiaansen, M. C., Posthuma, D., Groot, P. F., & de Geus, E. J. (2002). Event-related alpha and theta responses in a visuo-spatial working memory task. *Clinical Neurophysiology*, 113, 1882–1893.

Baving, L., Laucht, M., & Schmidt, M. H. (1999). Atypical frontal brain activation in ADHD: Preschool and elementary school boys and girls. *Journal of American Academy of Child and Adolescent Psychiatry*, 38, 1363–1371.

Baving, L., Laucht, M., & Schmidt, M. H. (2002). Frontal brain activity in anxious school children. *Journal of Child Psychology and Psychiatry*, 43, 265–274.

Begleiter, H., Porjesz, B., Bihari, B., & Kissin, B. (1984). Event-related brain potentials in boys at risk for alcoholism. *Science*, 225, pp. 1493–1496.

Begleiter, H., Porjesz, B., Reich, T., Edenberg, H. J., Goate, A., Blangero, J., ... & Bloom, F. E. (1998). Quantitative trait loci analysis of human event-related brain potentials: P3 voltage. *Electroencephalography and Clinical Neurophysiology/Evoked Potentials Section, 108*(3), 244–250.

Beste, C., Baune, B. ., Domschke, K., Falkenstein, M., Konrad, C. (2010). Paradoxical association of the brain-derived-neurotrophic-factor val66met genotype with response inhibition. *Neuroscience*, 166, 178–184.

Birbaumer, N., Elbert, T., Canavan, A. G., Rockstroh, B. (1990). Slow potentials of the cerebral cortex and behavior. *Physiological Reviews*, 70, 1–41.

Bismark, A. W., Moreno, F. A., Stewart, J. L., Towers, D. N., Coan, J. A., Oas, J., ... & Allen, J. J. (2010). Polymorphisms of the HTR1a allele are linked to frontal brain electrical asymmetry. *Biological psychology, 83*(2), 153–158.

Blokland, G. A. M., McMahon, K. L., Thompson, P. M., Martin, N. G., de Zubicaray, G. I., & Wright, M. J. (2011). Heritability of working memory brain activation. *The Journal of Neuroscience*, 31, 10882–10890.

Blokland, G. A. M., de Zubicaray, G. I., McMahon, K. L., Wright, M. J. (2012). Genetic and Environmental Influences on Neuroimaging phenotypes: a meta-analytical perspective on twin imaging studies. *Twin Research and Human Genetics*, 15, 351–371.

Blokland, G. A. M., McMahon, K. L., Jan Hoffman, J., Zhu, G., Meredith, M., Martin, N. G., Paul, M., Thompson, P. M., de Zubicaray, G. I., Margaret, J., & Wright, M. J. (2008). Quantifying the heritability of task-related brain activation during the N-back working memory task: A twin fMRI study. *Biological Psychology*, 79, 70–79

Boomsma, D. I., & Somsen, R. J. (1991). Reaction times measured in a choice reaction time and a double task condition: A small twin study. *Personality and Individual Differences*, *12*(6), 519–522.

Botvinick, M. M., Braver, T. S., Barch, D. M., Carter, C. S., & Cohen, J. D. (2001). Conflict monitoring and cognitive control. *Psychological Review*, 108, 624–652.

Botvinick, M., Nystrom, L. E., Fissell, K., Carter, C. S., & Cohen, J. D. (1999). Conflict monitoring versus selection-for-action in anterior cingulate cortex. *Nature*, 402, 179–181.

Boutros, N. N., Overall, J., & Zouridakis, G. (1991). Test-retest reliability of the P50 mid-latency auditory evoked response. *Psychiatry research*, *39*(2), 181–192.

Brans, R. G., van Haren, N. E., van Baal, G. C. M., Schnack, H. G., Kahn, R. S., & Pol, H. E. H. (2008). Heritability of changes in brain volume over time in twin pairs discordant for schizophrenia. *Archives of General Psychiatry*, *65*(11), 1259–1268.

Brun, C. C., Leporé, N., Pennec, X., Lee, A. D., Barysheva, M., Madsen, S. K., Avedissian, C., Chou, Y. Y., de Zubicaray, G. I., McMahon, K. L., Wright, M. J., Toga, A. W., & Thompson, P. M. (2009). Mapping the regional influence of genetics on brain structure variability – a tensor-based morphometry study. *NeuroImage*, *48*, 37–49.

Cameron, A. M., Geffen, G. M., Kavanagh, D. J., Wright, M. J., Mc-Grath, J. J., & Geffen, L. B. (2003). Event-related potential correlates of impaired visuospatial working memory in schizophrenia. *Psychophysiology*, *40*, 702–715.

Cannon, T. D., Thompson, P. M., van Erp, T. G., Huttunen, M., Lonnqvist, J., Kaprio, J., & Toga, A. W. (2006). Mapping heritability and molecular genetic associations with cortical features using probabilistic brain atlases: Methods and applications to schizophrenia. *Neuroinformatics*, 4, 5–19.

Cardenas, V. A., Gerson, J., & Fein, G. (1993). The reliability of P50 suppression as measured by the conditioning/testing ratio is vastly improved by dipole modeling. *Biological Psychiatry*, *33*, 335–344.

Carlson, S. R., & Iacono, W. G. (2006). Heritability of P300 amplitude development from adolescence to adulthood. *Psychophysiology*, *43*, 470–480.

Carmelli, D., DeCarli, C., Swan, G. E., Jack, L. M., Reed, T., Wolf, P. A., et al. (1998). Evidence for genetic variance in white matter hyperintensity volume in normal elderly male twins. *Stroke*, *29*, 1177–1181

Carmelli, D., Swan, G. E., DeCarli, C., & Reed, T. (2002). Quantitative genetic modeling of regional brain volumes and cognitive performance in older male twins. *Biological Psychology*, *61*(1–2), 139–155.

Carter, C. S., Braver, T. S., Barch, D. M., Botvinick, M. M., Noll, D., & Cohen, J. D. (1998). Anterior cingulate cortex, error detection, and the online monitoring of performance. *Science*, *280*, 747–749.

Chen, A. C., Tang, Y., Rangaswamy, M., Wang, J. C., Almasy, L., Foroud, T., ... & Porjesz, B. (2009). Association of single nucleotide polymorphisms in a glutamate receptor gene (GRM8) with theta power of event-related oscillations and alcohol dependence. *American Journal of Medical Genetics Part B: Neuropsychiatric Genetics, 150*(3), 359–368.

Chen, A. C., Manz, N., Tang, Y., Rangaswamy, M., Almasy, L., Kuperman, S., ... & Porjesz, B. (2010). Single-Nucleotide Polymorphisms in Corticotropin Releasing Hormone Receptor 1 Gene (CRHR1) Are Associated With Quantitative Trait of Event-Related Potential and Alcohol Dependence. *Alcoholism: Clinical and Experimental Research, 34*(6), 988–996.

Chiang, M. C., Barysheva, M., Shattuck, D. W., Lee, A. D., Madsen, S. K., Avedissian, C., ... & Thompson, P. M. (2009). Genetics of brain fiber architecture and intellectual performance. *The Journal of Neuroscience, 29*(7), 2212–2224.

Chiang, M. C., McMahon, K. L., de Zubicaray, G. I., Martin, N. G., Hickie, I., Toga, A. W., et al. (2011). Genetics of white matter development: a DTI study of 705 twins and their siblings aged 12 to 29. *NeuroImage, 54*, 2308–2317.

Chorlian, D. B., Tang, Y., Rangaswamy, M., O'Connor, S., Rohrbaugh, J., Taylor, R., & Porjesz, B. (2007). Heritability of EEG coherence in a large sib-pair population. *Biological Psychology, 75*(3), 260–266.

Clementz, B. A., Geyer, M. A., Braff, D. L. (1997). P50 suppression among schizophrenia and normal comparison subjects: a methodological analysis. *Biological Psychiatry, 41*, 1035–1044.

Cohen, H. L., Wang, W., Porjesz, B., Bauer, B. O., Kuperman, S., O'Connor, S. J., Rohrbaugh, J. and Begleiter, H. (1994). Visual P300: An interlaboratory consistency study. *Alcohol, 11*(6), 583–587.

Cohen, H. L., Wang, W., Porjesz, B., & Begleiter, H. (1995). Auditory P300 in young alcoholics: regional response characteristics. *Alcoholism: Clinical and Experimental Research, 19*(2), 469–475.

Cohen, R. M., Small, C., Lalonde, F., Friz, J., Sunderland, T. (2001). Effect of apolipoprotein E genotype on hippocampal volume loss in aging healthy women. *Neurology, 57*, 2223–2228.

Côté, C., Beauregard, M., Girard, A., Kauffmann, C., Mensour, B., Mancini-Marïe, A., Boivin, M., Pérusse, D. (2007). Individual variation in neural correlates of sadness in children: A twin fMRI study. *Human Brain Mapping, 28*, 482–487.

Courchesne, E. (1987). A neurophysiological view of autism. In E. Schopler & G. Mesibow (Eds), *Neurobiological issues in autism*. New York: Plenum Press.

Courchesne, E. (1978). Neurophysiological correlates of cognitive development: changes in long-latency event-related potentials from childhood to adulthood. *Electroencephalography and clinical neurophysiology, 45*(4), 468–482.

Davidson, R. J. (1998). Anterior electrophysiological asymmetries, emotion, and depression: conceptual and methodological conundrums. *Psychophysiology, 35*(5), 607–614.

Davidson, R. J. (2003). Affective neuroscience and psychophysiology: toward a synthesis. *Psychophysiology, 40*(5), 655–665.

Deary, I. J., Spinath, F. M., & Bates, T. C. (2006). Genetics of intelligence. *European Journal of Human Genetics, 14*, 690–700.

De Geus, E. J. C. (2010). From genotype to EEG endophenotype: a route for post-genomic understanding of complex psychiatric disease? *Genome Medicine*, 2, 63.

den Heijer, T., Oudkerk, M., Launer, L. J., van Duijn, C. M., Hofman, A., & Breteler, M. M. (2002). Hippocampal, amygdalar, and global brain atrophy in different apolipoprotein E genotypes. *Neurology, 59*, 746–748.

Desmedt, J. E., Debecker, J., & Manil, J. (1965). Mise en évidence d'un signe électrique cérébral associé à la détection par le sujet d'un stimulus sensoriel tactile [Demonstration of a cerebral electric sign associated with the detection by the subject of a tactile sensorial stimulus. The analysis of cerebral evoked potentials derived from the scalp with the aid of numerical ordinates]. *Bulletin de l'Académie Royale de Médecine de Belgique*, 5, 887–936.

Dien, J., Spencer, K. M., & Donchin, E. (2004). Parsing the late positive complex: Mental chronometry and the ERP components that inhabit the neighborhood of the P300. *Psychophysiology, 41*, 665–678.

Dikman, Z. V., & Allen, J. J. (2000). Error monitoring during reward and avoidance learning in high- and low-socialized individuals. *Psychophysiology, 37*, 43–54.

Donchin, E., & Coles, M. G. (1988). Is the P300 component a manifestation of context updating? *Behavioral and Brain Sciences, 11*(03), 357–374.

Doppelmayr, M., Klimesch, W., Sauseng, P., Hodlmoser, K., Stadler, W., & Hanslmayr, S. (2005). Intelligence related differences in EEG band power. *Neuroscience Letters, 381*, 309–313.

Ducci, F., Enoch, M. A., Yuan, Q., Shen, P. H., White, K. V., Hodgkinson, C., Albaugh, B., Virkkunen, M., Goldman, D. (2009). HTR3B is associated with alcoholism with antisocial behavior and alpha EEG power—an intermediate phenotype for alcoholism and co-morbid behaviors. *Alcohol, 43*, 73–84.

Duffy, F., & McAnulty, G. (1990). Neurophysiological heterogeneity and the definition of dyslexia: preliminary evidence for plasticity. *Neurophysiologica, 28*, 555–571.

Eason, R. G., Aiken Jr, L. R., White, C. T., & Lichtenstein, M. (1964). Activation and behavior: II. Visually evoked cortical potentials in man as indicants of activation level. *Perceptual and motor skills, 19*(3), 875–895.

Eischen, S. E., & Polich, J. (1994). P300 from families. *Electroencephalography and Clinical Neurophysiology/Evoked Potentials Section, 92*(4), 369–372.

Ethridge, L., Malone, S., Iacono, W., Clementz, B. (2013). Genetic influences on composite neural activations supporting visual target identification. *Biological Psychology, 92*(2), 329–341.

Fabiani, M., & Friedman, D. (1995). Changes in brain activity patterns in aging: the novelty oddball. *Psychophysiology, 32*, 579–594.

Falkenstein, M., Hohnsbein, J., Hoormann, J., & Blanke, L. (1991). Effects of cross-modal divided attention on late ERP components. II. Error processing in choice reaction tasks. *Electroencephalography and Clinical Neurophysiology, 78*, 447–455.

Falkenstein, M., Hoormann, J., Christ, S., & Hohnsbein, J. (2000). ERP components on reaction errors and their functional significance: A tutorial. *Biological Psychology, 51*, 87–107.

Fallgatter, A. J., Herrmann, M. J., Roemmler, J., Ehlis, A. C., Wagener, A., Heidrich, A., Ortega, G., Zeng, Y., Lesch, K.P ., 2004. Allelic variation of serotonin transporter function modulates the brain electrical response for error processing. *Neuropsychopharmacology, 29*, 1506–1511.

Filipovic, S. R., Jahanshahi, M., & Rothwell, J. C. (2001). Uncoupling of contingent negative variation and alpha band event-related desynchronization in a go/no-go task. *Clinical Neurophysiology, 112*, 1307–1315.

Fjell, A. M., Walhovd, K. B. (2004). Life-span changes in P3a. *Psychophysiology*, 41, 575–583.

Forbes, E., Shaw, D., Fox, N., Cohn, J., Silk, J., & Kovacs, M. (2006). Maternal depression, child frontal asymmetry, and child affective behavior as factors in child behavior problems. *Journal of Child Psychology and Psychiatry, 47*(1), 79–87.

Fornito, A., Zalesky, A., Bassett, D. S., Meunier, D., Ellison-Wright, I., Yücel, M., ... & Bullmore, E. T. (2011). Genetic influences on cost-efficient organization of human cortical functional networks. *The Journal of Neuroscience, 31*(9), 3261–3270.

Freedman, R., Coon, H., Myles-Worsley, M., Orr-Urtreger, A., Olincy, A., Davis, A., ... & Byerley, W. (1997). Linkage of a neurophysiological deficit in schizophrenia to a chromosome 15 locus. *Proceedings of the National Academy of Sciences, 94*(2), 587–592.

Freedman, R., Adams, C. E., Adler, L. E., Bickford, P. C., Gault, J., Harris, J. G., ... & Leonard, S. (2000). Inhibitory neurophysiological deficit as a phenotype for genetic investigation of schizophrenia. *American Journal of Medical Genetics, 97*(1), 58–64.

Friedman, D., Cornblatt, B., Vaughun, H., Erlenmeyer-Kimling, L. (1986). Event-related potentials in children at risk for schizophrenia during two continuous performance tests. *Psychiatry Research, 18*, 161–177.

Gale, A. (1983). Electroencephalographic studies of extraversion-introversion: a case study in the psychophysiology of individual differences. *Personality and Individual Differences*, 4, 371–380.

Gao, Y., Raine, A. (2009). P3 event-related potential impairments in antisocial and psychopathic individuals: a meta-analysis. *Biological Psychology, 82*, 199–210.

Gao, Y., Tuvblad, C., Raine, A., Lozano, D. I., Baker, L. A. (2009). Genetic and environmental influences on frontal EEG asymmetry and alpha power in 9–10-year-old twins. *Psychophysiology*, 46, 787–796.

Gehring, W. J., & Fencsik, D. E. (2001). Functions of the medial frontal cortex in the processing of conflict and errors. *Journal of Neuroscience*, 21, 9430–9437.

Geschwind, D. H., Miller, B. L., DeCarli, C., & Carmelli, D. (2002). Heritability of lobar brain volumes in twins supports genetic models of cerebral laterality and handedness. *Proceedings of the National Academy of Sciences USA*, 99, 3176–3181.

Giannitrapani, D. (1985). *The Electrophysiology of Intellectual Functions*. Kargere Press, New York.

Gould, J., Austin, F., & Cook, P. (1978). A genetic analysis of sleep stage organization in newborn twins. *Sleep Research, 7*, 132.

Hall, M. H., Schulze, K., Bramon, E., Murray, R. M., Sham, P., & Rijsdijk, F. (2006). Genetic overlap between P300, P50, and duration mismatch negativity. *American Journal of Medical Genetics Part B: Neuropsychiatric Genetics, 141*(4), 336–343.

Hall, M. H., Rijsdijk, F., Picchioni, M., Psych, M. R. C., Schulze, K., Ettinger, U., ... & Sham, P. (2007). Substantial shared genetic influences on schizophrenia and event-related potentials. *The American journal of psychiatry, 164*(5), 804–812.

Hansell, N. K., Wright, M. J., Geffen, G. M., Geffen, L. B., Smith, G. A., & Martin, N. G. (2001). Genetic influence on ERP slow wave measures of working memory. *Behavior Genetics, 31*(6), 603–614.

Hansell, N. K., Wright, M. J., Luciano, M., Geffen, G. M., Geffen, L. B., & Martin, N. G. (2005). Genetic covariation between event-related potential (ERP) and behavioral non-ERP measures of working-memory, processing speed, and IQ. *Behavior genetics, 35*(6), 695–706.

Hill, S. Y., Locke, J., Zezza, N., Kaplan, B., Neiswanger, K., Steinhauer, S. R., Wipprecht, G., Xu, J. (1998). Genetic association between reduced P300 amplitude and the DRD2 dopamine receptor A1 allele in children at high risk for alcoholism. *Biological Psychiatry*, 43, 40–51.

Hodgkinson, C. A., Enoch, M. A., Srivastava, V., Cummins-Oman, J. S., Ferrier, C., Iarikova, P., ... & Goldman, D. (2010). Genome-wide association identifies candidate genes that influence the human electroencephalogram. *Proceedings of the National Academy of Sciences, 107*(19), 8695–8700.

Hulshoff Pol, H. E., Schnack, H. G., Mandl, R. C., Brans, R. G., van Haren, N. E., Baare, W. F., et al. (2006a). Gray and white matter density changes in monozygotic and same-sex dizygotic twins discordant for schizophrenia using voxel-based morphometry. *NeuroImage, 31*(2), 482–488.

Hulshoff Pol, H. E., Schnack, H. G., Posthuma, D., Mandl, R. C., Baare, W. F., van Oel, C., et al. (2006b). Genetic contributions to human brain morphology and intelligence. *Journal of Neuroscience*, 26, 10235–10242.

Jahanshad, N., Zhan, L., Bernstein, M., Borowski, B. J., Jack Jr, C. R., Toga, A. W., & Thompson, P. M. (2010, April). Diffusion tensor imaging in seven minutes: determining trade-offs between spatial and directional resolution. In *Biomedical Imaging: From Nano to Macro, 2010 IEEE International Symposium on* (pp. 1161–1164). IEEE.

Jaušovec, N., & Jaušovec, K. (2000). EEG activity during the performance of complex mental problems. *International Journal of Psychophysiology, 36*(1), 73–88.

Jausovec, N., & Jausovec, K. (2004). Differences in induced brain activity during the performance of learning and working-memory tasks related to intelligence. *Brain and Cognition*, 54, 65–74.

Jensen, O., Gelfand, J., Kounios, J., & Lisman, J. E. (2002). Oscillations in the alpha band (9–12 Hz) increase with memory load during retention in a short-term memory task. *Cerebral Cortex*, 12, 877–882.

John, E. R., Ahn, H., Prichep, L., Trepetin, M., Brown, D., & Kaye, H. (1980). Developmental equations for the electroencephalogram. *Science, 210*(4475), 1255–1258.

Johnson, J. S., & Newport, E. L. (1989). Critical period effects in second language learning: The influence of maturational state on the acquisition of English as a second language. *Cognitive psychology, 21*(1), 60–99.

Johnstone, S. J., Barry, R. J., & Anderson, J. W. (2001). Topographic distribution and developmental timecourse of auditory event-related potentials in two subtypes of attention-deficit hyperactivity disorder. *International Journal of Psychophysiology, 42*(1), 73–94.

Jones, K. A., Porjesz, B., Almasy, L., Bierut, L., Goate, A., Wang, J. C., ... & Begleiter, H. (2004). Linkage and linkage disequilibrium of evoked EEG oscillations with CHRM2 receptor gene polymorphisms: implications for human brain dynamics and cognition. *International Journal of Psychophysiology, 53*(2), 75–90.

Joshi, A. A., Lepore, N., Joshi, S. H., Lee, A. D., Barysheva, M., Stein, J. L., ... & Thompson, P. M. (2011). The contribution of genes to cortical thickness and volume. *Neuroreport, 22*(3), 101–105.

Kang, S. J., Rangaswamy, M., Manz, N., Wang, J. C., Wetherill, L., Hinrichs, T., ... & Porjesz, B. (2012). Family-based genome-wide association study of frontal theta oscillations identifies potassium channel gene KCNJ6. *Genes, Brain and Behavior, 11*(6), 712–719.

Katsanis, J., Iacono, W. G., McGue, M. K., & Carlson, S. R. (1997). P300 event-related potential heritability in monozygotic and dizygotic twins. *Psychophysiology, 34*(1), 47–58.

Katsanis, J., Iacono, W. G., & McGue, M. K. (1996). The association between P300 and age from preadolescence to early adulthood. *International Journal of Psychophysiology, 24*(3), 213–221.

Kemner, C., van der Gaag, R. J., Verbaten, M., & van Engeland, H. (1999). ERP differences among subtypes of pervasive developmental disorders. *Biological Psychiatry, 46*(6), 781–789.

Kendler, K. S., & Baker, J. H. (2007). Genetic influences on measures of the environment: A systematic review. *Psychological Medicine*, 37, 615–626.

Kennedy, D. P., Redcay, E., & Courchesne, E. (2006). Failing to deactivate: resting functional abnormalities in autism. *Proceedings of the National Academy of Sciences, 103*(21), 8275–8280.

Kesslak, J. P., So, V., Choi, J., Cotman, C. W., & Gomez-Pinilla, F. (1998). Learning upregulates brain-derived neurotrophic factor messenger ribonucleic acid: a mechanism to facilitate encoding and circuit maintenance? *Behavioral neuroscience, 112*(4), 1012.

Kirchhoff, B. A., Buckner, R. L. (2006). Functional-anatomic correlates of individual differences in memory. *Neuron*, 51, 263–274.

Klimesch, W. (1999). EEG alpha and theta oscillations reflect cognitive and memory performance: a review and analysis. *Brain research reviews, 29*(2), 169–195.

Klimesch, W., Doppelmayr, M., Schwaiger, J., Auinger, P., & Winkler, T. (1999). "Paradoxical" alpha synchronization in a memory task. *Cognitive Brain Research*, 7, 493–501.

Klimesch, W., Sauseng, P., Hanslmayr, S., Gruber, W., & Freunberger, R. (2007). Event-related phase reorganization may explain evoked neural dynamics. *Neuroscience & Biobehavioral Reviews, 31*(7), 1003–1016.

Klimesch, W., Schimke, H. Pfurtscheller, G. (1993). Alpha frequency, cognitive load and memory performance. *Brain topography, 5*(3), 241–251.

Knott, V. J. (2001). Electroencephalographic characterization of cigarette smoking behavior. *Alcohol, 24*(2), 95–97.

Kochunov, P., Glahn, D. C., Lancaster, J. L., Winkler, A. M., Smith, S., Thompson, P. M., et al. (2010). Genetics of microstructure of cerebral white matter using diffusion tensor imaging. *NeuroImage*, 53, 1109–1116.

Kok, A. (2001). On the utility of P3 amplitude as a measure of processing capacity. *Psychophysiology, 38*(3), 557–577.

Koten, J. W., Wood, G., Hagoort, P., Goebel, R., Propping, P., Willmes, K., & Boomsma, D. I. (2009). Genetic contribution to variation in cognitive function: an FMRI study in twins. *Science, 323*(5922), 1737–1740.

Krause, C. M., Lang, A. H., Laine, M., Kuusisto, M., & Pörn, B. (1996). Event-related EEG desynchronization and synchronization during an auditory memory task. *Electroencephalography and clinical neurophysiology, 98*(4), 319–326.

Kremen, W. S., Prom-Wormley, E., Panizzon, M. S., Eyler, L. T., Fischl, B., Neale, M. C., et al. (2009). Genetic and environmental influences on the size of specific brain regions in midlife: the VETSA MRI study. *NeuroImage, 49*(2), 1213–1223.

Lazzaro, I., Gordon, E., Whitmont, S., Plahn, M., Li, W., Clarke, S., Dosen, A., Meares, R. (1998). Quantified EEG activity in adolescent attention deficit hyperactivity disorder. *Clinical EEG (Electroencephalography)*, 29, 37–42.

Lee, A. D., Leporé, N., Brun, C., Chou, Y. Y., Barysheva, M., Chiang, M. C., Madsen, S. K., de Zubicaray, G. I.,McMahon, K. L., Wright, M. J., Toga, A. W., & Thompson, P. M. (2009). Tensor-based analysis of genetic influences on brain integrity using DTI in 100 twins. *Medical Image Computing and Computer-Assisted Intervention*, 12, 967–974.

Lenroot, R. K., Schmitt, J. E., Ordaz, S. J., Wallace, G. L., Neale, M. C., Lerch, J. P., Kendler, K. S., Evans, A. C., & Giedd, J. N. (2009). Differences in genetic and environmental influences on the human cerebral cortex associated with development during childhood and adolescence. *Human Brain Mapping*, 30, 163–174.

Leonard, S., Gault, J., Hopkins, J., Logel, J., Vianzon, R., Short, M., ... & Freedman, R. (2002). Association of promoter variants in the α7 nicotinic acetylcholine receptor subunit gene with an inhibitory deficit found in schizophrenia. *Archives of General Psychiatry, 59*(12), 1085–1096.

Linkenkaer-Hansen, K., Monto, S., Rytsälä, H., Suominen, K., Isometsä, E., & Kähkönen, S. (2005). Breakdown of long-range temporal correlations in theta oscillations in patients with major depressive disorder. *The Journal of Neuroscience, 25*(44), 10131–10137.

Linkowski, P., Kerkhofs, M., Hauspie, R., Mendlewicz, J. (1991). Genetic determinants of EEG sleep: a study in twins living apart. *Electroencephalography and Clinical Neurophysiology, 79*(2), 114_/118.

Lu, P. H., Thompson, P. M., Leow, A., Lee, G. J., Lee, A., Yanovsky, I., et al. (2011). Apolipoprotein E genotype is associated with temporal and hippocampal atrophy rates in healthy elderly adults: a tensor-based morphometry study. *Journal of Alzheimer's disease*, 23, 433–442.

Lykken, D. T., Bouchard, T. J., Jr., McGue, M., & Tellegen, A. (1992). Emergenesis: Genetic traits that may not run in families. *American Psychologist*, 47, 1565–1577.

Lykken, D. T., Tellegen, A., Iacono, W. G. (1982). EEG spectra in twins: evidence for a neglected mechanism of genetic determination. *Physiological Psychology*, 10, 60–65.

Malykh, S. B., Egorova, M. S., Meshkova, T. A. (1998). *Osnovy psihogenetiki [Foundations of Behavior Genetics]*. Moscow, Epidaurus.

Makeig, S., Westerfield, M., Jung, T.-P., Enghoff, S., Townsend, J., Courchesne, E., Sejnowski, T. (2002). Dynamic Brain Sources of Visual Evoked Responses. *Science, 295*, 690–694.

Matthews, S. C., Simmons, A. N., Strigo, I., Jang, K., Stein, M. B., Paulus, M. P. (2007). Heritability of anterior cingulate response to conflict: an fMRI study in female twins. *NeuroImage*, 38, 223–227.

Mawuenyega, K., Sigurdson, W., Ovod, V., Munsell, L., Kasten, T., Morris, J.,Yarasheski, K., Bateman, R. (2010). Decreased Clearance of CNS β-Amyloid in Alzheimer's Disease. *Science, 330*(6012), 1774–1777.

McLoughlin, G., Makeig, S., & Tsuang, M. T. (2014). In search of biomarkers in psychiatry: EEG-based measures of brain function. *American Journal of Medical Genetics Part B: Neuropsychiatric Genetics, 165*(2), 111–121.

McLoughlin, G., Palmer, J. A., Rijsdijk, F., & Makeig, S. (2014). Genetic overlap between evoked frontocentral theta-band phase variability, reaction time variability, and attention-deficit/hyperactivity disorder symptoms in a twin study. *Biological psychiatry, 75*(3), 238–247.

Micheloyannis, S., Pachou, E., Stam, C. J., Breakspear, M., Bitsios, P., Vourkas, M., Zervakis, M. (2006). Small world networks and disturbed functional connectivity in schizophrenia. *Schizophrenia Research*, 87, 60–66.

Min, B. K., Busch, N. A., Debener, S., Kranczioch, C., Hanslmayr, S., Engel, A. K., et al. (2007). The best of both worlds: Phase-reset of human EEG alpha activity and additive power contribute to ERP generation. *International Journal of Psychophysiology*, 65, 58–68.

Mullis, R. J., Holcomb, P. J., Diner, B. C., & Dykman, R. A. (1985). The effects of aging on the P3 component of the visual event-related potential. *Electroencephalography and Clinical Neurophysiology*, 62, 141–149.

Mundy-Castle, A. (1957). The electroencephalogram and mental activity. *Electroencephalography and Clinical Neurophysiology, 9*(4), 643–655.

Myles-Worsley, M., Coon, H., Byerley, W., Waldo, M., Young, D., Freedman, R. (1996). Developmental and genetic influences on the P50 sensory gating phenotype. *Biological Psychiatry*, 39, 289–295.

Nikulin, V. V., Linkenkaer-Hansen, K., Nolte, G., Lemm, S., Muller,K. R., Ilmoniemi, R. J., et al. (2007). A novel mechanism for evoked responses in the human brain. *European Journal of Neuroscience*, 25, 3146–3154.

O'Connor, S., Morzorati, S., Christian, J. C., & Li, T. K. (1994). Heritable features of the auditory oddball event-related potential: peaks, latencies, morphology and topography. *Electroencephalography and Clinical Neurophysiology/Evoked Potentials Section, 92*(2), 115–125.

Olbrich, S., Arns, M. (2013). EEG biomarkers in major depressive disorder: Discriminative power and prediction of treatment response. *International Review of Psychiatry, 25*(5), 604–618.

Olvet, D. M., Hatchwell, E., Hajcak, G. (2010). Lack of association between the 5-HTTLPR and the error-related negativity (ERN). *Biological Psychology*, 85, 504–508.

Onton, J., Delorme, A., & Makeig, S. (2005). Frontal midline EEG dynamics during working memory. *NeuroImage*, 27, 341–356.

Orekhova, E. V., Stroganova, T. A., Posikera, I. N., Malykh, S. B. (2003). Heritability and "environmentability" of electroencephalogram in infants: the twin study. *Psychophysiology*, 40, 727–741.

Park, J., Shedden, K., & Polk, T. A. (2012). Correlation and heritability in neuroimaging datasets: A spatial decomposition approach with application to an fMRI study of twins. *NeuroImage*, 59, 1132–1142.

Patterson, B. W., Williams, H. L., Mclean, G. A., Smith, L. T. and Schaeffer, K. W. (1987). Alcoholism and family history of alcoholism: Effects on visual and auditory event-related potentials. *Alcohol*, 4, 265–274.

Pennington, B. F., Filipek, P. A., Lefly, D., Chhabildas, N., Kennedy, D. N., Simon, J. H., et al. (2000). A twin MRI study of size variations in human brain. *Journal of Cognitive Neuroscience, 12*(1), 223–232.

Peper, J. S., Brouwer, R. M., Boomsma, D. I., Kahn, R. S., & Hulshoff Pol, H. E. (2007). Genetic influences on human brain structure: a review of brain imaging studies in twins. *Human Brain Mapping, 28*(6), 464–473.

Peper, J. S., Zwiers, M. P., Boomsma, D. I., Kahn, R. S., & Hulshoff Pol, H. E. (2009). Human brain volume: What's in the genes? In Y.-K. Kim (Ed.), *Handbook of Behavior Genetics* (pp. 137–158). New York: Springer.

Perez-Edgar, K., Fox, N. A., Cohn, J. F., & Kovacs, M. (2006). Behavioral and electrophysiological markers of selective attention in children of parents with a history of depression. *Biological Psychiatry*, 60, 1131–1138.

Pfefferbaum, A., Sullivan, E. V., Swan, G. E., & Carmelli, D. (2000). Brain structure in men remains highly heritable in the seventh and eighth decades of life. *Neurobiology of aging, 21*(1), 63–74.

Plomin, R., & Spinath, F. M. (2004). Intelligence: genetics, genes, and genomics. *Journal of personality and social psychology, 86*(1), 112.

Pogarell, O., Juckel, G., Mavrogiorgou, P., Mulert, C., Folkerts, M., Hauke, W., Zaudig, M., Moller, H. J., & Hegerl, U. (2006). Symptom-specific EEG power correlations in patients with obsessive-compulsive disorder. *International Journal of Psychophysiology*, 62, 87–92.

Polk, T. A., Park, J., Smith, M. R., & Park, D. C. (2007). Nature versus nurture in ventral visual cortex: A functional magnetic resonance imaging study of twins. *The Journal of Neuroscience*, 27, 13921–13925.

Polich, J., & Burns, T. (1987). P300 from identical twins. *Neuropsychologia, 25*(1B), 299–304.

Polich, J., Pollock, V. E., & Bloom, F. E. (1994). Meta-analysis of P300 amplitude from males at risk for alcoholism. *Psychological Bulletin*, 115, 55–73.

Polich, J. (1997). On the relationship between EEG and P300: individual differences, aging, and ultradian rhythms. *International Journal of Psychophysiology*, 26, 299–317.

Pollock, V. E., Schneider, L. S., & Lyness, S. A. (1991). Reliability of topographic quantitative EEG amplitude in healthy late-middle-aged and elderly subjects. *Electroencephalography and Clinical Neurophysiology*, *79*(1), 20–26.

Porjesz, B., & Begleiter, H. (1996). Effects of alcohol on electrophysiological activity of the brain. In H. Begleiter & B. Kissin (Eds), *Alcohol and Alcoholism, The Pharmacology of Alcohol and Alcohol Dependence* (Vol. 2, pp. 207–247). New York: Oxford University Press.

Porjesz, B., Begleiter, H., Reich, T., Van Eerdewegh, P., Edenberg, H. J., Foroud, T., ... & Rohrbaugh, J. (1998). Amplitude of visual P3 event-related potential as a phenotypic marker for a predisposition to alcoholism: preliminary results from the COGA project. *Alcoholism: Clinical and Experimental Research*, *22*(6), 1317–1323.

Porjesz, B., Begleiter, H., Wang, K., Almasy, L., Chorlian, D. B., Stimus, A. T., ... & Reich, T. (2002). Linkage and linkage disequilibrium mapping of ERP and EEG phenotypes. *Biological Psychology*, 61, 229–248.

Posthuma, D. (2002). *Genetic variation and cognitive ability*. Unpublished doctoral dissertation. Vrije Universiteit Amsterdam, The Netherlands.

Posthuma, D., de Geus, E., Mulder, E., Smit, D., Boomsma, D., Stam, C. (2005). Genetic Components of Functional Connectivity in the Brain: The Heritability of Synchronization Likelihood. *Human Brain Mapping*, 26, 191–198.

Posthuma, D., De Geus, E. J. C., & Boomsma, D. I. (2001). Perceptual speed and IQ are associated through common genetic factors. *Behavior Genetics*, *31*(6), 593–602.

Posthuma, D., de Geus, E. J., Neale, M. C., Hulshoff Pol, H. E., Baare, W. E. C., Kahn, R. S., & Boomsma, D. (2000). Multivariate genetic analysis of brain structure in an extended twin design. *Behavior Genetics*, 30, 311–319.

Potts, G. F., George, M. R., Martin, L. E., & Barratt, E. S. (2006). Reduced punishment sensitivity in neural systems of behavior monitoring in impulsive individuals. *Neuroscience Letters*, 397, 130–134.

Pritchard, W. S., & Duke, D. W. (1995). Measuring chaos in the brain – a tutorial review of EEG dimension estimation. *Brain and Cognition*, *27*(3), 353–397.

Reiser, E., Schulter, G., Elisabeth, M., Weiss, E., Fink, A., Rominger, C., & Papousek, I. (2012). Decrease of prefrontal–posterior EEG coherence: Loose control during social–emotional stimulation. *Brain and Cognition*, 80, 144–154.

Reveley, A. M., Reveley, M. A., Chitkara, B., & Clifford, C. (1984). The genetic basis of cerebral ventricular volume. *Psychiatry research*, *13*(3), 261–266.

Rijsdijk, F. V., Viding, E., De Brito, S., Forgiarini, M., Mechelli, A., Jones, A. P., & McCrory, E. (2010). Heritable variations in gray matter concentration as a potential endophenotype for psychopathic traits. *Archives of General Psychiatry*, 67, 406–413.

Rijsdijk, F. V., van Haren, N. E., Picchioni, M. M., McDonald, C., Toulopoulou, T., Hulshoff Pol, H. E., Kahn, R. S., Murray, R., & Sham, P. C. (2005). Brain MRI abnormalities in schizophrenia: Same genes or same environment? *Psychological Medicine*, 35, 1399–1409.

Richter-Schmidinger, T., Alexopoulos, P., Horn, M., Maus, S., Reichel, M., Rhein, C., ... & Kornhuber, J. (2011). Influence of brain-derived neurotrophic-factor and apolipoprotein E genetic variants on hippocampal volume and memory performance in healthy young adults. *Journal of neural transmission, 118*(2), 249–257.

Rimol, L. M., Agartz, I., Djurovic, S., Brown, A. A., Roddey, J. C.,Kahler, A. K., et al. (2010). Sex-dependent association of common variants of microcephaly genes with brain structure. *Proceedings of the National Academy of Sciences of the United States of America*, 107, 384–388.

Rockstroh, B., Elbert, T., Canavan, A., Lutzenberger,W., & Birbaumer, N. (1989). Slow cortical potentials and behavior (2nd ed.). Muenchen: Urban & Schwarzenberg.

Rogers, T. D., & Deary, I. (1991). The P300 component of the auditory event-related potential in monozygotic and dizygotic twins. *Acta Psychiatrica Scandinavica, 83*(5), 412–416.

Ruchsow, M., Spitzer, M., Gron, G., Grothe, J., & Kiefer, M. (2005). Error processing and impulsiveness in normals: Evidence from event-related potentials. *Cognitive Brain Research*, 24, 317–325.

Ruchkin, D. S., Canoune, H. L., Johnson, R., & Ritter, W. (1995). Working memory and preparation elicit different patterns of slow wave event-related brain potentials. *Psychophysiology*, 32, 399–410.

Salinsky, M. C., Oken, B. S., Morehead, L. (1991). Test/retest reliability in EEG frequency analysis. *Electroencephalography and Clinical Neurophysiology, 79*(5), 382–392.

Santesso, D. L., Segalowitz, S. J., & Schmidt, L. A. (2005). ERP correlates of error monitoring in 10-year olds are related to socialization. *Biological Psychology*, 70, 79–87.

Santesso, D. L., Segalowitz, S. J., & Schmidt, L. A. (2006). Error-related electrocortical responses in 10-year-old children and young adults. *Developmental Science*, 9, 473–481.

Satterfield, J. H., Schell, A. M., Backs, R. W., & Hidaka, K. C. (1984). A cross-sectional and longitudinal study of age effects of electrophysiological measures in hyperactive and normal children. *Biological Psychiatry*, 19, 973–990.

Sauseng, P., Klimesch, W., Doppelmayr, M., Pecherstorfer, T., Freunberger, R., & Hanslmayr, S. (2005). EEG alpha synchronization and functional coupling during top–down processing in a working memory task. *Human Brain Mapping*, 26, 148–155.

Sauseng, P., Klimesch, W., Schabus, M., & Doppelmayr, M. (2005). Fronto-parietal EEG coherence in theta and upper alpha reflect central executive functions of working memory. *International Journal of Psychophysiology*, 57, 97–103.

Scamvougeras, A., Kigar, D. L., Jones, D., Weinberger, D. R., & Witelson, S. F. (2003). Size of the human corpus callosum is genetically determined: An MRI study in mono and dizygotic twins. *Neuroscience Letters*, 338, 91–94.

Schmitt, J. E., Lenroot, R. K., Wallace, G. L., Ordaz, S., Taylor, K. N., Kabani, N., ... & Giedd, J. N. (2008). Identification of genetically mediated cortical networks: A multivariate study of pediatric twins and siblings. *Cerebral Cortex*, 18, 1737–1747.

Schutte, N. M., Hansell, N. K., de Geus, E. J., Martin, N. G., Wright, M. J., & Smit, D. J. (2013). Heritability of resting state EEG functional connectivity patterns. *Twin Research and Human Genetics, 16*(05), 962–969.

Silverman, M. H., Krueger, R. F., Iacono, W. G., Malone, S. M., Hunt, R. H., & Thomas, K. M. (2014). Quantifying familial influences on brain activation during the monetary incentive delay task: An adolescent monozygotic twin study. *Biological psychology, 103*, 7–14.

Schmid, R., Tirsch, W., Scherb, H. (2002). Correlation between spectral EEG parameters and intelligence test variables in school-age children. *Clinical Neurophysiology, 13*(10), 1647–1656.

Smit, D. J. A., Posthuma, D., Boomsma, D. I., & de Geus, E. D. (2005). Heritability of background EEG across the power spectrum. *Psychophysiology, 42*(6), 691–697.

Smit, D. J., Posthuma, D., Boomsma, D. I., & de Geus, E. J. (2007). Genetic contribution to the P3 in young and middle-aged adults. *Twin Research and Human Genetics, 10*(02), 335–347.

Smit, D. J., Stam, C. J., Posthuma, D., Boomsma, D. I., & de Geus, E. J., (2008). Heritability of "small-world" networks in the brain: a graph theoretical analysis of resting-state EEG functional connectivity. *Human Brain Mapping, 29*, 1368–1378.

Smit, D. J., Posthuma, D., Boomsma, D. I., & de Geus, E. J. C. (2009). Phenotypic and genetic correlations between evoked EEG/ERP measures during the response anticipation period of a delayed response task. *Psychophysiology*, 46, 344–356.

Smit, D. J. A., Boersma, M., van Beijsterveldt, C. E. M., Posthuma, D., Boomsma, D. I., Stam, C. J., & De Geus, E. J. C. (2010). Endophenotypes in a dynamically connected brain. *Behavior Genetics, 40*(2), 167–177.

Smith, D. A., Boutros, N. N., Schwarzkopf, S. B. (1994). Reliability of P50 auditory event-related potential indices of sensory gating. *Psychophysiology*, 31, 495–502.

Steriade, M. (2000). Corticothalamic resonance, states of vigilance and mentation. *Neuroscience*, 101, 243–276.

Stroganova, T., Posikera, I., Orekhova, E., Tsetlin, M., Novikova, S., & Kuravsky, L. (2009). The nature of individual differences in EEG parameters during homogeneous visual field in 5-and 6-year-old twins: A pilot study. *Clinical Neurophysiology, 120*(8), 1492–1500.

Surwillo, W. W. (1980). Cortical evoked potentials in monozygotic twins and unrelated subjects: comparisons of exogenous and endogenous components. *Behavior Genetics, 10*(2), 201–209.

Sutton, S., Braren, M., Zubin, J., & John, E. R. (1965). Evoked-potential correlates of stimulus uncertainty. *Science*, 150, 1187–1188.

Taylor, M. J. (1993). Maturational changes in ERPs to orthographic and phonological tasks. *Electroencephalography and Clinical Neurophysiology*, 88, 494–507.

Tesche, C., & Karhu, J. (2000). Theta oscillations index human hippocampal activation during a working memory task. *Proceedings of the National Academy of Sciences*, 97, 919.

Thatcher, R. W. (1991). Maturation of the human frontal lobes: Physiological evidence for staging. *Developmental Neuropsychology, 7*(3), 370–394.

Thatcher, R. W. (1994). Cyclic cortical reorganization, origins of human cognitive development. In G. Dawson & K. Fischer (Eds), *Human behavior and the developing brain* (pp. 232–266). New York: Guilford Press.

Thatcher, R., McAlaster, R., Lester, M., Horst, R., & Cantor, D. (1983). Hemispheric EEG asymmetries related to cognitive functioning in children. In E. Perecman (Ed.), *Cognitive processing in the right hemisphere* (pp. 125–146). Orlando: Academic Press.

Thatcher, R. W., North, D. M., & Biver, C. J. (2008). Development of cortical connections as measured by EEG coherence and phase delays. *Human Brain Mapping, 29*, 1400–1415.

Thatcher, R., Walker, R., & Guidice, S. (1987). Human cerebral hemispheres develop at different rates and ages. *Science*, 236, 1110–1113.

Thompson, P. M., Cannon, T. D., Narr, K. L., van Erp, T., Poutanen, V. P., Huttunen, M., et al. (2001). Genetic influences on brain structure. *Nature Neuroscience*, 4, 1253–1258.

Thompson, P. M., Stein, J. L., Medland, S. E., Hibar, D. P. et al., (2014). The ENIGMA Consortium: large-scale collaborative analyses of neuroimaging and genetic data. *Brain Imaging and Behavior*, 8, 153–182.

Toga, A. W., & Thompson, P. M. (2005). Genetics of brain structure and intelligence. *Annual Review of Neuroscience*, *28*, 1–23.

Tomarken, A. J., Davidson, R. J., & Henriques, J. B. (1990). Resting frontal brain asymmetry predicts affective responses to films. *Journal of Personality and Social Psychology*, *59*(4), 791–801.

Tran, Y., Craig, A., Boord, P., Connell, K., Cooper, N., & Gordon, E. (2006). Personality traits and its association with resting regional brain activity. *International Journal of Psychophysiology*, 60, 215–224.

van Baal, G. C. M., De Geus, E. J. C., & Boomsma, D. I. (1996). Genetic architecture of EEG power spectra in early life. *Electroencephalography and clinical Neurophysiology*, *98*(6), 502–514.

van Baal, G. C. M., De Geus, E. J. C., & Boomsma, D. I. (1998a). Genetic influences on EEG coherence in 5-year-old twins. *Behavior genetics*, *28*(1), 9–19.

van Baal, G. C. M., de Geus, E. J., & Boomsma, D. I. (1998b). Longitudinal study of genetic influences on ERP-P3 during childhood. *Developmental Neuropsychology*, *14*(1), 19–45.

van Baal, G. C., Boomsma, D. I., & de Geus, E. J. (2001). Longitudinal genetic analysis of EEG coherence in young twins. *Behavioral Genetics*, 31, 637–651.

Van Beijsterveldt, C. E., & van Baal, G. C. (2002). Twin and family studies of the human electroencephalogram: a review and a meta-analysis. *Biological Psychology*, 61, 111–138.

Van Beijsterveldt, C. E. M., & Boomsma, D. I. (1994). Genetics of the human electroencephalogram (EEG) and event-related brain potentials (ERPs): A review. *Human Genetics*, 94, 319–330.

Van Beijsterveldt, C. E., van Baal, G. C., Molenaar, P. C., Boomsma, D. I., & de Geus, E. J. (2001). Stability of genetic and environmental influences on P300 amplitude, a longitudinal study in adolescent twins. *Behavior Genetics*, 31, 533–543.

Van Beijsterveldt, C. E., Molenaar, P. C., De Geus, E. J., & Boomsma, D. I. (1996). Heritability of human brain functioning as assessed by electroencephalography. *American journal of human genetics*, *58*(3), 562.

Van Beijsterveldt, C. E. M., Molenaar, P. C. M., De Geus, E. J. C., & Boomsma, D. I. (1998a). Genetic and environmental influences on EEG coherence. *Behavior genetics*, *28*(6), 443–453.

Van Beijsterveldt, C. E., Molenaar, P. C., de Geus, E. J., & Boomsma, D. I., (1998b). Individual differences in P300 amplitude: a genetic study in adolescent twins. *Biological Psychology*, 47, 97–120.

Van Erp, T. G., Cannon, T. D., Tran, H. L., Wobbekind, A. D., Huttunen, M., Lönnqvist, J., ... & Thompson, P. M. (2004, April). Genetic influences on human brain morphology. In *Biomedical Imaging: Nano to Macro, 2004. IEEE International Symposium on* (pp. 583–586). IEEE.

Van den Heuvel, M. P., van Soelen, I. L., Stam, C. J., Kahn, R. S., Boomsma, D. I., Hulshoff Pol, H. E., (2013). Genetic control of functional brain network efficiency in children. *European Neuropsychopharmacology*, 23, 19–23.

Van der Schot, A. C., Vonk, R., Brans, R. G., van Haren, N. E., Koolschijn, P. C., Nuboer, V., ... & Kahn, R. S. (2009). Influence of genes and environment on brain volumes in twin pairs concordant and discordant for bipolar disorder. *Archives of General Psychiatry*, 66, 142–151.

Van Soelen, I. L., Brouwer, R. M., Peper, J. S., van Leeuwen, M., Koenis, M. M., van Beijsterveldt, T. C., ... & Boomsma, D. I. (2012). Brain SCALE: brain structure and cognition: an adolescent longitudinal twin study into the genetic etiology of individual differences. *Twin Research and Human Genetics*, *15*(03), 453–467.

Van Veen, V., & Carter, C. S. (2002). The anterior cingulate as a conflict monitor: fMRI and ERP studies. *Physiology & behavior*, *77*(4), 477–482.

Verleger, R. (1988). Event-related potentials and cognition: A critique of the context updating hypothesis and an alternative interpretation of P3. *The Behavioral and Brain Sciences*, 11, 343–427.

Vernon, P. A., & Mori, M. (1989). Intelligence, reaction times, and nerve conduction velocity. *Behavior Genetics*, 19, 779.

Wallace, G. L., Schmitt, J. E., Lenroot, R. K., Viding, E., Ordaz, S., Rosenthal, M. A., et al. (2006). A pediatric twin study of brain morphometry. *Journal of Child Psychology and Psychiatry*, 47, 987–993.

Watson, N. F. (2011). Genetics of electroencephalography during wakefulness and sleep. *Sleep Medicine Clinics*, *6*(2), 155–169.

Webb, W. B., & Campbell, S. S. (1983). Relationships in sleep characteristics of identical and fraternal twins. *Archives of general psychiatry*, *40*(10), 1093–1095.

Wheeler, R. E., Davidson, R. J., & Tomarken, A. J. (1993). Frontal brain asymmetry and emotional reactivity: a biological substrate of affective style. *Psychophysiology*, *30*(1), 82–89.

Whitfield-Gabrieli, S., Thermenos, H. W., Milanovic, S., Tsuang, M. T., Faraone, S. V., McCarley, R. W., ... & Seidman, L. J. (2009). Hyperactivity and hyperconnectivity of the default network in schizophrenia and in first-degree relatives of persons with schizophrenia. *Proceedings of the National Academy of Sciences*, *106*(4), 1279–1284.

Winkler, A. M., Kochunov, P., Blangero, J., Almasy, L., Zilles, K., Fox, P. T., et al. (2010). Cortical thickness or grey matter volume? The importance of selecting the phenotype for imaging genetics studies. *NeuroImage*, *53*(3), 1135–1146.

Wright, M. J., Hansell, N. K., Geffen, G. M., Geffen, L. B., Smith, G. A., & Martin, N. G. (2001). Genetic influence on the variance in P3 amplitude and latency. *Behavior Genetics*, 31, 555–565.

Wright, I. C., Sham, P., Murray, R. M., Weinberger, D. R., & Bullmore, E. T. (2002). Genetic contributions to regional variability in human brain structure: Methods and preliminary results. *NeuroImage*, 17, 256–271.

Yoon, U., Fahim, C., Perusse, D., & Evans, A. C. (2010). Lateralized genetic and environmental influences on human brain morphology of 8-year-old twins. *NeuroImage*, 53, 1117–1125.

Yoon, U., Perusse, D., Lee, J. M., & Evans, A. C. (2011). Genetic and environmental influences on structural variability of the brain in pediatric twin: Deformation based morphometry. *Neuroscience Letters*, 493, 8–13.

Young, D. A., Waldo, M., Rutledge III, J. H., & Freedman, R. (1996). Heritability of inhibitory gating of the P50 auditory-evoked potential in monozygotic and dizygotic twins. *Neuropsychobiology*, 33, 113–117.

Zietsch, B. P., Hansen, J. L., Hansell, N. K., Geffen, G. M., Martin, N. G., & Wright, M. J. (2007). Common and specific genetic influences on EEG power bands delta, theta, alpha, and beta. *Biological Psychology*, 75, 154–164.

Zlojutro, M., Manz, N., Rangaswamy, M., Xuei, X., Flury-Wetherill, L., Koller, D., ... & Almasy, L. (2011). Genome-wide association study of theta band event-related oscillations identifies serotonin receptor gene HTR7 influencing risk of alcohol dependence. *American Journal of Medical Genetics Part B: Neuropsychiatric Genetics*, *156*(1), 44–58.

13
Conclusion: Behavioural Genomics and Education

Yulia Kovas and Sergey Malykh

This book presents a systematic review of investigations that provide evidence for the importance of genetics for education. This new area of research on the intersection of genetics and education – educational genomics – has already made a significant contribution to our understanding of the processes underlying individual differences in educationally relevant traits. In recent years, our view of genetic and environmental influences on child development has changed significantly, from a dichotomous view (genes versus environment) to the idea of complex processes of gene–environment interplay during development. Current research has moved towards studying the co-action between molecular genetic mechanisms and the environment, such as environmental influences on gene expression (epigenetic processes) and genetic influences on perceptions and choices of environment (gene–environment correlations).

Contemporary investigations in education assimilate new technological and conceptual approaches to the study of the aetiology of individual differences in educationally relevant traits. Today, quantitative genetic research involves large representative twin and adoption samples and utilizes the latest analytic and statistical methodology. Recently, a new "adoption at conception design" has been developed, where children are conceived through IVF technology and can be divided into several groups stratified by different degrees of genetic relatedness between parents and children (Harold et al., 2012). This method allows us to separate genetic influences from the influences of the prenatal environment on child development. Another new addition to the quantitative genetic methodology toolbox – Genome-wide Complex Trait Analysis (GCTA) – estimates genetic influences on complex traits using genome-wide genotypes in large samples of unrelated individuals (Plomin & Deary, 2015). Comparing GCTA results to the results of family studies provides important insights into the genetic architecture of complex traits.

Intuitive views of how genes affect psychological development often differ paradoxically from reality. In contrast to erroneous views of deterministic effects of genes, genetic effects on complex behavioural traits are probabilistic. Moreover, genetic influences on behaviour are not static and can change in the course of development. For example, genetic effects on individual differences in intelligence in early childhood are significantly weaker than environmental effects. However, with age they gradually increase, with individual differences in intelligence among adults largely explained by genetic differences. This phenomenon may be explained by the processes of gene–environment correlation, whereby people actively select, modify and react to the environment under the influence of their genes. At the same time, academic achievement has been shown to be highly heritable from the beginning of school education. This may be explained by the unified environment provided by schools, such as standardized curricula and standard teaching methods. It is possible that when the environment is unified, the observed differences among people are largely explained with genetic differences (Kovas et al., 2013). If schools did not exist, or if schools differed dramatically from one another, genetic influences on academic achievement would be much weaker, and the observed differences would be under stronger environmental control. Indeed, there is some evidence that with greater variation in school type and quality, genetic factors explain a smaller proportion of individual differences in learning ability and achievement (e.g., Petrill et al., 2007).

Even in unified educational environment there are differences in children's average level of academic achievement among different classes, schools, regions and countries. For example, according to the Programme for International Student Assessment (PISA) report, the average level of mathematical achievement of school children differs across counties (PISA, 2009). These average differences are very small and explain only a small proportion of the existing individual differences. Similarly, average differences among regions, cities, schools and classes explain only a small proportion of the variation in ability and achievement within each country. In the absence of selection (e.g., selection of pupils into specialized schools), the effect of common environment (e.g., unifying effect of curriculum, school environment, etc.) is weak. In other words, almost the whole spectrum of the existing individual variation can be observed within each class (Kovas et al., 2007; Byrne et al., 2010). See Box 3 for an explanation of the relevant concepts.

These weak effects of common environment may seem paradoxical and at first sight contradictory to the importance of a teacher, school and educational system for children's academic achievement. However, there are several potential explanations for this phenomenon. For example, this phenomenon may reflect achievements of contemporary education. If the society provides equal access to quality education for all children, most children can reach a certain level of knowledge. The differences among children above this level are largely explained by genetic differences and individual experiences. It is possible that in a less egalitarian society, in which access to quality education is reserved for privileged members of society, the influence of common educational environment on academic motivation and achievement would be stronger. However, in such a society the variation in academic achievement would probably be even greater. Another, less optimistic, explanation for the weak effects of common environment is also possible: the same environment may be suitable for one pupil (increasing his/her results) and not suitable for another pupil (decreasing his/her results). In other words, the influences are masked at the level of average class performance. Personalized educational methods can be more effective and lead to better results for all pupils (i.e. a higher average level of knowledge), and to a narrower range between different children's levels of knowledge (i.e. smaller variance).

Every experienced teacher knows that children within every class differ greatly (particularly in the absence of special selection into classes or schools). Accepting genetic differences as an important factor for individual differences in educationally relevant traits does not reduce the role of education, but helps to realize that different children require different methods, approaches and resources to achieve the same result. Research that focuses on environmental factors also suggests the importance of individualization in education. Numerous studies show that environment largely contributes to differences among family members, rather than to their similarities. For example, academic motivation is quite different even for monozygotic twins, who are genetically identical, live in the same home and study in the same class (Kovas et al., in press). This book has provided a number of examples that demonstrate the important role of non-shared environment in the development of children's behaviour.

Today genetically informative research focuses not only on the aetiology of different psychological characteristics separately, but also on the aetiology of their interrelationships. For example, multivariate genetic models have been applied to the study of associations between features of the environment and reading and mathematical abilities (Kovas et al., 2007; Hart, Petrill, & Kamp-Dush, 2010; Hart, Petrill, Thompson, & Plomin, 2009); attention and reading and mathematical abilities (Hart Petrill, & Kamp-Dush, 2010); academic motivation and academic achievement in the course of development (Luo et al., 2011); and reading abilities (Harlaar et al., 2011). If the same genes affect different traits (biological phenomenon called *pleiotropy*), a *genetic correlation* is observed between the traits. Pleiotropy has been found across many educationally relevant traits, meaning that many genetic effects are general rather than specific to any one trait. Substantial genetic overlap has been found not only for normal variation in traits, but also for learning disabilities. For example, genetic correlations are high between reading, mathematics and language disabilities; and are moderate between learning disabilities and other developmental difficulties, such as ADHD (Plomin & Kovas, 2005; Butterworth & Kovas, 2013).

Two recent multivariate twin studies examined the genetic architecture of academic achievement and reported several important for education results. First, heritability of achievement (measured by performance in the state exams) was moderate for all school subjects, including mathematics, language, science, art and business (Rimfeld et al., in press). Second, the observed moderate to high correlations in performance in different academic subjects were largely explained by overlapping genetic influences. Third, this large overlap in genetic influences on different academic subjects was not reduced (or was reduced by very little) once intelligence was controlled for. In other words, to a large extent the same genes affected exam performance in different subjects (e.g., mathematics and art), even after accounting for any shared effects with intelligence. Fourth, genetic influences on achievement in core school subjects partly overlap with genetic factors affecting intelligence, as well as a whole range of other traits, including self-efficacy, personality, psychopathology, behavioural problems, health, wellbeing and even perceptions of home and school environment (Krapohl et al., 2014).

Our understanding of gene–environment co-action is continuously improving. However, in itself it cannot influence educational policy. The same scientific knowledge may lead to completely different educational strategies, depending on societal values. For example, the society may distribute educational resources aiming to achieve a certain level of knowledge for all children. Alternatively, it may spend the resources focusing on the development of gifted children. Of course, an optimal strategy for any society would be to increase the budget spent on education, as it is justified both socially and economically. For example, it has been shown that in the UK poor numeracy affects not only the life chance of individuals, but costs society about £2.4 billion per year (Butterworth, Varma & Laurillard, 2011), with a causal link between mathematical competence and economic performance of the society (according to the Organisation for Economic Co-operation and Development, OECD report, 2010; Butterworth & Walsh, 2011).

However, if increasing the educational budget is not immediately possible, in conditions of limited economic resources the most promising strategy is to distribute resources in such a way that guarantees that every child can reach a level of knowledge necessary for successful functioning in today's technologically and numerically demanding society. This strategy can succeed only if education is personalized, in other words if every child receives the necessary educational resources. Every child is not a passive object of instruction, but an active participant in the educational process. Educational models that are oriented towards an "average" pupil do not take into account gene–environment interplay processes described in this book. Children differ not only in their learning abilities, but in their interest, motivation and other characteristics important for education. In future, we hope, children will have more opportunities to participate actively in developing their personal educational trajectories. In this process, adaptive informational technologies can serve as an important resource. These technologies allow teachers to adapt the level of difficulty, content and the sequence of learning – preventing boredom in some children and repeated failure in others, and optimizing learning for all children.

Progress in genomic research can also make an important contribution to the education of children with learning disabilities, such as dyslexia and dysgraphia. Such disabilities are often heterogeneous in nature. Better understanding of genetic and neurobiological mechanisms will lead to better diagnoses, correction and possibly prevention of these disabilities through the use of the most suitable environments.

As mentioned in the preceding chapters of this book, progress in genomics is very fast. It is likely that in the near future, scientists will discover many genes, as well as epigenetic and neurobiological mechanisms, through which these genes contribute to educationally relevant traits. Although the effect of each genetic variant is weak, a constellation of a large number of such genetic factors (polygenic risk score; Plomin & Deary, 2015) can be used for early diagnosis of learning disabilities, and for finding the most suitable educational environments. Moreover, it is possible that such probabilistic diagnostics, based on polygenic risk scores, will be used not only for learning disabilities, but also for the whole spectrum of individual variation and educational needs. From this point of view, learning and behavioural problems are considered not as "disorders", but as part of a quantitative continuum; and genetic variants are considered not as risk factors but as factors of common variation (Plomin & Deary, 2014). Indeed, many twin studies have shown that learning disabilities (for example, very low mathematical performance) lie on the same continuum as ability. In other words, the same genetic (and environmental) influences are involved in placing someone at the very low end of the continuum, as are involved in placing one person just slightly below another in terms of achievement at the high end. It is the number and combination of such factors that determine each person's particular position on this continuum (Plomin & Kovas, 2005; Plomin et al., 2009; Butterworth & Kovas, 2013). It is however possible that different factors operate at the very low or very high extremes of the variation. These possibilities can only be tested in large-scale genetically informative investigations (Plomin & Deary, 2014).

Quantitative genetic designs remain valuable, providing a more refined investigation of educationally relevant traits, and paving the way for new molecular genetic investigations. The rapid advances in molecular genetics bring closer the aim of identifying the actual genetic variants that are involved in variation in traits. Today behavioural geneticists have at their disposal a whole range of molecular genetic tools. These include new and continuously improving technologies, such as *microarrays*, that allow genotyping of hundreds of thousands of *DNA* markers and evaluation of epigenetic processes of all genes simultaneously; and increasingly accessible *whole-genome sequencing* that establishes the entire DNA sequence of a person. *Molecular genetics* is possibly the fastest developing area in the history of science. The area is still in its infancy, but has already provided many important insights into the origins and mechanisms of individual differences.

It is possible that the information on our complete genomic profiles will be routinely available long before we can truly utilize this knowledge. The process of tracing a path from each genetic variant to behaviour may take a long time. Research into the exact mechanisms by which each gene affects a trait is complex and involves multiple levels: from gene expression profiles, to specific protein functions, to physiology and to the structure and function of the brain (Plomin, Haworth, & Davis, 2009). We hope that these important insights will make a positive contribution to the development of education. This can only be possible with active support from all participants of the educational process: teachers, parents, administration, policy makers and students themselves. Their support, in turn, depends on their understanding of the complex processes of gene–environment interplay in child development. Unfortunately, the relevant information is often not accessible to them. This leads to misconceptions, distorted views on real processes, such as the mistaken view of heritability as relating to a specific person. For example, the substantial heritability of height (approximately 90%) is often understood in a way that 90% of an individual's height is determined by the person's genes. This is totally wrong. The substantial heritability of height means that approximately 90% of the individual differences in height in the population at a particular time is explained by genetic differences among the people in this population. Therefore, the concept of heritability refers to differences among people, and not to a specific individual (see Box 3). In this book we have tried to clarify the main concepts, definitions and methodology used in genetic research. We hope that this will help with bridging the gap between scientific knowledge and its practical applications in education. Universal access to education is one of the most important achievements of contemporary society. However, today society faces a new global challenge – to provide each child with optimal conditions for learning.

References

Butterworth, B., & Kovas, Y. (2013). Understanding neurocognitive developmental disorders can improve education for all. *Science*, 340, 300–305.

Butterworth, B., Varma, S., & Laurillard, D. (2011). Dyscalculia: from brain to education. *Science, 332*, 1049–1053.

Butterworth, B., & Walsh, V. (2011).Neural basis of mathematical cognition. *Current Biology, 21*(16), R618–R621.

Byrne, B., Coventry, W. L., Olson, R. K., Wadsworth, S., Samuelsson, S., Petrill, S. A., Willcutt, E. G., & Corley, R. (2010). "Teacher effects" in early literacy development: evidence from a study of twins. *Journal of Educational Psychology, 102*(1), 32–42.

Harlaar, N., Deater-Deckard, K., Thompson, L., DeThorne, L., Petrill, S. (2011). Associations between reading achievement and independent reading in early elementary school: a genetically informative cross-lagged study. *Child Development, 82*(6), 2123–2137.

Harold, G. T., Elam, K. K., Lewis, G., Rice, F., & Thapar, A. (2012). Interparental conflict, parent psychopathology, hostile parenting, and child antisocial behavior: examining the role of maternal versus paternal influences using a novel genetically sensitive research design. *Development and Psychopathology, 24*(4), 1283–1295.

Hart, S., Petrill, S., Thompson, L., Plomin, R. (2009). The ABCs of math: a genetic analysis of mathematics and its links with reading ability and general cognitive ability. *Journal of Educational Psychology, 101*(2), 388–402.

Hart, S., Petrill, S., & Kamp Dush, C. (2010). Genetic influences on language, reading, and mathematics skills in a national sample: an analysis using the national longitudinal survey of youth. *Language, Speech, and Hearing Services in Schools, 41*, 118–128.

Kovas, Y., Garon-Carrier, et al. (2015). Why children differ in motivation to learn: insights from 13,000 twins from 6 countries. *Personality and Individual Differences, 80*, 51–63.

Kovas, Y., Haworth, C. M. A., Dale, P. S., & Plomin, R. (2007). The genetic and environmental origins of learning abilities and disabilities in the early school years. *Monographs of the Society for Research in Child Development, 72*, whole number 3, Serial No. 188, 1–144.

Kovas, Y., Voronin, I., Kaydalov, A., Malykh, S. B., Dale, P. S., & Plomin, R. (2013). Literacy and numeracy are more heritable than intelligence in primary school. *Psychological Science, 24*(10), 2048–2056.

Krapohl, E., Rimfeld, K., Shakeshaft, N. G., Trzaskowski, M., McMillan, A., Pingault, J. B., ... & Plomin, R. (2014). The high heritability of educational achievement reflects many genetically influenced traits, not just intelligence. *Proceedings of the National Academy of Sciences, 111*(42), 15273–15278.

Luo, L. L. Y., Kovas, Y., Haworth, C. M. A., & Plomin, R. (2011). The etiology of mathematical self-evaluation and mathematics achievement: understanding the relationship using a cross-lagged twin study from ages 9 to 12. *Learning and Individual Differences, 21*(6), 710–718.

OECD (2010). The High Cost of Low Educational Performance: The Long-Run Economic Impact of Improving Educational Outcomes. *OECD*, Paris.

Petrill, S. A., Deater-Deckard, K., Thompson, L., Schatschneider, C., & DeThorne, L. (2007). Longitudinal genetic analysis of early reading: The Western Reserve Reading Project. *Reading and Writing, 20*(1–2), 127–246.

PISA (2009). Results: Executive summary.

Plomin, R., & Deary, I. (2015). Genetics and intelligence differences: five special findings. *Molecular Psychiatry, 20*, 98–108.

Plomin, R., Haworth, C. M., & Davis, O. S. (2009). Common disorders are quantitative traits. *Nature Reviews Genetics, 10*(12), 872–878.

Plomin, R., & Kovas, Y. (2005). Generalist genes and learning disabilities. *Psychological Bulletin, 131*, 592–617.

Rimfeld, et al. (in review). From mathematics to art – pleiotropy across academic subjects at the end of compulsory education.

Appendix

Box A.1 Twin method

People differ tremendously in all complex traits, from physical appearance to psychological characteristics. These individual differences (variance) are called phenotypic differences or phenotypic variance. All the variance (100%, or 1) in a particular complex trait (e.g., height) in a particular population (e.g., adult population of Russia in 2015) can be explained by genetic and environmental influences. The DNA sequence (~3 billion base pairs of the human genome) is practically identical for all people, with only ~1% difference on average. Some unrelated people are more genetically similar than others, by chance. This usually means that they are more similar physically and/or psychologically. Physical and psychological similarities among unrelated people can also be explained by more similar environments. For example, playing chess may have a positive effect on cognitive ability. In this example, playing chess would be a shared environment (and have a shared effect on cognitive ability) for all those who engage in this activity, and would be a non-shared environment for those who do and those who do not engage in it.

In behavioural genetics, 100% of phenotypic variance (P) in a trait in a particular population is decomposed into genetic (h^2), shared environmental (c^2) and non-shared environmental (e^2) components (effects):

$$P = h^2 + c^2 + e^2$$

Biological relatives are genetically more similar than unrelated individuals. For example, siblings share ~50%, whereas first cousins share ~25%, of the ~1% of variable DNA (segregating DNA). These different degrees of genetic similarity are reflected in various degrees of phenotypic (observed, measured) similarity across people. For example, first degree relatives (e.g., siblings; parents and their children) are usually more similar than second degree relatives (e.g., cousins; aunts and nieces) on physical and behavioural traits.

Studying family members with different degrees of genetic relatedness can help to evaluate the relative contributions of genetic and environmental factors to individual differences (variance) in physical and behavioural characteristics. A particularly informative approach is comparing monozygotic (identical, MZ) and dizygotic (non-identical, DZ) twins (see Figure A.1).

MZ twins result from the division of the same fertilized egg (zygote). Therefore, they are 100% genetically similar in most cases. DZ twins develop at the same time from two, separately fertilized, zygotes. Like other siblings they share about 50% of the segregating DNA on average.

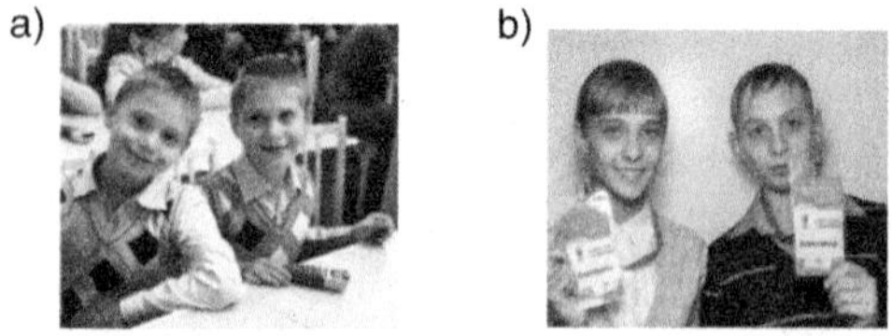

Figure A.1 (a) MZ twins and (b) DZ twins

MZ twin resemblance (twin correlation; rMZ) stems from common genetic and shared environmental factors. DZ twin resemblance (rDZ) also stems from common genetic and shared environmental factors, but as mentioned above, DZ twins share only half of their segregating DNA:

$$r_{MZ} = h^2 + c^2$$

$$r_{DZ} = \frac{1}{2}h^2 + c^2$$

If genetic factors play a role in individual differences in a trait (phenotypic variance), MZ twins should be more similar than DZ twins in that trait. Estimated genetic contributions to variance are called *heritability* (h^2). Heritability of a trait can be computed as twice the difference between MZ and DZ twin correlations in this trait. Environment contributing to twin similarity in a trait are called *shared* (c^2) and are computed as the difference between MZ twin correlation and heritability. *Non-shared environment* (e^2) is the extent to which MZ twins differ, and is estimated as the difference between 100% (the whole phenotypic variance) and the MZ twin correlation. Non-shared environment includes a whole range of environmental factors that contribute to dissimilarities between twins, but also measurement error:

$$h^2 = 2 \times (r_{MZ} - r_{DZ})$$
$$c^2 = r_{MZ} - h^2$$
$$e^2 = 1 - r_{MZ}$$

For example, if the correlation in mathematical achievement for MZ twins is 0.68 (68%) and for DZ twins is 0.41 (41%) (see Figure A.2), heritability of mathematical achievement is estimated as: 2 × (0.68 – 0.41) = 0.54. In other words, 54% of the phenotypic variance in mathematical achievement is explained by genetic factors. The effect of shared environment is: 0.68 – 0.54 = 0.14 (14% of the phenotypic variance); and of non-shared environment is: 1 – 0.68 = 0.32 (32% of the phenotypic variance).

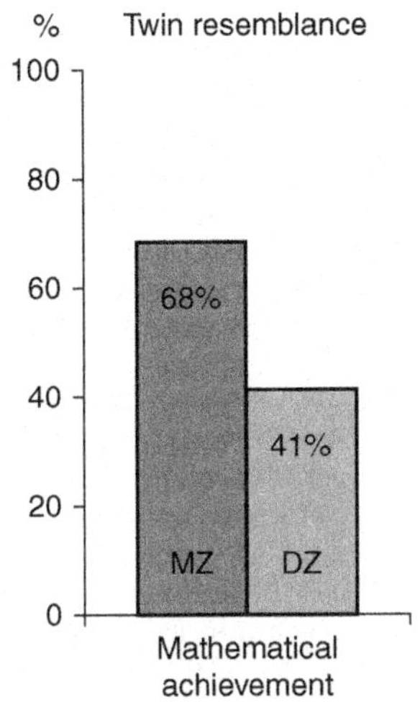

Figure A.2 Twin correlations in mathematical achievement

Source: Based on Luo et al. (2011).

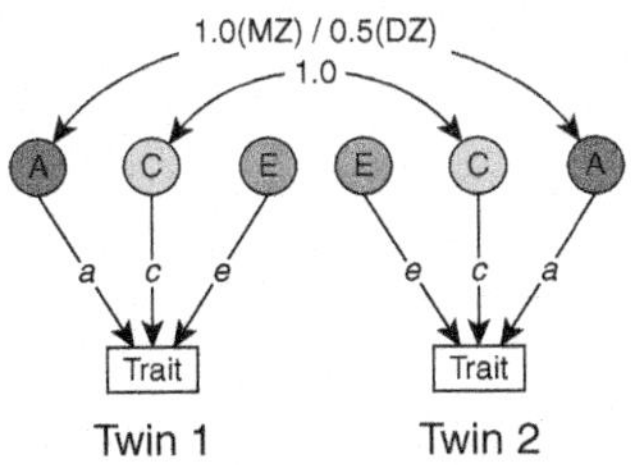

Figure A.3 Univariate (one trait) ACE model (path diagram). Traits (observed variables) are represented by rectangles. A, C and E (latent, unmeasured variables) are represented by circles. MZ twins share all the genetic variance, A (1.0), DZ twins share 0.5 of A. Both twins have the same common environment, C; non-shared environments, E, are unique to each twin. The paths, drawn as arrows, reflect the statistical influence of one variable on another. Note that **h** is denoted as **a** in ACE models

Source: Plomin et al. (2013, pp. 392–393) for more detail.

The components h^2, c^2, and e^2 can also be estimated using structural model fitting analyses with specialist software (e.g., OpenMx). This model, called an ACE model, can be presented as a path diagram (see Figure A.3). In these analyses, genetic and environmental effects are modelled as latent (unmeasured) variables, where A represents (Additive) genetic effects, C represents shared (Common) environmental effects, and E represents non-shared Environmental effects. The latent variables, A, C, E explain all the variance in the trait.

The twin method can also be used to assess why different traits correlate with each other. Multivariate models are used to decompose the variance between two or more traits (see Figure A.4). The latent variables A_1, C_1, E_1 explain all the variance in Trait 1 and some of the variance in Trait 2. The residual variance in Trait 2 (specific to Trait 2) is explained by the latent variables A_2, C_2, E_2, independent from influences on Trait 1. The paths a_1, c_1, e_1 and a_2, c_2, e_2 represent genetic, shared and non-shared environmental influences specific to Trait 1 and Trait 2, respectively. The paths x, y and z represent genetic (x), shared environmental (y) and non-shared environmental (z) influences that are common for these two traits.

The correlation between the two latent variables A_1 and A_2 (r_g, genetic correlation) represents the amount of overlapping genetic influences on Trait 1 and Trait 2. The absolute value of genetic correlation can vary from 0 (no overlapping genetic influences) to 1 (all genetic influences contributing to the twin similarity in Trait 1 are the same as the genetic influences contributing to the twin similarity in Trait 2). Similarly, the shared environmental correlation (r_c) indexes environments common for Trait 1 and Trait 2 contributing to twin similarity in both traits; and the non-shared environmental correlation (r_e) indexes environmental effects that make twins less similar in both traits, Trait 1 and Trait 2.

Multivariate genetic analysis is also used to estimate the relative contribution of genetic and environmental factors to the phenotypic correlation between two traits. Bivariate heritability is the proportion of the correlation between Trait 1 and Trait 2 that is explained by common genetic factors. Similarly, it is possible to estimate the proportion of the phenotypic correlation explained by shared and non-shared environmental factors. These bivariate heritability and environmentalities add up to 1, as they explain all the covariance between Trait 1 and Trait 2.

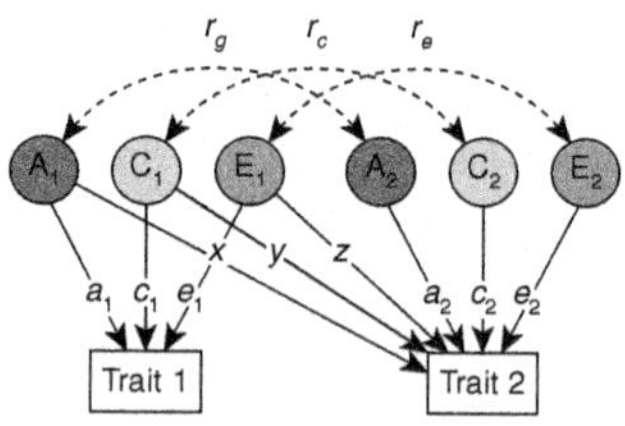

Figure A.4 Bivariate (two variables) ACE model (path diagram). The arrows represent standardized paths: the sum of the squared values of paths contributing to each trait accounts for all (100% or 1) the variance in that trait. For Trait 1 all the variance is: $a_1^2 + c_1^2 + e_1^2$; all variance in Trait 2 is: $a_2^2 + c_2^2 + e_2^2 + x^2 + y^2 + z^2$. The heritability of Trait 2 is the sum of all genetic influences exercised on it: $a_2^2 + x^2$. The same logic applies to shared and non-shared environment: $c_2^2 + y^2$ and $e_2^2 + z^2$, respectively. *r*g, *r*c and *r*e, shown in the diagram can also be derived

Source: Plomin et al. (2013, pp. 393–399) for more detail.

Multivariate genetic analyses have been particularly useful for understanding the aetiology of comorbidity across psychopathologies; co-variation across learning abilities; and performance on cognitive or behavioural tasks at different ages. For example, a study that employed this method, found that genetic influences, rather than environments, were largely responsible for the reciprocal relationship between children's mathematics performance and their motivation from age 9 to age 12 years (Luo et al., 2011). Another study found a genetic correlation of 0.55 between hyperactivity and inattention components of ADHD, suggesting aetiological heterogeneity within the ADHD construct (Greven et al., 2011).

Box A.2 Neuroscience methods

Electroencephalogram (EEG) measures the voltage of electrical activity in the brain produced by active nerve cells. The electrical activity propagates to the scalp, from which it can be measured with EEG sensors. The EEG activity is described in terms of its frequency and amplitude. Frequency is measured in Hertz (Hz) and amplitude is measured in microvolts (μV). There are several types of EEG rhythms (alpha, beta, theta, gamma), which are categorized according to their frequency. For example, alpha waves oscillate up and down around 10 times per second. This type of activity is usually observed at the back of the head and when participants have their eyes closed or are drowsy. EEG recordings can be used to monitor brain activity associated with specific stimuli or events. There are several widely studied EEG parameters, including power, coherence and asymmetry (see the Glossary for more information).

Magnetoencephalogram (MEG) is closely related to the EEG method but, instead of electrical activity, MEG measures changes in the magnetic field. Owing to the differences in how electrical and magnetic currents spread, the MEG signal is related only to activity in the cortex.

Several types of EEG and MEG analyses are used:

Stimulus-related activity or event related potential (ERP) is obtained by averaging EEG recordings across many trials. This allows one to observe the activity related to a stimulus by removing spontaneous EEG activity not associated with this stimulus. An averaged ERP waveform consists of a series of positive and negative peaks, which are also referred to as components (Figure A.5). Several ERP components have been associated with different aspects of cognition and behaviour.

EEG spectral analysis (Fourier-transform) is a method of EEG analysis that allows us to explore the complex EEG waves frequency structure (EEG spectrum). EEG spectrum is divided into a number of frequency bands (delta, theta, alpha, beta, gamma), called brain rhythms.

Event Related Spectral Perturbation (ERSP) is a method of EEG analysis used to quantify the brain activity related to the stimulus presentation (Figure A.6). ERSP is a measure of the average changes over time in the amplitude of the EEG frequency (EEG amplitude spectrum) that is related to a certain event. ERSP analyses are often conducted on more than one EEG frequency band.

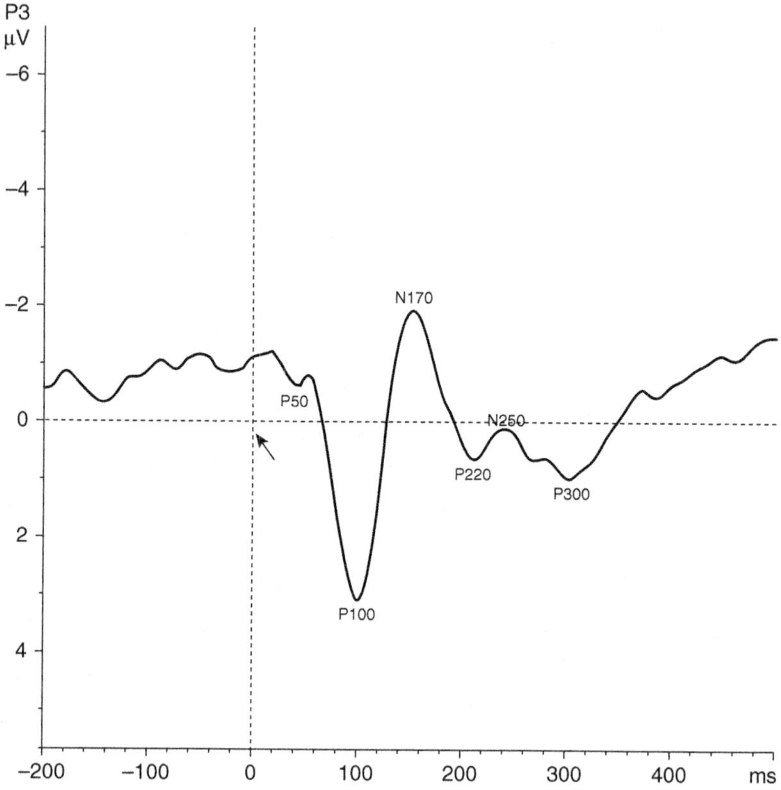

Figure A.5 ERP recorded during presentation of a visual stimulus. The *x*-axis shows the timeline in milliseconds (ms) from before the stimulus presentation (negative values) and afterwards (positive values). The moment of the stimulus presentation is indicated by an arrow pointing at 0 on the *x*-axis. The *y*-axis indicates voltage of the brain electrical activity, the ERP plotted with negative voltage upward on the *y*-axis. An ERP waveform consists of a series of positive and negative voltage deflections, which are related to a set of underlying components. Most components are referred to by a letter (N/P) indicating polarity (negative/positive), followed by a number indicating either the latency in ms or the ordinal position of the component in the waveform. For instance, a negative peak that is the first substantial peak in the waveform and often occurs about 100 ms after a stimulus is presented is often called the N100 (indicating its latency is 100 ms after the stimulus and that it is negative) or N1 (indicating that it is the first peak and is negative); it is often followed by a positive peak, usually called the P200 or P2

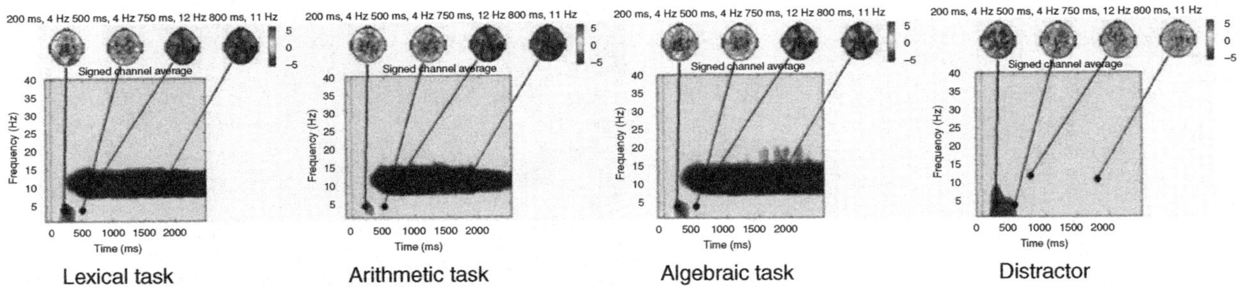

Figure A.6 ERSP activity related to three error recognition tasks (lexical, arithmetic and algebraic) and to the presentation of an auditory distractor. For the error recognition tasks, the figure shows early short-term synchronization in the delta and theta bands rhythms, and more long-lasting desynchronization in the alpha and beta bands (synchronization refers to increase, whereas desynchronization refers to decrease, in oscillations in the particular band; see Chapter 12 for more detail). The distractor is shown to be associated with synchronization in the high-amplitude slow-wave range

Magnetic resonance imaging (MRI) is a non-invasive method that produces images of the inner structures of the body. Participants are placed in a special chamber with a strong, stable magnetic field, and are exposed to an additional transient magnetic field. This makes atoms resonate and this resonance can be used to compute images of the body structures.

There are several main types of MRI:

Structural MRI (sMRI) can be used to examine the anatomical properties and structures of the brain. Structural MRI allows volumes of the brain tissues to be calculated and it is largely used in medicine as a diagnostic tool (for example in the case of tumours, demyelination, and cerebrovascular diseases). Neuroscience research uses sMRI data mostly to explore the relationship between brain volume and behavioural phenotypes, such as cognitive abilities, neurodevelopmental disorders, etc.

Functional MRI (fMRI) is based on the principle that when a brain area is processing information, it requires more oxygen. Oxygen is brought to neurons via haemoglobin in the capillary red blood cells. Functional MRI measures the changes in blood flow and oxygenation related to neural activity. For example, fMRI can detect brain areas that show greater activation during cognitive processing, such as reading, counting, etc. (Figure A.7).

Diffusion Tensor Imaging (DTI) is an MRI technique used to visualize the restricted diffusion of molecules, mainly water, in biological tissues of the brain. Through DTI it is possible to obtain images of the fibres of the white matter (Figure A.8) and their rate or preferred direction of diffusion (anysothropy). This

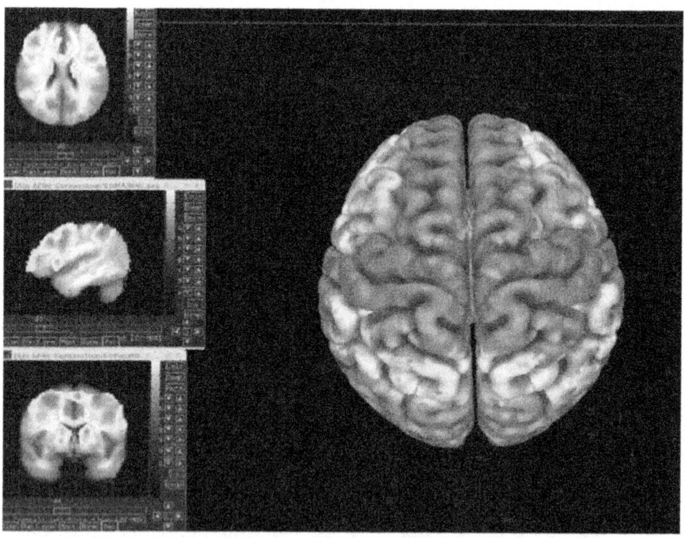

Figure A.7 Localization of brain activity obtained using fMRI: increased activation (usually indicated by warmer yellow-orange colours) in the frontal and parietal areas

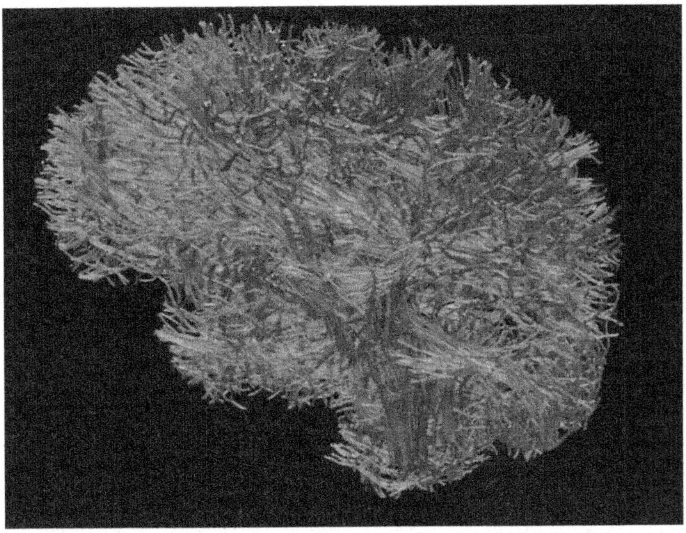

Figure A.8 The whole brain view of the DTI-based tractography. Different tracts are represented with separate fibres; different colours refer to different directions of the diffusion of water molecules

technique is mostly used to study microscopic anatomical details *in vivo*. In clinical settings, it is largely used to study neurological conditions. This method also provides visual models of the brain connectivity. The reconstruction of neural connectivity via DTI is called tractography.

Brain connectivity refers to the pattern of anatomical links (anatomical connectivity), statistical dependencies (functional connectivity), or causal interactions (effective connectivity) between neurons or groups of neurons. Brain connectivity allows the observation of how neuronal networks process information. It can be measured using different methods such as EEG, fMRI or DTI.

Box A.3 Individual and group differences (height as an example)

An example of variation and averages (means) of height in men and women in three countries is shown in Figure A.9. Both country and sex differences in height can be observed. For example, the average height of people in Russia is greater than in India. Overall, the height of men is greater than of women.

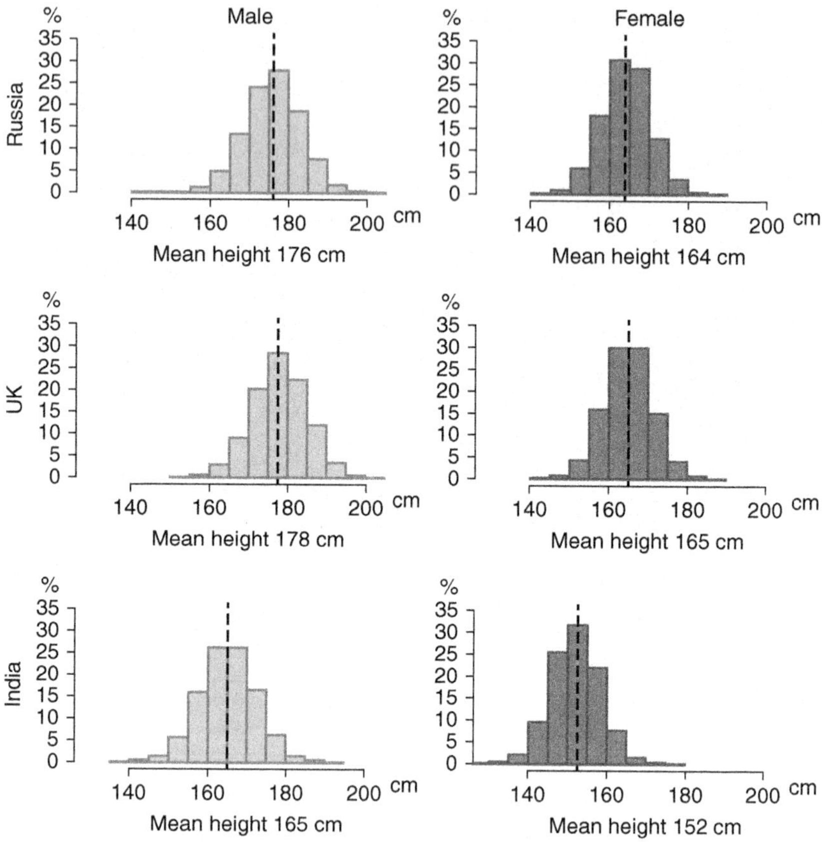

Figure A.9 Variation and averages of height in men and women

Source: Data obtained from: http://en.wikipedia.org/wiki/Human_height#Average_height_around_the_world.

In addition, there are differences in average height between different generations in the same country (i.e., cohort effect). An example of average height of men in the United Kingdom (UK) between 1880 and 1980 is shown in Figure A.10. During this period (100 years), the average height of men in the UK increased by approximately 10 cm. A similar tendency is observed in many countries (Silventoinen et al., 2003).

However, the average differences among the groups (sex, population, cohort differences) are significantly smaller than differences among people within these groups (individual differences). For example, as demonstrated in Figure A.9, the maximum difference in average height between India and the UK is 13 cm, whereas the maximum difference within the UK population is more than 50 cm.

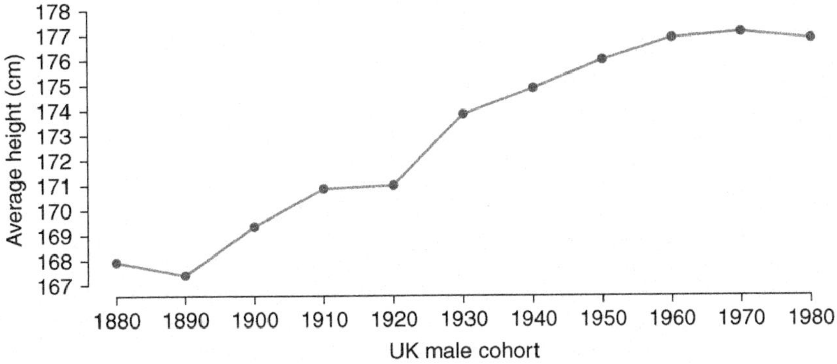

Figure A.10 Average height of men in the UK between 1880 and 1980
Source: Data obtained from: https://www.clio-infra.eu/datasets/indicators.

Factors that contribute to average differences may be the same or different from factors which contribute to individual differences. For example, cross-population differences in average height may be partly explained by such environmental factors as Gross Domestic Product (GDP) and availability of quality nutrition. Average sex differences in height are largely explained by genetic factors that lead to the whole range of biological differences between men and women (sexual dimorphism).

Quantitative genetic studies in different populations have shown that individual differences in height largely result from genetic factors. For men, genetic factors explain approximately 90% of individual differences in height, with the remaining part of the variation explained by non-shared environmental factors. For women, between 70 and 80% of individual differences in height are explained by genetic factors, with the remaining part explained by both shared and non-shared environmental factors (Silventoinen et al., 2003). Interestingly, an increase in heritability of height has been shown from the beginning of the 20th century to the present time (Silventoinen et al., 2000). A number of molecular genetic investigations have shown that a large number of genetic variants of small effects contribute to individual differences in height (polygenic trait). For example, a recent large-scale study has demonstrated that thousands of genetic variants together explain approximately 30% of individual differences in height (Wood et al., 2014).

References

Greven, C. U., Rijsdijk, F. V., & Plomin, R. (2011). A twin study of ADHD symptoms in early adolescence: hyperactivity-impulsivity and inattentiveness show substantial genetic overlap but also genetic specificity. *Journal of Abnormal Child Psychology, 39*(2), 265–275.

Luo, Y. L., Kovas, Y., Haworth, C. M., & Plomin, R. (2011). The etiology of mathematical self-evaluation and mathematics achievement: understanding the relationship using a cross-lagged twin study from age 9 to 12. *Learning and Individual Differences*, *21*(6), 710–718.

Plomin, R., DeFries, J. C., Knopik, V. S., & Neiderhiser, J. M. (2013). Behavioral Genetics, 6th edition, Worth Publishers.

Silventoinen, K., Kaprio, J., Lahelma, E., & Koskenvuo, M. (2000). Relative effect of genetic and environmental factors on body height: differences across birth cohorts among Finnish men and women. *American Journal of Public Health, 90*(4), 627–630.

Silventoinen, K., Sammalisto, S., Perola, M., Boomsma, D. I., Cornes, B. K., Davis, C., ... & Kaprio, J. (2003). Heritability of adult body height: a comparative study of twin cohorts in eight countries. *Twin Research, 6*(05), 399–408.

Wood, A. R., Esko, T., Yang, J., Vedantam, S., Pers, T. H., Gustafsson, S., ... & Lim, U. (2014). Defining the role of common variation in the genomic and biological architecture of adult human height. *Nature genetics, 46*(11), 1173–1186.

Glossary

Many terms provided in this Glossary are associated with different scientific disciplines. Here, we provide definitions of the concepts in the context of psychological science.

ACE model: A quantitative genetic model that assumes that variation in a phenotype results from additive genetic (A), shared environmental (C), and non-shared environmental (E) factors.

Active gene–environment correlation: A type of gene–environment correlation whereby a person actively seeks, modifies or creates an environment that corresponds with his/her genetic propensities. Different perceptions of the same environment by different people can also reflect this type of gene–environment correlation.

Additive genetic factors: Genetic factors whose independent effects are summed up in their influence on a trait.

Adoption method: A method used for estimating genetic and environmental influences on individual differences in a trait by comparing similarities on this trait: 1) between adopted children and their biological and adoptive parents; 2) between siblings from biological families and siblings from adoptive families.

Allele: An alternative form (variant) of a gene (or a genetic locus).

Amplification theory: An increased effect of genetic factors on a phenotype during ontogenesis as a result of gene–environment correlation.

Behavioural genetics: An interdisciplinary research area that studies the contribution of genetic and environmental factors to individual differences in psychological (behavioural) traits. There are two major branches of behavioural genetics: quantitative and molecular.

Bivariate twin analysis: Analysis based on cross-trait twin correlations that allows an estimation of the contribution of genetic and environmental factors to covariation between two traits.

Brain connectivity: The pattern of anatomical links in the brain (anatomical connectivity) and statistical dependencies between neurons or groups of neurons (functional connectivity). Analysis of these links within brain structures and among them allows to define neuronal networks involved in specific information processing. See Box A.2.

Candidate gene: A gene encoding a protein that is involved in a biological process thought to be related to individual differences in a particular complex trait.

Causality: A direct effect of one trait on another. For example, reading problems may cause problems with learning mathematical material that is presented in a text form.

Chromosome: A nucleoprotein structure that is located inside the nucleus of a cell. There are 23 chromosome pairs in a human cell that contain most of the DNA and are used to store, process and transfer DNA.

Chromosome abnormality: A pathology that is caused by changes in a structure or a number of chromosomes.

Common polymorphism: A variable DNA locus whose minor allele frequency in a population is more than 1%.

Common (shared) teacher effect: Teacher effect at the level of all pupils in the same class whereby each pupil's performance is better (or worse) than if they studied with a different teacher (the so called 'good' or 'bad' teacher effect).

Comorbidity: Co-occurrence of two or more psychological and/or physical disorders/ illnesses.

Concordance: Similarity on a trait among family members, for example, when both twins in a pair have a diagnosis of dyslexia.

Cross-cultural behavioural genetics: A scientific discipline that focuses on investigating the role of culture (e.g., language, traditions, social norms) in the aetiology of individual differences in human behaviour.

Cross-lagged analysis: Statistical analysis of longitudinal data on two or more traits used to investigate causal links among them. The analysis is based on regressing variables at one time point on variables at preceding time points in order to evaluate the extent to which preceding variables explain variation in later variables.

Cross-sectional method (design): A research method where data are collected at one time point.

Deletion: A type of DNA polymorphism characterized by deletion of one or several nucleotides.

Diffusion tensor imaging (DTI): An MRI technique that uses the effects of magnetic resonance for visualization of direction of water molecule diffusion in the body tissues, including the brain tissues. This method allows one to obtain images of the white matter tracts as well as indicators of diffusion directions (anisotropy). See Box A.2.

Discordance: Dissimilarity in a trait among family members, for example, when one twin in a pair has a diagnosis of dyslexia, and another twin does not.

Discordant MZ twin design: A method used for evaluating the role of environmental factors in trait variation. For example, research using this method has shown that if twins from an MZ pair differ in the level of conflict relationship with parents, then a twin with a higher level of conflict relationship is at a higher risk for behavioural problems. This suggests that child–parent conflict is an environmental risk factor for behavioural problems.

Dizygotic twins (fraternal twins, DZ twins): Twins, both same sex and opposite sex, born as a result of one pregnancy, from two zygotes (two independently fertilized ova). Accordingly, genetic similarity between twins of this type is the same as for non-twin brothers and sisters, who share approximately 50% of their segregating DNA.

DNA chip: Technology that is used for genotyping multiple (hundreds of thousands) DNA markers at the same time.

DNA code/Genetic code: A nucleotide sequence of a DNA (RNA) molecule that codes the sequence of amino acids forming different proteins. A DNA molecule is built up with four types of nucleotide bases: adenine (A), guanine (G), cytosine (C) and thymine (T). These letters of genetic code specify 20 amino acids. A sequence of three nucleotides (called a triplet codon) is used to code a specific amino acid or to signal a termination of translation (a stop codon).

DNA marker/Genetic marker: A DNA locus. **Single nucleotide polymorphism (SNP)** and **variable number of tandem repeats (VNTR)** are examples of a polymorphic DNA marker.

DNA sequencing: Methods used to identify nucleotide sequence of the DNA of a particular individual.

Educational genomics: A research area on the intersection of genetics and education that aims to make a contribution to our understanding of genetic, environmental, neural and psychological processes underlying individual differences in educationally relevant traits.

EEG spectral analysis (Fourier transform): A method of EEG analysis that allows exploration of the complex frequency structure of EEG waves (EEG spectrum). The EEG spectrum is divided into a number of frequency bands (i.e., delta, theta, alpha, beta, and gamma), called brain rhythms. See Box A.2.

Electroencephalography (EEG): A non-invasive method used to record electrical activity of the brain from the scalp. EEG measures voltage fluctuations resulting from ionic current flows within the neurons of the brain. See Box A.2.

Environmental correlation: Amount of overlapping environmental influences on two traits or on one trait at two different ages. Shared and non-shared environmental correlations can be estimated with **multivariate model fitting**. See Box A.1.

Epigenetic mechanisms: Mechanisms that influence gene expression without changes in the DNA sequence.

Epistasis/Gene–gene interaction: A phenomenon whereby activity of one gene is influenced by other genes.

Estimation of variance components: In quantitative genetics, statistical evaluation of the relative contribution of genetic and environmental factors to variation of a trait (phenotype).

Event-related potential (ERP): A method of EEG analysis based on averaging of bioelectric activity of the brain in response to a number of repeated stimuli. See Box A.2.

Event related spectral perturbation (ERSP): A method of EEG analysis used to quantify the brain activity related to stimulus presentation. ERSP provides measures of amplitude, temporal and frequency characteristics of signal. See Box A.2.

Familiality: Similarity among family members on a particular trait which can be explained by common genetic and/or common environmental factors.

Familial risk: The relative risk of a particular trait in a family in comparison to the risk of this trait in a population.

Functional magnetic resonance imaging (fMRI): A non-invasive neuroimaging method used to measure brain activity by detecting associated changes in blood flow. For example, during visual stimulation there is increased activation in the occipital lobes of the brain, whereas during planning tasks there is increased activation in the frontal lobes of the brain. See Box A.2.

GCSE (General Certificate of Secondary Education): Unified state examination in different subjects taken at the end of the formal secondary education (at the end of Year 11, age 15–16) according to the national curriculum of the United Kingdom.

Gene: A basic unit of heredity; a region of the DNA that encodes for a protein or RNA, and therefore is involved in the development of a phenotype.

Gene–environment correlation (rGE): A type of **gene–environment interplay** that refers to the situation when a person's exposure to environment depends on his/her genetic characteristics. As a result, some genotypes can be observed more often in one environment than in another. There are three types of rGE: **passive, reactive** and **active.**

Gene–environment interaction (GxE): A type of **gene–environment interplay** that reflects interdependence of genetic and environmental influences on a trait. In this case, genetic influences on a trait depend on the presence of a specific environment; or environmental influences depend on the presence of a specific genetic factor.

Gene–environment interplay: A concept that refers to different types of combined influences of genetic and environmental factors on complex traits (phenotypes), including **gene–environment correlation** and **gene–environment interaction**.

Gene expression: A process by which information from a gene is used to synthesize RNA and protein.

Generalist genes: A phenomenon describing largely general effects of genes on different traits. For example, largely the same set of genetic factors affects mathematical and reading abilities.

Genetically informative designs: Research designs that study complex traits taking into account genetic influences.

Genetic association: A statistical link between a **DNA polymorphism** and individual differences in a trait.

Genetic association study: A study used to detect associations between one or multiple DNA polymorphisms and a quantitative trait (e.g., reading ability) or a categorical diagnosis (e.g., dyslexia).

Genetic correlation: Amount of overlapping genetic influences on two traits or on one trait at two different ages. Genetic correlation can be estimated with **multivariate model fitting**. See Box A.1.

Genetic linkage: A link between two DNA loci that are located close on a chromosome making them more likely to be inherited together than in independent inheritance.

Genetic linkage analysis: A method that uses multiple genetic markers in pedigree (also called family tree) to locate genomic regions that are linked with a trait. See **Genetic linkage**.

Genetic profile: A unique genome of an individual, or a unique pattern of the genome expression: a product of gene–gene and gene–environment co-action.

Genome: All inherited material (DNA) in an organism's cell (i.e., nuclear and mitochondrial genome). Human nuclear genome consists of 23 **chromosome** pairs (22 pairs of autosomal chromosomes and one pair of sex chromosomes) and is built up of 3 billion **nucleotide** pairs.

Genome mapping: Creation of a detailed physical map of the human genome of maximum resolution, i.e., the description of its complete nucleotide sequence. This term is also used to describe functional mapping, i.e., identification of associations between DNA polymorphisms and individual differences in traits.

Genome-Wide Association Study (GWAS): Association analysis of many (hundreds of thousands) DNA loci across the whole genome with a particular trait.

Genomic "address": A position of a genetic marker in the genome sequence, such as its position on a chromosome, in a gene.

Genomic imprinting: An epigenetic process by which certain genes are expressed in a parent-of-origin-specific manner: for example, an allele inherited from the father is imprinted, (i.e., silenced) and an allele inherited from the mother is expressed.

Genotype: a) a combination of all genetic factors of one individual that contribute to his/her **phenotype**; b) a combination of alleles at a specific DNA locus of a person.

Genotyping: Molecular genetic techniques used to identify variants (alleles and genotypes) in the DNA sequence of an individual.

g factor/general intelligence/general cognitive ability: A common factor derived from the different measures of cognitive abilities correlated among each other.

Haplotype: A combination of alleles of different DNA loci that are located on the same chromosome and are linked with each other (i.e., inherited together). See **Linkage disequilibrium.**

HapMap international project: The project to develop the human haplotype map. Specifically, the project aims to define the locations of single nucleotide polymorphisms (SNPs) in the human genome, their allele and genotype frequencies, and the linkage disequilibrium between them (haplotypes), in different populations and to provide open access to these data. The data for more than 10 million SNPs are now available for European, African and Asian populations.

Heritability (h^2, A): A proportion of the phenotypic variation in a particular trait that is explained by genetic factors. It can be estimated as twice the difference between MZ and DZ twin correlations on a particular trait. See Box A.1.

Individual-specific teacher effect: Teacher effect whereby performance of different pupils in the same class can both increase and decrease. Therefore, the same teacher can have a positive influence on performance of some pupils and a negative influence on performance of others.

Insertion: A type of DNA polymorphism characterized by insertion of one or several nucleotides.

Intraparietal sulcus: A sulcus on the lateral surface of the parietal lobe; it separates the superior parietal lobule and inferior parietal lobule.

Linkage disequilibrium (LD): Non-random association of alleles of different DNA loci that are located on the same chromosome and are more likely to be inherited together (linked). In association studies, this phenomenon allows one to impute alleles of DNA loci that have not been directly genotyped in this sample.

Magnetic resonance imaging (MRI): A non-invasive method of visualization of inner structures of the body, including the brain. This method uses a strong, stable magnetic field, and the studied part of the body is also exposed to a high frequency transient magnetic field. This combination of stable and transient magnetic fields makes atoms in molecules resonate, and this resonance can be used to compute images of the body structures under investigation.

Major allele: The most common allele of a particular DNA locus in a population. For example, for a DNA locus with two alleles (e.g., SNP), the frequency of allele 1 can be 70%, and the frequency of allele 2 is thus 30%. The allele with the highest frequency (70% in this case) is a major allele for this locus.

Microsatellite marker: See **Variable number of tandem repeats (VNTR).**

Minor allele: The least common allele of a particular DNA locus in a population. For example, for a DNA locus with two alleles (e.g., SNP), the frequency of allele 1 can be 70%, and the frequency of allele 2 is thus 30%. The allele with the lowest frequency (30% in this case) is a minor allele for this locus.

"Missing heritability" problem: The discrepancy between relatively high heritability estimates for complex traits obtained with quantitative genetic methods and small effects of DNA markers identified with molecular genetic methods.

Molecular genetic study: A study used to link individual differences in DNA sequence with individual differences in a phenotype.

Monogenic disease: A disease that is caused by a single gene mutation(s).

Monozygotic twins (identical, MZ twins): Twins born as a result of one pregnancy, from a single spontaneously divided zygote (a fertilized ovum). Accordingly, MZ twins share 100% of the segregating DNA.

Multivariate twin analysis: Analysis based on cross-trait twin correlations that allows estimation of the contribution of genetic and environmental factors to covariation among multiple traits.

MZ (DZ) twin correlation: Intra-class correlation coefficient for MZ and DZ twins. See Box A.1.

Neurogenomics (imaging genomics): A research area that investigates individual differences in structures and functions of the brain using genetically informative designs.

Non-additive genetic factors: Genetic factors with a non-liner effect on a phenotype (dominance, **epistasis**). For example, in the case of dominance, the effect of one allele of a particular DNA locus may supress (weaken) the effect of another allele of the same locus.

Non-shared environment (unique, individual-specific environment): In quantitative genetics, any non-genetic factors that contribute to dissimilarity among family members. See Box A.1.

Nuclear family: A family that consists of two partners with or without children, or one parent with a child/children.

Nucleotide: A molecule that consists of nitrogenous base, a five-carbon sugar (ribose or deoxyribose), and at least one phosphate group. Nucleotides are building blocks of nucleic acids, such as DNA and RNA.

Passive gene–environment correlation: A type of gene–environment correlation whereby the rearing environment that parents provide to their children is influenced by the parents' own behavioural characteristics, and these characteristics are partly under genetic influence. For example, parents can provide both genetic predisposition to musical giftedness and a "musical" environment.

Phenotype: An observed or measured variable characteristic (trait), for example, height, weight, eye colour, personality traits, learning abilities.

Pleiotropy: Influence of the same genetic factor on multiple phenotypes. For example, a gene that influences hair colour is also related to skin colour and the presence/absence of freckles.

Polygenic index (score): Summing genotypes/alleles of a number of DNA markers (e.g., SNPs). For example, if 50 SNPs (each with two alleles) have been found to be involved in a particular trait (e.g., dyslexia), a polygenic risk index/score can be calculated in the following way: for one SNP with two alleles (A1 and A2), with the A1 allele associated, and the A2 allele not associated with dyslexia, values can be assigned for these alleles: A1=1 and A2=0. A person will have a score for this SNP based on his/her genotype: A1A1=2, A1A2=1, and A2A2=0. Therefore, for all 50 SNPs, a person can have a score from 0 (no risk alleles) to 100 (all alleles are risk). This polygenic score/index can be used for estimating genetic risk for dyslexia. See **Quantitative trait locus.**

Polygenicity: Influence of many genetic factors on the same phenotype.

Polygenic trait: A trait whose variation is influenced by multiple genetic factors (DNA loci).

Polymorphism/polymorphic locus: A DNA locus with two or more variants (alleles) in a population.

Probability of phenotypic expression: Effect of each genetic factor at the level of phenotypic expression is probabilistic (not deterministic) as there are many genetic factors each of which makes only a small contribution to a trait. In addition, it is possible for a genetic factor to have a positive or negative effect on a trait depending on other genetic and environmental factors.

Proband: In quantitative genetics, a person studied because of their position on a particular trait continuum beyond a **threshold.**

Qualitative sex differences: In quantitative genetics, (partial) differences in genetic and environmental factors that affect variation in a particular trait in men and women. For example, research has shown that genetic factors involved in depression are partially different in men and women.

Quantitative sex differences: In quantitative genetics, differences in the relative contribution of genetic and environmental factors to variation in a particular trait in men and women. For example, research has shown that heritability of depression is higher for women than for men.

Quantitative trait locus (QTL): A DNA locus (e.g., SNP) in the polygenic system that contributes to quantitative variation of a trait. See **Polygenic trait.**

Rapid Eye Movement (REM) sleep: One of the stages of sleep characterized by the electrical activity of the brain similar to that in waking state. A person is completely immobile as a result of a sharp decline in muscle tone, whereas his/her eyes move rapidly under the close eyelids.

Rare polymorphism/mutation: A variable DNA locus whose minor allele frequency in a population is less than 1%.

Reactive (evocative) gene–environment correlation: A type of gene–environment correlation whereby a person's behaviour, which is partly influenced by genetic factors, evokes a reaction from other people. For example, a child's aggressive behaviour (which is partly under genetic influences) may evoke negative attitudes from peers.

Representative sample: In statistics, a sub-set of individuals from a general population in which all basic characteristics are presented with the same frequency as in the population of interest (i.e., a population to which the results are being generalized).

Segregating DNA: Variable DNA; DNA loci whose alleles/genotypes differ among people. Segregating DNA accounts for only ~1% of the whole DNA.

Shared environment (common environment, C): In quantitative genetics, any non-genetic factors that increase similarity between members of one family.

Siblings: Children of the same parents; brothers and sisters.

Single Nucleotide Polymorphism (SNP): A type of DNA sequence variation that results from a replacement of one nucleotide for another at a particular locus (for example, adenine (A) for thymine (T)), and is characterized by the presence of different alleles (usually two) in a population (for example, allele A and allele T).

Specialist genes: A phenomenon describing a relatively specialized effect of genetic factors on specific traits. For example, although most genetic factors that influence mathematical ability also influence general cognitive ability and reading ability, there are also genetic factors that affect mathematical, but not other cognitive, abilities.

Structural magnetic resonance imaging (sMRI): A non-invasive neuroimaging method used to examine anatomical structures of the brain. See Box A.2.

Threshold: A critical value on a trait continuum, scores beyond which are considered outside the norm. For example, a threshold for reading disability may be a score on a particular test (e.g., below two standard deviations from the population mean on this test), which indicates severe learning problems and is classified as a disorder (dyslexia).

Twin registry: A database of twins created to study the role of genetic and environmental factors in individual differences in human traits.

Twin method: See Box A.1.

Twins: See Box A.1.

Variable Number of Tandem Repeats (VNTR): A type of DNA polymorphism characterized by a replication of a short sequence of nucleotides (2–3 nucleotide pairs) that are located together (tandem repeat). Alleles of a VNTR polymorphism have different numbers of tandem repeats.

X-linked genes: Genes located on a sex chromosome (X). Women have two X-chromosomes (XX); men have one X-chromosome and one Y-chromosome (XY). Although women have two X-chromosomes, in somatic cells one of them is inactivated; therefore only genes of one of the X-chromosomes are expressed.

Index

Page numbers in *italics* refer to glossary entries

CPSIA information can be obtained at www.ICGtesting.com
Printed in the USA
BVOW06*1538160616

452334BV00015B/15/P

9 781137 437310